Particle Swarm Optimization and Intelligence:

Advances and Applications

Konstantinos E. Parsopoulos
University of Ioannina, Greece

Michael N. Vrahatis
University of Patras, Greece

A volume in the Advances in
Computational Intelligence and Robotics
(ACIR) Book Series

Director of Editorial Content:	Kristin Klinger
Director of Book Publications:	Julia Mosemann
Development Editor:	Joel Gamon
Publishing Assistant:	Sean Woznicki
Typesetter:	Deanna Zombro
Quality Control:	Jamie Snavely
Cover Design:	Lisa Tosheff

Published in the United States of America by
Information Science Reference (an imprint of IGI Global)
701 E. Chocolate Avenue
Hershey PA 17033
Tel: 717-533-8845
Fax: 717-533-8661
E-mail: cust@igi-global.com
Web site: http://www.igi-global.com

Copyright © 2010 by IGI Global. All rights reserved. No part of this publication may be reproduced, stored or distributed in any form or by any means, electronic or mechanical, including photocopying, without written permission from the publisher. Product or company names used in this set are for identification purposes only. Inclusion of the names of the products or companies does not indicate a claim of ownership by IGI Global of the trademark or registered trademark.

Library of Congress Cataloging-in-Publication Data

Particle swarm optimization and intelligence : advances and applications / Konstantinos E. Parsopoulos and Michael N. Vrahatis, editors.
    p. cm.
  Summary: "This book presents the most recent and established developments of Particle swarm optimization (PSO) within a unified framework by noted researchers in the field"--Provided by publisher.
  Includes bibliographical references and index.
  ISBN 978-1-61520-666-7 (hardcover) -- ISBN 978-1-61520-667-4 (ebook) 1. Mathematical optimization. 2. Particles (Nuclear physics) 3. Swarm intelligence. I. Parsopoulos, Konstantinos E., 1974- II. Vrahatis, Michael N., 1955-
  QC20.7.M27P37 2010
  519.6--dc22
                                        2009041376

This book is published in the IGI Global book series Advances in Computational Intelligence and Robotics (ACIR) Book Series (ISSN: 2327-0411; eISSN: 2327-042X)

British Cataloguing in Publication Data
A Cataloguing in Publication record for this book is available from the British Library.

All work contributed to this book is new, previously-unpublished material. The views expressed in this book are those of the authors, but not necessarily of the publisher.

Advances in Computational Intelligence and Robotics (ACIR) Book Series

ISSN: 2327-0411
EISSN: 2327-042X

MISSION

While intelligence is traditionally a term applied to humans and human cognition, technology has progressed in such a way to allow for the development of intelligent systems able to simulate many human traits. With this new era of simulated and artificial intelligence, much research is needed in order to continue to advance the field and also to evaluate the ethical and societal concerns of the existence of artificial life and machine learning.

The **Advances in Computational Intelligence and Robotics (ACIR) Book Series** encourages scholarly discourse on all topics pertaining to evolutionary computing, artificial life, computational intelligence, machine learning, and robotics. ACIR presents the latest research being conducted on diverse topics in intelligence technologies with the goal of advancing knowledge and applications in this rapidly evolving field.

COVERAGE

- Adaptive & Complex Systems
- Agent Technologies
- Artificial Intelligence
- Cognitive Informatics
- Computational Intelligence
- Natural Language Processing
- Neural Networks
- Pattern Recognition
- Robotics
- Synthetic Emotions

IGI Global is currently accepting manuscripts for publication within this series. To submit a proposal for a volume in this series, please contact our Acquisition Editors at Acquisitions@igi-global.com or visit: http://www.igi-global.com/publish/.

The Advances in Computational Intelligence and Robotics (ACIR) Book Series (ISSN 2327-0411) is published by IGI Global, 701 E. Chocolate Avenue, Hershey, PA 17033-1240, USA, www.igi-global.com. This series is composed of titles available for purchase individually; each title is edited to be contextually exclusive from any other title within the series. For pricing and ordering information please visit http://www.igi-global.com/book-series/advances-computational-intelligence-robotics-acir/73674. Postmaster: Send all address changes to above address. Copyright © 2010 IGI Global. All rights, including translation in other languages reserved by the publisher. No part of this series may be reproduced or used in any form or by any means – graphics, electronic, or mechanical, including photocopying, recording, taping, or information and retrieval systems – without written permission from the publisher, except for non commercial, educational use, including classroom teaching purposes. The views expressed in this series are those of the authors, but not necessarily of IGI Global.

Titles in this Series

For a list of additional titles in this series, please visit: www.igi-global.com

Intelligent Technologies and Techniques for Pervasive Computing
Kostas Kolomvatsos (University of Athens, Greece) Christos Anagnostopoulos (Ionian University, Greece) and Stathes Hadjiefthymiades (University of Athens, Greece)
Information Science Reference ● copyright 2013 ● 349pp ● H/C (ISBN: 9781466640382) ● US $195.00 (our price)

Mobile Ad Hoc Robots and Wireless Robotic Systems Design and Implementation
Raul Aquino Santos (University of Colima, Mexico) Omar Lengerke (Universidad Autónoma de Bucaramanga, Colombia) and Arthur Edwards-Block (University of Colima, Mexico)
Information Science Reference ● copyright 2013 ● 347pp ● H/C (ISBN: 9781466626584) ● US $190.00 (our price)

Intelligent Planning for Mobile Robotics Algorithmic Approaches
Ritu Tiwari (ABV – Indian Institute of Information, India) Anupam Shukla (ABV – Indian Institute of Information, India) and Rahul Kala (School of Systems Engineering, University of Reading, UK)
Information Science Reference ● copyright 2013 ● 320pp ● H/C (ISBN: 9781466620742) ● US $195.00 (our price)

Simultaneous Localization and Mapping for Mobile Robots Introduction and Methods
Juan-Antonio Fernández-Madrigal (Universidad de Málaga, Spain) and José Luis Blanco Claraco (Universidad de Málaga, Spain)
Information Science Reference ● copyright 2013 ● 497pp ● H/C (ISBN: 9781466621046) ● US $195.00 (our price)

Prototyping of Robotic Systems Applications of Design and Implementation
Tarek Sobh (University of Bridgeport, USA) and Xingguo Xiong (University of Bridgeport, USA)
Information Science Reference ● copyright 2012 ● 321pp ● H/C (ISBN: 9781466601765) ● US $195.00 (our price)

Cross-Disciplinary Applications of Artificial Intelligence and Pattern Recognition Advancing Technologies
Vijay Kumar Mago (Simon Fraser University, Canada) and Nitin Bhatia (DAV College, India)
Information Science Reference ● copyright 2012 ● 784pp ● H/C (ISBN: 9781613504291) ● US $195.00 (our price)

Handbook of Research on Ambient Intelligence and Smart Environments Trends and Perspectives
Nak-Young Chong (Japan Advanced Institute of Science and Technology, Japan) and Fulvio Mastrogiovanni (University of Genova, Italy)
Information Science Reference ● copyright 2011 ● 770pp ● H/C (ISBN: 9781616928575) ● US $265.00 (our price)

Particle Swarm Optimization and Intelligence Advances and Applications
Konstantinos E. Parsopoulos (University of Ioannina, Greece) and Michael N. Vrahatis (University of Patras, Greece)
Information Science Reference ● copyright 2010 ● 328pp ● H/C (ISBN: 9781615206667) ● US $180.00 (our price)

www.igi-global.com

701 E. Chocolate Ave., Hershey, PA 17033
Order online at www.igi-global.com or call 717-533-8845 x100
To place a standing order for titles released in this series, contact: cust@igi-global.com
Mon-Fri 8:00 am - 5:00 pm (est) or fax 24 hours a day 717-533-8661

Dedication

To my wife Anastasia and our sons Vangelis and Manos

K.E. Parsopoulos

To my wife Irene

M.N. Vrahatis

Table of Contents

Chapter 3

Theoretical Derivations and Application Issues

Chapter 4

Chapter 5

Section 2
Applications of Particle Swarm Optimization

Chapter 6

Chapter 12

Foreword

Swarm intelligence is an exciting new research field still in its infancy compared to other paradigms in artificial intelligence. With many successful applications in a wide variety of complex problems, swarm-based algorithms have shown to have much promise, being efficient and robust, yet very simple to implement. A number of computational swarm-based systems have been developed in the past decade, where the approach is to model the very simple local interactions among individuals, from which complex problem-solving behaviors emerge. One of the research areas within computational swarm intelligence is particle swarm optimization (PSO), which has its origins in bird flocking models. Each individual, referred to as a particle, follow two very simple behaviors, i.e., to follow the best performing individual, and to move towards the best conditions found by the individual itself. In terms of optimization, each particle moves towards two attractors, with the result that all particles converge on one solution.

Since its inception in 1995, research and application interest in PSO have increased, resulting in an exponential increase in the number of publications and applications. Research in PSO has resulted in a large number of new PSO algorithms that improves the performance of the original PSO and enables application of PSO to different optimization problem types (e.g., unconstrained optimization, constrained optimization, multiobjective optimization, optimization in dynamic environments, and finding multiple solutions). Elaborate theoretical studies of PSO dynamics have been done, and PSO parameter sensitivity analyses have resulted in a better understanding of the influence of PSO control parameters. PSO applications vary in complexity and cover a wide range of application areas. To date, the total number of PSO publications counts to approximately 1500 since 1995.

It should be evident to the reader that the published knowledge available on PSO is vast. This then provides motivation for a dedicated, up-to-date book on particle swarm optimization. However, such a task is not an easy one. These authors have succeeded in the daunting task of sifting through the large volumes of PSO literature to produce a text that focuses on the most recent and significant developments in PSO. The authors have also succeeded in conveying their significant experience in PSO development and application to the benefit of the reader. It should be noted that the intention of this book was not to produce an encyclopedia of PSO research and applications, but to provide both the novice and the experienced PSO user and researcher with an introductory as well as expert level overview of PSO. As such the authors provide the reader with a compact source of information on PSO, and a foundation for the development of new PSO algorithms and applications.

The book is very well organized, starting with an overview of optimization, evolutionary computation, and swarm intelligence in general. This is followed by a detailed development of the original PSO and first improvements. A concise summary of theoretical analyses is given followed by detailed discussions of state-of-the-art PSO models. An excellent contribution made by the book is the coverage of a wide range of real-world applications, and of different optimization problem types. Throughout, the authors have provided a book which is hands-on, making the book accessible to first-time PSO users. Another positive of the book is the collection of benchmark problems given in the appendix, and the list of resources provided.

The authors have succeeded in their objective to produce a book which covers the main trends in PSO research and applications, while still producing text that is accessible to a wide range of authors. I have no second thoughts of recommending the book and making the statement that this book will be a valuable resource to the PSO practitioner and researcher.

Andries P. Engelbrecht,
University of Pretoria, South Africa

Andries Engelbrecht *is a professor in Computer Science at the University of Pretoria, South Africa. He also holds the position as South African Research Chair in Artificial Intelligence, and leads the Computational Intelligence Research Group at the University of Pretoria, consisting of 50 Masters and PhD students. He obtained his Masters and PhD degrees in Computer Science from the University of Pretoria in 1994 and 1999 respectively. His research interests include swarm intelligence, evolutionary computation, artificial neural networks, artificial immune systems, and the application of these CI paradigms to data mining, games, bioinformatics, and finance. He has published over 130 papers in these fields in journals and international conference proceedings, and is the author of the two books, "Computational Intelligence: An Introduction" and "Fundamentals of Computational Swarm Intelligence". In addition to these, he is a co-editor of the upcoming books, "Applied Swarm Intelligence" and "Foundations on Computational Intelligence". He is very active in the international community, annually serving as a reviewer for over 20 journals and 10 conferences. He is an associate-editor of the IEEE Transactions on Evolutionary Computation, Journal of Swarm Intelligence, and the recent IEEE Transactions on Computational Intelligence and AI in Games. Additionally, he serves on the editorial board of 3 other international journals, and is co-guest-editor of special issues of the IEEE Transactions on Evolutionary Computation and the Journal of Swarm Intelligence. He served on the international program committee and organizing committee of a number of conferences, organized special sessions, presented tutorials, and took part in panel discussions. As member of the IEEE CIS, he is a member of the Games technical committee and chair of its Swarm Intelligence for Games task force. He also serves as a member of the Computational Intelligence and Machine Learning Virtual Infrastructure Network.*

Preface

Optimization is the procedure of detecting attributes, configurations or parameters of a system, to produce desirable responses. For example, in structural engineering, one is interested in detecting the best possible design to produce a safe and economic structure that adheres to specific mechanical engineering rules. Similarly, in computer science, one is interested in designing high-performance computer systems at the lowest cost, while, in operations research, corporations struggle to identify the best possible configuration of their production lines to increase their operational flexibility and efficiency. Numerous other systems (including human) can be mentioned, where there is a need or desire for explicit or implicit improvement.

All these problems can be scientifically resolved through modeling and optimization. Modeling offers a translation of the original (physical, engineering, economic, etc.) problem to a mathematical structure that can be handled through algorithmic optimization procedures. The model is responsible for the proper representation of all key features of the original system and its accurate simulation. Concurrently, it offers a mathematical means of identifying and modifying the system's properties to produce the most desirable outcome without requiring its actual construction, thereby saving time and cost.

The produced models are usually formulated as functions, called *objective functions*, in one or several variables that correspond to adaptable parameters of the system. The model is built in such a way that, based on the particular optimality criteria per case, the most desirable system configurations correspond to the extremal values of the objective function. Thus, the original system optimization problem is transformed to an equivalent function minimization or maximization problem. The difficulty in solving this problem is heavily dependent on the form and mathematical properties of the objective function.

It is possible that a solution of the optimization problem can be achieved through an analytic approach that involves minimum effort. Unfortunately, this case is rather an exception. In most problems, complicated systems are modeled with complicated multi-dimensional functions that cannot be easily addressed. In such cases, algorithmic procedures that take full advantage of modern computer systems can be implemented to solve the underlying optimization problems numerically. Of course, only approximations of the original solutions can be obtained under this scope. Thus, computation accuracy, time criticality, and implementation effort become important aspects of the numerical optimization procedure.

To date, a multitude of algorithms that exploit favorable mathematical properties of the objective function, such as differentiability and Lipschitz continuity, have been developed. These approaches use first-order and second-order derivatives and achieve high convergence rates. However, the necessary assumptions for their application are not usually met in practice. Indeed, the blossoming of technological research and engineering has introduced a plethora of optimization problems with minimum available information regarding their form and inherent attributes. Typical properties of such problems are the existence of discontinuities, the lack of analytical representation of the objective function, and noise dissemination.

In these circumstances, the applicability and efficiency of classical optimization algorithms are questionable, giving rise to the need for the development of different optimization methods. Early attempts towards this direction were focused on stochastic algorithms that employ only function values. Pure random search is the most trivial approach, although its performance is rapidly degenerating with problem dimension and complexity, since it does not exploit information gained in previous steps of the algorithm. On the other hand, combinations of random and classical algorithms have offered better results; nevertheless, the necessity for strong mathematical assumptions on the objective function was still inevitable.

However, researchers were gradually realizing that several systems observed in nature were able to cope efficiently with similar optimization problems. Thus, the trend to study and incorporate models of natural procedures in optimization algorithms gradually gained ground, becoming an appealing alternative. Early approaches, such as *simulated annealing*, offered the potential of solving problems that were laborious for other algorithms. However, for a number of years, they remained in the margin of relative literature, due to their limited theoretical developments at that time.

At the same time, a new type of algorithms was slowly but steadily emerging. The inspiration behind their development stemmed from the study of adaptation mechanisms in natural systems, such as DNA. The underlying operations that support evolutionary mechanisms according to the Darwinian biological theory were modeled and used to evolve problem solutions, based on user-defined optimality criteria. Although these optimization approaches were initially supported by limited theoretical analyses, their promising results on complex problems previously considered as intractable offered a boost to research, especially in the engineering community. Research groups in the USA and Europe attained to refine early variants of these algorithms, introducing a set of efficient approaches under the general name of *evolutionary algorithms*.

Theoretical studies were soon conducted, rendering these algorithms promising alternatives in cases where classical approaches were not applicable. The new type of algorithms concentrated all desirable optimization features as well as novel concepts. For example, a search was not performed by one but rather a population of interacting search agents without central control. Stochasticity, communication, information exchange, and adaptation became part of their operation, while the requirements on objective function were restricted to the least possible, namely the ability to perform function evaluations.

The success recognized by evolutionary approaches sparked off research all over the world. As a result, in the mid-90's a new category of algorithms appeared. Instead of modeling evolutionary procedures in microscopic (DNA) level, these methods model populations in a macroscopic level, i.e., in terms of social structures and aggregating behaviors. Once again, nature was offering inspiration and motivation to scientists. Hierarchically organized societies of simple organisms, such as ants, bees, and fish, with a very limited range of individual responses, exhibit fascinating behaviors with identifiable traits of intelligence as a whole. The lack of a central tuning and control mechanism in such systems has triggered scientific curiosity. Simplified models of these systems were developed and studied through simulations. Their dynamics were approximated by mathematical models similar to those used in particle physics, while probability theory and stochastic processes offered a solid theoretical background for the development of a new category of algorithms under the name of *swarm intelligence*.

Particle swarm optimization (PSO) belongs to this category and constitutes the core subject of the book at hand. Its early precursors were simulators of social behavior that implemented rules such as nearest-neighbor velocity matching and acceleration by distance, to produce swarming behavior in groups of simple agents. As soon as the potential of these models to serve as optimization algorithms was recognized, they were refined, resulting in the first version of PSO, which was published in 1995 (Eberhart & Kennedy, 1995; Kennedy & Eberhart, 1995).

Since its development, PSO has gained wide recognition due to its ability to provide solutions efficiently, requiring only minimal implementation effort. This is reflected in Fig. 1, which illustrates the number of journal papers with the term "particle swarm" in their titles published by three major publishers, namely Elsevier, Springer, and IEEE, during the years 2000-2008. Also, the potential of PSO for straightforward parallelization, as well as its plasticity, i.e., the ability to adapt easily its components and operators to assume a desired form implied by the problem at hand, has placed PSO in a salient position among intelligent optimization algorithms.

The authors of the book at hand have contributed a remarkable amount of work on PSO and gained in-depth experience on its behavior and handling of different problem types. This book constitutes an attempt to present the most recent and established developments of PSO within a unified framework and share their long experience, as described in the following section, with the reader.

WHAT THIS BOOK IS AND IS NOT ABOUT

This book is not about numerical optimization or evolutionary and swarm intelligence algorithms in general. It does not aim at analyzing general optimization procedures and techniques or demonstrating

Figure 1. Number of journal papers with the term "particle swarm" in their titles, published by three major publishers, namely Elsevier, Springer, and IEEE, during the years 2000-2008.

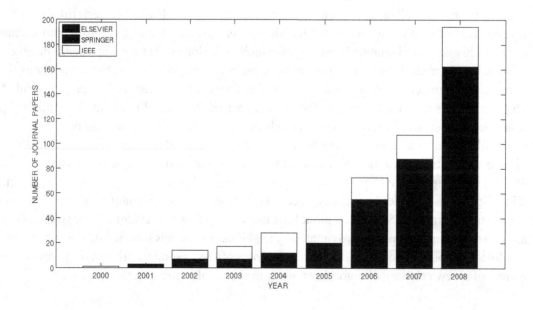

the application of evolutionary algorithms. The book at hand is completely devoted to the presentation of PSO. Its main objectives are to:

1. Provide a unified presentation of the most established PSO variants, from early precursors to concurrent state-of-the-art approaches.
2. Provide a rough sketch of the established theoretical analyses and their impact on the established variants of the algorithm.
3. Provide complementary techniques that enhance its performance on particular problem types.
4. Illustrate its workings on a plethora of different problem types, providing guidelines and experimental settings along with representative results that equip the reader with the necessary background and offer a starting point for further investigation in similar problems.

All these objectives were meant to be achieved using the simplest possible notation and descriptions to render non-expert readers capable of taking full advantage of the presented material.

In order to achieve these ambitious goals, the authors relied on their accumulated experience of more than ten years working on PSO. The majority of the presented material is based on their own work, which corresponds to more than 60 papers in international refereed scientific journals, book chapters, edited volumes, and conference proceedings, and more than a thousand citations from other researchers. A wide variety of applications is presented together with details on published results. Also, pseudocode is provided for the presented methodologies and procedures when applicable, while special mathematical manipulations and transformations are analyzed whenever needed. Besides that, many recent developments of other researchers are reported, always striving to keep the presentation simple and understandable. The provided material and references can serve as an Ariadne's thread for further acquisition and development of more sophisticated approaches.

We tried to avoid the pitfall of rendering this book another literature review on PSO. Nevertheless, there are excellent survey papers for this purpose (Al Rashidi & El-Hawary, in press; Banks et al., 2007, 2008; Hendtlass & Randall, 2001; Parsopoulos & Vrahatis, 2002; Reyes-Sierra & Coello Coello, 2006; Yang & Li, 2004). Instead, only the most important developments with potential for further extension and improvement are analyzed. We did not aim at deluging the reader with a huge number of minor developments or reproductions of established sound results, but rather offer the most essential concepts and considerations as the amalgam of our experience and personal viewpoint in the ongoing progress of PSO research all these years.

The target audience of the book embraces undergraduate and graduate students with a special interest in PSO and relative approaches, as well as researchers and scientists that employ heuristic algorithms for problem solving in science and engineering. The level of mathematics was intentionally kept low to make descriptions comprehensible even to a multidisciplinary (non-expert) audience. Thus, elementary calculus is adequate for the comprehension of most of the presented algorithms in the first part of the book, while essential specialized knowledge on machine learning, dynamical systems, and operations research may be useful in specific chapters of the second part.

This book can also serve as an essential reference guide of established advances on PSO, as well as a stepping stone for further developments. It can be distinguished by relevant books by its specialization solely on PSO, without however restricting its algorithmic material to the narrow personal achievements and considerations of the authors. Thus, it can be valuable to a wide scientific audience of both novice and expert researchers.

ORGANIZATION OF THE BOOK

The book is divided in two sections. Section 1 consists of Chapters 1 to 5 and presents basic developments in PSO along with theoretical derivations, state-of-the-art variants, and performance-enhancing techniques. Section 2 consists of Chapters 6 to 12, where various applications of PSO are presented and trends for future research are reported. The book also has two appendices. Appendix A contains descriptions of the test problems used throughout the text, especially in Section 2. Appendix B contains an indicative simple implementation of PSO in Matlab©, as well as further web resources on PSO. A brief description of each chapter is provided in the following paragraphs.

Chapter 1 introduces the reader to basic concepts of global optimization, evolutionary computation, and swarm intelligence, outlines the necessity for solving optimization problems and identifies various problem types. Also, a rough classification of established optimization algorithms is provided, followed by the historical development of evolutionary computation. The three fundamental evolutionary approaches, namely genetic algorithms, evolutionary programming, and evolution strategies are briefly presented, along with their basic features and operations. Then, swarm intelligence is presented followed by short descriptions of its three main algorithms, namely ant colony optimization, stochastic diffusion search, and particle swarm optimization. Finally, reference is made to the no-free-lunch theorem to justify the necessity for further development of intelligent optimization algorithms.

Chapter 2 is devoted to the presentation of particle swarm optimization (PSO). The description begins with the main inspiration source that led to the development of its early precursors. Severe deficiencies of these approaches are pointed out and addressed by introducing new concepts, such as inertia weight, velocity clamping, and the concept of neighborhood. Finally, the reader is brought to the present day by exposing the standard contemporary developments, which are considered the state-of-the-art PSO variants nowadays.

Chapter 3 briefly exposes the fundamental theoretical derivations and issues of PSO. Emphasis is given to developments that offered new insight in configuring and tuning its parameters. The chapter begins with a discussion on initialization procedures, followed by the first attempts to investigate particle trajectories. These studies opened the way for the stability analysis of PSO, which resulted in contemporary sophisticated variants and rules for parameter setting. We then present a useful technique for the optimal tuning of PSO on specific problems, based on computational statistics. The workings of this technique are illustrated in detail in two problems. The chapter closes with a short discussion on the most common termination conditions.

Chapter 4 presents established and recently proposed PSO variants. The presented methods were selected from the extensive PSO literature according to various criteria, such as sophisticated inspiration source, close relationship to the standard PSO form, wide applicability in problems of different types, satisfactory performance and theoretical properties, number of reported applications, and potential for further development and improvement. Thus, the unified PSO, memetic PSO, composite PSO, vector evaluated PSO, guaranteed convergence PSO, cooperative PSO, niching PSO, TRIBES, and quantum PSO are described and their fundamental concepts are exposed.

Chapter 5 closes the first section of the book by presenting performance-enhancing techniques. These techniques consist of transformations of either the objective function or the problem variables, enabling PSO to alleviate local minimizers, detect multiple global minimizers, handle constraints, and solve integer programming problems. The chapter begins with a short discussion of the filled functions approach, followed by the stretching technique as an alternative for alleviating local minimizers. Next, deflection and repulsion techniques are presented as means for detecting multiple minimizers with PSO. The penalty function approach for handling constraints is discussed and two recently proposed approaches are reported. The chapter closes with the description of two rounding schemes that enable the real-valued PSO to solve integer programming problems. The techniques are thoroughly described and, wherever possible, graphically illustrated.

Chapter 6 focuses on the application of PSO on machine learning problems. Two representative cases, namely the training of artificial neural networks and learning in fuzzy cognitive maps, are first defined within a general framework, followed by illustrative examples that familiarize the reader with the main procedures and nominate possible obstacles of the underlying optimization procedure.

Chapter 7 presents the application of PSO in dynamical systems and more specifically in the detection of periodic orbits. The transformation of the original fixed-point problem to the corresponding optimization problem is analyzed and indicative examples on widely used nonlinear mappings are reported in the first part of the chapter. The second part thoroughly describes an important application on the detection of periodic orbits in 3-dimensional galactic potentials, illustrating the ability of PSO to detect previously unknown solutions.

Chapter 8 consists of three representative applications of PSO in operations research. Similarly to previous chapters, attention is focused on the presentation of essential aspects of the applications rather than reviewing the existing literature. Thus, we present the formulation of the optimization problem from the original one, along with the efficient treatment of special problem requirements that cannot be handled directly by PSO. Applications from the fields of scheduling, inventory optimization, and game theory are given, accompanied by recently published results to provide a flavor of PSO's efficiency per case.

Chapter 9 presents two interesting applications of PSO in bioinformatics and medical informatics. The first one consists of the adaptation of probabilistic neural network models for medical classification tasks. The second application considers PSO variants for tackling two magnetoencephalography problems, namely source localization and refinement of approximation models. The main points where PSO interferes with the employed computational models are analyzed, and details are provided regarding the formulation of the corresponding optimization problems and their experimental settings. Indicative results are also reported as representative performance samples.

Chapter 10 discusses the workings of PSO on two intimately related research fields with special focus on real world applications, namely noisy and dynamic environments. Noise simulation schemes are presented and experimental results on benchmark problems are reported. Also, the application of PSO in a simulated real world problem, namely the particle identification by light scattering, is presented. Moreover, a hybrid scheme that incorporates PSO in particle filtering methods to estimate system states online is analyzed and representative experimental results are provided. Finally, the combination of noisy and continuously changing environments is shortly discussed, providing illustrative graphical representations of performance for different PSO variants.

Chapter 11 essentially closes the second section of the book by presenting applications of PSO in three very interesting problem types, namely multiobjective, constrained, and minimax optimization problems. The largest part of the chapter is devoted to the multiobjective case, which is supported by an extensive bibliography with a rich assortment of PSO approaches developed to date. Different algorithm types are presented and briefly discussed, insisting on the most influential approaches for the specific problem types.

Chapter 12 closes the book by providing current trends and future directions in PSO research. This information can be beneficial to new researchers with an interest in conducting PSO-related research, since it enumerates open problems and active research topics.

Appendix A contains all test problems employed throughout the text. Thus, widely used unconstrained optimization problems, nonlinear mappings, inventory optimization problems, game theory problems, data sets for classification tasks in bioinformatics, multiobjective problems, constrained benchmark and engineering design problems, as well as minimax problems are reported. Citations for each problem are also provided.

Finally, Appendix B contains an indicative simple implementation of the PSO algorithm in Matlab©, as well as references to web resources for further information on developments and implementations of PSO.

Each chapter is written in a self-contained manner, although several references to the first section of the book are made in chapters devoted to applications. We hope that readers will find our approach interesting and help us improve it by providing their comments, considerations, and suggestions.

REFERENCES

AlRashidi, M. R., & El-Hawary, M. E. (2009). A survey of particle swarm optimization applications in electric power systems. *IEEE Transactions on Evolutionary Computation, 13*(4), 913-918.

Banks, A., Vincent, J., & Anyakoha, C. (2007). A review of particle swarm optimization. Part I: background and development. *Natural Computing, 6* (4), 476-484.

Banks, A., Vincent, J., & Anyakoha, C. (2008). A review of particle swarm optimization. Part II: hybridization, combinatorial, multicriteria and constrained optimization, and indicative applications. *Natural Computing, 7* (1), 109-124.

Eberhart, R. C., & Kennedy, J. (1995). A new optimizer using particle swarm theory. In *Proceedings of the 6th Symposium on Micro Machine and Human Science, Nagoya, Japan* (pp. 39-43). Piscataway, NJ: IEEE Service Center.

Hendtlass, T., & Randall, M. (2001). A survey of ant colony and particle swarm meta-heuristics and their application to discrete optimisation problems. In *Proceedings of the Inaugural Workshop on Artificial Life, Adelaide, Australia (AL 2001)* (pp. 15-25).

Kennedy, J., & Eberhart, R. C. (1995). Particle swarm optimization. In *Proceedings of the IEEE International Conference on Neural Networks, Perth, Australia* (Vol. IV, pp. 1942-1948). Piscataway, NJ: IEEE Service Center.

Parsopoulos, K. E., & Vrahatis, M. N. (2002). Recent approaches to global optimization problems through particle swarm optimization. *Natural Computing, 1* (2-3), 235-306.

Reyes-Sierra, M., & Coello Coello, C. A. (2006). Multi-objective particle swarm optimizers: a survey of the state-of-the-art. *International Journal of Computational Intelligence Research, 2* (3), 287-308.

Yang, W., & Li, Q. Q. (2004). Survey on particle swarm optimization algorithm. *Engineering Science, 5* (6), 87–94.

Acknowledgment

The authors wish to thank all those who offered their support during the elaboration of this book. Special thanks are due to our collaborators over all these years of research on Particle Swarm Optimization. Also, we feel obligated to thank the pioneers, Professor Russell C. Eberhart at the Purdue School of Engineering and Technology, Indiana University Purdue University Indianapolis (IUPUI), Indianapolis (IN), USA, and Dr. James Kennedy at the Bureau of Labor Statistics, US Department of Labor, Washington (DC), USA, who created a new and fascinating research direction by introducing Particle Swarm Optimization, back in 1995.

Special thanks are also due to the anonymous reviewers, whose constructive criticism and recommendations helped us enhance the quality of this book. Our sincere gratitude also goes to Professor Andries P. Engelbrecht at Department of Computer Science, University of Pretoria, South Africa, for writing the foreword for this book. One of us (K.E.P.) wishes to thank also the State Scholarships Foundation (IKY) of Greece for partially supporting his efforts.

The authors wish to thank their wives for their patience, understanding, and affection all these days of endless work. Additionally, Konstantinos E. Parsopoulos wishes to thank his two little sons for putting the pressure on him to eventually finalize this project, as well as his parents for their unconditional support and encouragement.

K.E. Parsopoulos and M.N. Vrahatis
Patras, Greece
June 2009

Section 1
Theory and Methods

Chapter 1
Introduction

In this chapter, we provide brief introductions to the basic concepts of global optimization, evolutionary computation, and swarm intelligence. The necessity of solving optimization problems is outlined and various problem types are reported. A rough classification of established optimization algorithms is provided, followed by the historical development of evolutionary computation. The three fundamental evolutionary approaches are briefly presented, along with their basic features and operations. Finally, the reader is introduced to the field of swarm intelligence, and a strong theoretical result is concisely reported to justify the necessity for further development of global optimization algorithms.

WHAT IS OPTIMIZATION?

Optimization is a scientific discipline that deals with the detection of optimal solutions for a problem, among alternatives. The optimality of solutions is based on one or several criteria that are usually problem- and user-dependent. For example, a structural engineering problem can admit solutions that primarily adhere to fundamental engineering specifications, as well as to the aesthetic and operational expectations of the designer. Constraints can be posed by the user or the problem itself, thereby reducing the number of prospective solutions. If a solution fulfills all constraints, it is called a *feasible solution*. Among all feasible solutions, the *global optimization* problem concerns the detection of the optimal one. However, this is not always possible or necessary. Indeed, there are cases where suboptimal solutions are acceptable, depending on their quality compared to the optimal one. This is usually described as *local optimization*, although the same term has been also used to describe local search in a strict vicinity of the search space.

DOI: 10.4018/978-1-61520-666-7.ch001

Copyright © 2010, IGI Global. Copying or distributing in print or electronic forms without written permission of IGI Global is prohibited.

A modeling phase always precedes the optimization procedure. In this phase, the actual problem is modeled mathematically, taking into account all the underlying constraints. The building blocks of candidate solutions are translated into numerical variables, and solutions are represented as numerical vectors. Moreover, a proper mathematical function is built, such that its global minimizers, i.e., points where its minimum value is attained, correspond to optimal solutions of the original problem. This function is called the *objective function*, and the detection of its global minimizer(s) is the core subject of global optimization. Instead of minimization, an optimization problem can be equivalently defined as maximization by inverting the sign of the objective function. Without loss of generality, we consider only minimization cases in the book at hand.

The objective function is accompanied by a domain, i.e., a set of feasible candidate solutions. The domain is delimited by problem constraints, which need to be quantified properly and described mathematically using equality and inequality relations. In the simplest cases, constraints are limited to bounding boxes of the variables. In harder problems, complex relations among the variables must hold in the final solution, rendering the minimization procedure rather complicated.

Analytical derivation of solutions is possible for some problems. Indeed, if the objective function is at least twice continuously differentiable and has a relatively simple form, then its minimizers are attained by determining the zeros of its gradient and verifying that its Hessian matrix is positive definite at these points. Apparently, this is not possible for functions of high complexity and dimensionality or functions that do not fulfill the required mathematical assumptions. In the latter case, the use of algorithms that approximate the actual solution is inevitable. Such algorithms work iteratively, producing a sequence of search points that has at least one subsequence converging to the actual minimizer.

Optimization has been an active research field for several decades. The scientific and technological blossoming of the late years has offered a plethora of difficult optimization problems that triggered the development of more efficient algorithms. Real-world optimization suffers from the following problems (Spall, 2003):

a.   Difficulties in distinguishing global from local optimal solutions.
b.   Presence of noise in solution evaluation.
c.   The "curse of dimensionality", i.e., exponential growth of the search space with the problem's dimension.
d.   Difficulties associated with the problem's constraints.

The different nature and mathematical characteristics of optimization problems necessitated the specialization of algorithms to specific problem categories that share common properties, such as nonlinearity, convexity, differentiability, continuity, function evaluation accuracy etc. Moreover, the inherent characteristics of each algorithm may render it more suitable either for local or global optimization problems. Such characteristics include, among others, stochasticity, potential for parallelization in modern computer systems and limited computational requirements.

Today, there is a rich assortment of established algorithms for most problem types. Nevertheless, even different instances of the same problem may have different computational requirements, leaving space for development of new algorithms and the improvement of established ones. Consequently, there will be an ongoing need for new and more sophisticated ideas in optimization theory and applications.

In the next section, we put the optimization problem into a mathematical framework, which allows the distinction between different problem types, and we identify major categories of optimization algorithms, related to the topics of the book at hand.

TYPES OF OPTIMIZATION PROBLEMS

An optimization (minimization) problem can be defined mathematically in several ways, depending on the underlying application. In general, any function, $f{:}A{\rightarrow}Y$, defined over a domain, A, also called the *search space*, and with range, Y, can be subjected to optimization given a total ordering relation over Y. In literature, the most common optimization problems consist of the minimization of functions whose domain is a subset of the n-dimensional Euclidean space, $\mathbf{R}^n$, and their range is a subset of the real numbers. Moreover, the problem may have constraints in the form of inequality relations. Thus, the minimization problem can be formally described as:

$$\min_{x \in A} f(x), \quad \text{subject to} \quad C_i(x) \le 0, \quad i = 1, 2, ..., k, \tag{1}$$

where, $A \subseteq \mathbf{R}^n$, is a subset of the n-dimensional Euclidean space; $Y \subseteq \mathbf{R}$, is a subset of the real numbers; and, k, is the number of constraints. The form of constraints in relation (1) is not restrictive, since different forms can be represented equivalently as follows:

$$C_i(x) \ge 0 \Leftrightarrow -C_i(x) \le 0,$$

$$C_i(x) = 0 \Leftrightarrow C_i(x) \le 0 \text{ and } -C_i(x) \le 0.$$

Relation (1) defines a *constrained optimization* problem, and its constraints usually restrict the search space. A point, $x \in A$, is called *feasible point* if it satisfies all the constraints; the set of all feasible points of A is called the *feasible set*. Obviously, in constrained problems, only solutions that satisfy all constraints (or, in some cases, slightly violate them) are acceptable. In cases where constraints are absent, we have an *unconstrained optimization* problem. This can be defined with relation (1) by simply omitting the constraints. By definition, in unconstrained problems, the whole domain of the objective function is the feasible set.

Figure 1 depicts the contour lines of the following constrained optimization problem.

$$f(x,y) = x^2 + y^2, \text{ subject to } C(x,y) = 2x - y^2 + 1 \le 0, \tag{2}$$

for $(x,y) \in [-3,3]^2$. Although the domain of $f(x,y)$ is the set $A = [-3,3]^2$, the constraint $C(x)$, which is depicted as the parabola on the right part of Fig. 1, confines the search space by excluding the shadowed region. Note that the origin, where the global minimizer of the unconstrained problem lies, becomes infeasible for the constrained problem, while the new global minimizer lies on the parabolic boundary of the constrained feasible space. Therefore, even a relatively simple constraint may increase the complexity of a problem significantly.

Global optimization aims at the detection of a global minimizer:

$$x^* = \arg\min_{x \in A} f(x),$$

of the objective function, as well as its corresponding minimum, $f(x^*)$. This will be the core problem addressed by the algorithms presented in the book at hand. On the other hand, in *local optimization*, the detection of a local minimizer:

3

Figure 1. Contour lines of the problem defined in equation (2). The shadowed area is the region excluded from search due to the constraint

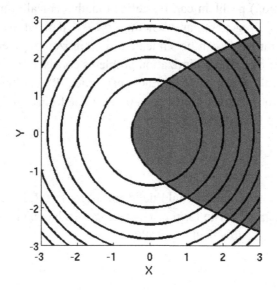

$$x' = \underset{x \in A' \subset A}{\arg \min} f(x),$$

along with its local minimum, $f(x\square)$, is adequate. Such problems will be considered only when local minimizers are acceptable for the studied problems. Figure 2 illustrates the local and the global minimum of the simple function, $f(x) = \sin(x)-x/2$, in the range [-10, 1].

Major optimization subfields can be further distinguished based on properties of the objective function, its domain, as well as the form of the constraints. A simple categorization is presented in the following paragraphs and summarized in Table 1. Some of the most interesting and significant subfields, with respect to the form of the objective function, are:

1. **Linear optimization (or linear programming):** It studies cases where the objective function and constraints are linear.
2. **Nonlinear optimization (or nonlinear programming):** It deals with cases where at least one nonlinear function is involved in the optimization problem.
3. **Convex optimization:** It studies problems with convex objective functions and convex feasible sets.
4. **Quadratic optimization (or quadratic programming):** It involves the minimization of quadratic objective functions and linear constraints.
5. **Stochastic optimization:** It refers to minimization in the presence of randomness, which is introduced either as noise in function evaluations or as probabilistic selection of problem variables and parameters, based on statistical distributions.

There is a multitude of comprehensive studies on the aforementioned optimization problems. Some of the most widely-used resources are (Horst & Pardalos, 1995; Horst & Tuy, 2003; Luenberger, 1989; Nocedal & Wright, 2006; Polak, 1997; Torn & Žilinskas, 1989; Zhigljavsky & Žilinskas, 2008).

Figure 2. The local and global minimum of f(x) = sin(x)-x/2 in the range [-10,1]

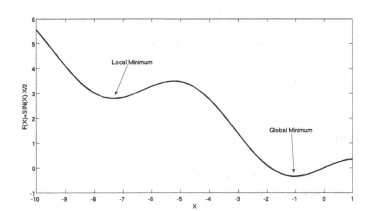

Usually, optimization problems are modeled with a single objective function, which remains unchanged through time. However, many significant engineering problems are modeled with one or a set of static or time-varying objective functions that need to be optimized simultaneously. These cases give rise to the following important optimization subfields:

1. **Dynamic optimization:** It refers to the minimization of time-varying objective functions (it should not be confused with *dynamic programming*). The goal in this case is to track the position of the global minimizer as soon as it moves in the search space. Also, it aims at providing *robust solutions*, i.e., solutions that will not require heavy computational costs for refinement in case of a slight change in the objective function.
2. **Multiobjective optimization, also known as multiple criteria optimization:** It refers to problems where two or more objective functions need to be minimized concurrently. In this case, the

Table 1. A categorization of optimization problems, based on different criteria

Classification Criterion	Type of Optimization Problem	Special Characteristics
Form of the objective function and/or constraints	Linear	Linear objective function and constraints
	Nonlinear	Nonlinear objective function and/or constraints
	Convex	Convex objective function and feasible set
	Quadratic	Quadratic objective function and linear constraints
	Stochastic	Noisy evaluation of the objective function or probabilistically determined problem variables and/or parameters
Nature of the search space	Discrete	Discrete variables of the objective function
	Continuous	Real variables of the objective function
	Mixed integer	Both real and integer variables
Nature of the problem	Dynamic	Time-varying objective function
	Multiobjective	Multitude of objective functions

optimality of solutions is redefined, since global minima of different objective functions are rarely achieved at the same minimizers.

Comprehensive books on these two subfields are (Branke, 2001; Chiang, 1991; Deb, 2001; Ehrgott & Gandibleux, 2002).

A different categorization can be considered with respect to the nature of the search space and problem variables:

1. **Discrete optimization:** In such problems, the variables of the objective function assume discrete values. The special case of integer variables is referred to as *integer optimization*.
2. **Continuous optimization:** All variables of the objective function assume real values.
3. **Mixed integer optimization:** Both integer and real variables appear in the objective function.

The reader can refer to (Boros & Hammer, 2003; Floudas, 1995) for specialized introductions to these topics.

There is a plethora of methods for solving efficiently problems in most of the aforementioned categories. However, many of these approaches are based on strong mathematical assumptions that do not hold in real-world applications. For example, there are very efficient deterministic algorithms for nonlinear optimization problems; however, they require that properties such as differentiability or convexity of the objective function hold. Unfortunately, these properties are not met in many significant applications. Indeed, there are problems where the objective function is not even analytically defined, with its values being obtained through complex procedures, computer programs, or measurements coming from observation equipment.

The aforementioned modeling deficiencies render the optimization problem very difficult, while solutions provided by mathematical approximations of the original problem are not always satisfactory. This effect gives rise to the concept of *black-box optimization*, where the least possible amount of information regarding the underlying problem is available, while function values of prospective solutions are obtained as the output of a hidden, complex procedure. The demand for efficient algorithms to tackle such problems is continuously increasing, creating a boost in relevant research.

The core of the book at hand is the *particle swarm optimization* algorithm and its applications, rather than optimization itself. Thus, in the rest of the book we will concentrate on problem categories on which particle swarm optimization has been applied successfully, providing significant results. In general, such problems include global optimization of nonlinear, discontinuous objective functions, and black-box optimization. Also, noisy, dynamic, and multiobjective problems will be considered. Continuous nonlinear optimization will be in the center of our interest; linear, discrete, and mixed integer problems will be considered only occasionally, since there are different approaches especially suited for these problem types (mainly for the linear case). In the next section, we provide a rough classification of optimization algorithms, based on their inherent properties.

CLASSIFICATION OF OPTIMIZATION ALGORITHMS

In general, two major categories of optimization algorithms can be distinguished: *deterministic* and *stochastic* algorithms. Although stochastic elements may appear in deterministic approaches to improve their performance, this rough categorization has been adopted by several authors, perhaps due to the similar inherent properties of the algorithms in each category (Archetti & Schoen, 1984; Dixon & Szegö, 1978). Deterministic approaches are characterized by the exact reproducibility of the steps taken by the algorithm, in the same problem and initial conditions. On the other hand, stochastic approaches produce samples of prospective solutions in the search space iteratively. Therefore, it is almost impossible to reproduce exactly the same sequence of samples in two distinct experiments, even in the same initial conditions.

Deterministic approaches include *grid search, covering methods*, and *trajectory-based methods*. Grid search does not exploit information of previous optimization steps, but rather assesses the quality of points lying on a grid over the search space. Obviously, the grid density plays a crucial role on the final output of the algorithm. On the other hand, trajectory-based methods employ search points that traverse trajectories, which (hopefully) intersect the vicinity of the global minimizer. Finally, covering methods aim at the detection and exclusion of parts of the search space that do not contain the global minimizer. All the aforementioned approaches have well-studied theoretical backgrounds, since their operation is based on strong mathematical assumptions (Horst & Tuy, 2003; Torn & Žilinskas, 1989).

Stochastic methods include *random search, clustering*, and methods based on *probabilistic models* of the objective function. Most of these approaches produce implicit or explicit estimation models for the position of the global minimizer, which are iteratively refined through sampling, using information collected in previous steps. They can be applied even in cases where strong mathematical properties of the objective function and search space are absent, albeit at the cost of higher computational time and limited theoretical derivations.

A more refined classification of global optimization algorithms is provided by Törn and Žilinskas (1989, p. 19):

1. Methods with guaranteed accuracy:
    a. Covering methods.
2. Direct search methods:
    b. Generalized descent methods.
    c. Clustering methods.
    d. Random search methods.
3. Indirect search methods:
    e. Level set approximation methods.
    f. Objective function approximation methods.

Their classification criterion differs from the previous categorization; it is rather based on the guaranteed (or not) accuracy provided by each algorithm. *Covering methods* represent the only category with guaranteed accuracy. They embrace methods that work in the general framework of bisection approaches, where the global minimizer is iteratively bounded within smaller intervals, omitting parts of the search space that have no further interest. In order to achieve theoretically sound results on covering methods, extensive information on the search space is required. Unfortunately, this requirement is rarely met in real world applications, while their performance usually becomes comparable with that of exhaustive search, rendering their applicability questionable (Törn & Žilinskas, 1989).

The category of *direct search* methods consists of algorithms heavily based on the fundamental computation element, namely function evaluation. *Generalized descent* methods additionally require first-order and second-order gradient information of the objective function to produce trajectories that intersect regions of attraction of the minimizers. For this reason, the combination of trajectory-based approaches, with random search schemes to provide their initial conditions, is preferable for the minimization of multimodal functions. This category also includes penalty function techniques, where the objective function is modified after the detection of a local minimizer, so that a new landscape is presented to the algorithm, where the detection of a lower minimum is possible.

Clustering approaches strive to address a crucial deficiency of the generalized descent techniques. More specifically, various iterative schemes may converge on the same minimizer, although initialized at different initial positions. This effect adds significant computational cost to the algorithm and reduces its performance. Clustering aims at initiating a single local search per detected local minimizer. This is possible by producing samples of points and clustering them in order to identify the neighborhoods of local minimizers. Then, a single local search can be conducted per cluster.

Random search algorithms are based on probability distributions to produce samples of search points. In pure random search, a new sample of points is generated at each iteration of the algorithm. Obviously, output accuracy increases as sample size approaches infinity. In practice, this approach is inefficient since it requires a vast number of function evaluations to produce acceptable results, even for problems of moderate dimensionality. For this reason, the performance of pure random search is considered as the worst acceptable performance for an algorithm. In improved variants, pure random search has been equipped with a local search procedure, applied either on the best or on several sampled points, giving rise to *singlestart* and *multistart* methods, respectively.

Significant efforts towards the improvement of random search methods have lead to the development of novel algorithms that draw inspiration from stochastic phenomena in nature. These algorithms are called *heuristics* or *metaheuristics* and they simulate fundamental elements and procedures that produce evolution and intelligent behaviors in natural systems. This category of algorithms is intimately related to modern research fields, such as *evolutionary computation* and *swarm intelligence*, and it will be henceforth in the center of our attention.

The last category, *indirect search*, consists of methods that build models of either the objective function or its level sets by exploiting local information. In the first case, Bayesian models are employed to approximate the objective function using random variables, while in the latter, polynomial approximations are used to fit its level sets. Although these approaches are very appealing theoretically, in practice there are several implementation problems that need to be addressed to render them applicable within an algorithmic framework (Törn & Žilinskas, 1989).

In the book at hand, we focus on variants of particle swarm optimization. These approaches belong rather to the category of direct search methods. However, regardless of different classifications, there is a general feeling that the most efficient global optimization methods combine features or algorithms of different types. This will also become apparent in the rest of the book. In the next section, we provide a brief flashback to the development of evolutionary computation, which constitutes the precursor of swarm intelligence and, consequently, of particle swarm optimization.

THE DEVELOPMENT OF EVOLUTIONARY COMPUTATION

The term *evolutionary computation* is used to describe a category of heuristic optimization methods that lie in the intersection of global optimization with computational intelligence. *Evolutionary algorithms* combine elements such as stochasticity, adaptation, and learning, in order to produce *intelligent optimization schemes*. Such schemes can adapt to their environment through evolution, even if it changes dynamically, while exploiting information from previous search steps.

The special structure of evolutionary algorithms is based on biological principles from the Darwinian theory of species. More specifically, they assume populations of potential solutions that evolve iteratively. Evolution occurs as the outcome of special operators on the population, such as *crossover* (also called *recombination*), *mutation*, and *selection*. All these operators work in direct analogy with their corresponding biological procedures. It is precisely these procedures that, according to Darwinian theory, promoted the biological evolution of humans and animals through natural selection.

The first applications of Darwinian principles on problem solving appeared in the mid-50's in the field of machine learning, and a few years later in optimization (Friedberg, 1958; Bremermann, 1962). However, it was not until the 90's that the term "evolutionary computation" was used to describe this new and promising scientific field. In the meantime, the three major evolutionary approaches, namely *evolutionary programming*, *genetic algorithms*, and *evolution strategies*, were developed in Europe and the USA. After the establishment of relevant annual meetings and journals, the evolutionary research community was able to exchange knowledge and present its latest developments to scientists and engineers from different disciplines. This has lead to a continuous growth of the field, as reflected in the number of published books, journals, and conferences organized all over the world.

In the following sections, we sketch the development of the three major evolutionary approaches over the last 50 years. Our aim is the exposition of basic principles and operations, also discussed later in the present chapter. A more detailed history of evolutionary algorithms is provided by De Jong *et al.* (1997).

Evolutionary Programming

Evolutionary programming was developed by L.J. Fogel in the mid-60's as a learning process for finite-state machines (Fogel, 1962; Fogel, 1964). More specifically, Fogel aimed at evolving an algorithm (a program) in an environment that consisted of symbols from a finite alphabet. The main goal was to produce an algorithm with traits of intelligence, i.e., the ability to predict its environment and trigger proper responses based on its predictions. This was possible by observing the sequence of passing symbols and producing an output that should predict the next symbol of the environment accurately.

A pay-off function was constructed to assess performance based on prediction quality. For this purpose, a prediction error (square root, absolute error etc.) was used. Thus, a population of finite-state machines was exposed to the environment, i.e., the symbol strings observed so far. For each input symbol, a predicted output symbol was produced by each machine of the population, and its prediction quality was assessed with the pay-off function. This procedure was repeated for all input symbols, and the average pay-off value per machine was assigned as its *fitness value*. Then, each machine was mutated by proper stochastic operations that changed one of its fundamental elements, such as states, state transitions, output symbols etc.

The mutated machines were exposed to the same environment and evaluated with the aforementioned procedure. Those with the highest fitness values were selected to comprise the population in the next iteration, also called *generation*, of the algorithm. This evolution procedure was repeated until a machine with exact predictions for all symbols was produced. Then, a new symbol was added to the known environment and the population was evolved anew. Detailed descriptions of these experiments, together with several applications, are provided by Fogel *et al.* (1964, 1965, 1966).

After the 80's, research on evolutionary programming matured and a plethora of different approaches and applications appeared in literature. The traveling salesman problem, neural networks training, scheduling, and continuous optimization, are just a fraction of the problems that were addressed by evolutionary programming (Fogel, 1988; Fogel & Fogel, 1988; Fogel *et al.*, 1990; McDonnell *et al.*, 1992), while studies on the adaptation of its parameters were also appearing (Fogel *et al.*, 1991). Today, there is such a huge amount of work on evolutionary programming and its relative field of *genetic programming*, that any attempt of a comprehensive presentation in the limited space of a book is condemned to failure. For further information, we refer the interested reader to specialized literature sources such as Eiben and Smith (1993), Fogel *et al.* (1966), Koza (1992), Langdon and Poli (2002).

Genetic Algorithms

Genetic algorithms were introduced by J.H. Holland in the mid-60's. By that time, Holland's research was focused on understanding adaptive systems capable of responding to environmental changes. Continuous competition and generation of offspring seemed to be the fundamental elements of long-term adaptive behaviors in natural systems. The idea of simulating these elements for designing artificial adaptive systems was very appealing. The core of Holland's ideas appeared in several dissertations of his students, establishing a new optimization methodology under the name of genetic algorithms (Bagley, 1967; Cavicchio, 1970; Hollstien, 1971). Holland also studied solution representation schemes and he developed a theoretical framework for the analysis of evolution in adaptive systems, producing a very strong theoretical result: the *schema theorem* (Goldberg, 1989; Holland, 1992). A further boost to genetic algorithms research was provided by the work of De Jong (1975), who produced convincing evidence that genetic algorithms could serve as an efficient optimization algorithm in difficult problems.

Today, genetic algorithms constitute one of the most popular heuristics for global optimization. They exploit populations of candidate solutions, similarly to the rest of evolutionary algorithms. The distinguishing characteristic of their essential variants is the binary representation of solutions, which requires a translation function between binary vectors and the actual variables of the problem. For example, in real-valued optimization problems, all real candidate solutions must be translated from and to binary vectors. In shape optimization problems, a proper representation between structures and binary vectors is required. The biological notation is retained in genetic algorithms; thus, the binary representation of a solution is called its *chromosome* or *genotype*, while the actual (e.g., real-valued) form is called its *phenotype*. Obviously, the size of chromosomes depends on the range of the variables and the required accuracy.

Evolution in genetic algorithms is achieved by three fundamental genetic operators: *selection*, *crossover*, and *mutation*. Based on a ranking scheme that promotes the best individuals of the population, i.e., those with the lowest function values, a number of *parents* are selected from the current population. The parents are stochastically combined and produce offspring that carry parts of their chromosomes. Then, a stochastic mutation procedure alters one or more digits in each chromosome, and the mutated offspring are evaluated. Finally, the best among all individuals (parents and offspring) are selected to comprise the population of the next generation. The algorithm is terminated as soon as a solution of desirable quality is found or the maximum available computational budget is exhausted.

The aforementioned procedures are subject to modifications, depending on the employed variant of the algorithm. For example, if real valued representation is used, the crossover operator can be defined as numerical (usually convex) combination of the real vectors. A more representative description of the main genetic operators is provided in a following section. The application domain of genetic algorithms is very wide, embracing almost all engineering disciplines. Detailed presentations of genetic algorithms as well as a plethora of references for further inquiry can be found in Falkenauer (1997), Goldberg (2002), Michalewicz (1999), Mitchell (1996), Vose (1999).

Evolution Strategies

Evolution strategies were developed in the 60's by Rechenberg (1965), Schwefel (1965) and Bienert (1967), for solving shape optimization engineering problems. Their main difference compared to previous evolutionary approaches was the lack of a crossover operator (only mutation and selection were used). In early implementations, discrete probability distributions such as the binomial one were used to mutate the shape of structures. Instead of a population, a single search point was used to produce one descendant per generation. In later implementations, continuous probability distributions (e.g., Gaussians) were also considered, especially for solving numerical optimization problems (Schwefel, 1965).

The parameters of the employed distributions have a severe impact on the quality of descendants produced by the mutation operator. For this purpose, evolution strategies incorporate a self-adaptation scheme for their parameters. More recent approaches also consider independent adaptation of each solution component, adding flexibility to the algorithm. These characteristics render evolution strategies a more appealing approach for solving numerical optimization problems than the evolutionary algorithms presented in previous sections (Beyer, 2001; Beyer & Schwefel, 2002; Hansen & Ostermeier, 2001; Schwefel, 1995).

Today, research in evolution strategies has been advanced mostly thanks to the work of several groups in Germany. The previously described scheme with one parent and one descendant is denoted as $(1+1)$-ES and uses only mutation. More generalized variants apply also recombination among parents, and they are denoted as:

$(\mu/\rho + \lambda)$-ES or $(\mu/\rho, \lambda)$-ES,

where μ stands for the total number of parents; $\rho \leq \mu$ is the number of parents selected to generate offspring; and λ is the number of offspring. The sign "+" stands for the *plus-strategy*, which selects the μ best individuals among both the μ parents and the λ offspring, to comprise the population in the next generation. On the other hand, the sign "," stands for the *comma-strategy*, which retains the μ best among the λ offspring in the next generation. Obviously, in this case the constraint, $\mu < \lambda$, must hold.

An interesting issue in recent implementations is the mutation of strategy parameters and their use to control the statistical properties of mutation for the actual variables. A relatively recent sophisticated approach, called *covariance matrix adaptation evolution strategy* (Hansen & Ostermeier, 2001; Ostermeier *et al.*, 1994), uses mutation distributions obtained through the adaptation of a covariance matrix, based on information from previous steps. This feature improves performance significantly, since it can capture the local shape of the objective function. This algorithm is typically denoted as $(\mu/\mu_I, \lambda)$-CMA. Further information on the theory and practice of evolution strategies can be found in Arnold (2002), Bäck (1996), Beyer (2001), Beyer and Schwefel (2002), Hansen and Ostermeier (2001).

FUNDAMENTAL EVOLUTIONARY OPERATIONS

A standard evolutionary algorithm can be described with the following steps:

```
Initialize population.
Evaluate population.
While (stopping condition not met)
   Apply selection.
   Apply crossover (recombination).
   Apply mutation.
   Evaluate generated individuals.
   Update population.
End While
```

Except from the evaluation procedure, which is typically based on the objective function, the rest of the procedures and operations constitute distinguishing features of an evolutionary algorithm. A plethora of different variants per operator have been developed to enhance efficiency in specific problem types. The form of these operators usually depends on solution representation. For example, mutation and recombination between individuals with binary representation differ significantly from their counterparts for real number representations. It is outside the scope of this book to present all available approaches. However, in the following paragraphs, we present the basic concepts, with brief discussions of their most common issues.

Population Initialization

Population initialization is the procedure where the first generation of the population is determined within the search space. Obviously, it is highly related to the overall available information on the problem at hand. For example, if there is prior knowledge of any special characteristics of the objective function implying a region that contains the global minimizer, then it would be reasonable to assign the whole or the largest part of the population into this region. On the other hand, if there is no such information, then treating each region of the search space equivalently would be the most appropriate choice.

Favoring or prohibiting regions of the search space without special reason can slow down convergence, in the best case, or get the algorithm trapped in local minima, in the worst. Furthermore, a necessary and sufficient condition for an algorithm to achieve convergence in optimization problems that lack favorable mathematical properties, is the ability to produce sequences of search points that are everywhere dense in the domain of the objective function, as stated in the following theorem.

Theorem (Zhigljavsky & Žilinskas, 2008, p. 13): Let the objective function, $f(x)$, be continuous in the neighborhood $D \subset \mathbf{R}^n$ of a global minimizer, x^*, and its feasible region, A, be compact. Then, a global minimization algorithm converges in the sense that, $f(x_n) \rightarrow f(x^*)$ as $n \rightarrow \infty$, if and only if it generates a sequence of points, x_i, which is everywhere dense in A.

The most common initialization procedure is the uniform dispersion of the population within the search space. Uniformity can be considered either deterministically or stochastically. In the first case, the population is initialized on the nodes of a grid that equidistantly covers the whole search space, while, in the latter and most common case, the population is produced through probabilistic sampling, following a uniform distribution over the search space. The final choice between the two approaches usually depends on the form of the specific search space.

More sophisticated initialization approaches employ deterministic algorithms and/or heuristics to provide the initial population. In this way, the local tendency of the objective function is revealed to the evolutionary algorithm in its first steps. Such an approach will be presented in a later chapter of the book at hand. According to a rule of thumb, the initialization scheme shall not be instrumental for the algorithm if there is no special information that justifies a biased assignment of the initial population.

Selection

Selection is the procedure of selecting individuals from the population to form a pool of parents that will be used to produce offspring through recombination. The criterion for selecting an individual can be either stochastic or deterministic. Nevertheless, it is always dependent on its function value. Thus, deterministic selection approaches directly select the best individuals, i.e., those with the lowest function values, while stochastic approaches perform a probabilistic selection, assigning higher probabilities to the best individuals. There are many different selection schemes reported in literature. We will describe two of the most common schemes, namely *tournament selection* and *roulette-wheel selection* (also called *fitness proportionate selection*).

Tournament selection consists of a series of tournaments among a randomly selected set of individuals, adding each time one of them into the parent pool. To put it more formally, let the population, P, consist of N individuals, m be a fixed integer from the set $\{2, 3,..., N\}$, and k be the number of parents to be selected. Then, tournament selection can be described with the following pseudocode:

```
Do (i = 1...k)
    Choose randomly m individuals from the population P.
    Select one among the m individuals.
    Add the selected individual into the parent pool.
End Do
```

Deterministic tournament always selects the best one among the m individuals, i.e., the one with the lowest function value, thereby promoting *elitism*. On the other hand, *stochastic tournament* uses function values to assign selection probabilities to individuals, and then performs a probabilistic selection among them. Thus, the overall best individual (among the m) has a selection probability, p, the second best, $p(1-p)$, the third best, $p(1-p)^2$, etc. The deterministic variant can be considered as a special case of the stochastic one, for $p = 1$. Also, the selected individual can be either replaced back into the population and probably reselected in the next tournament, or removed from the population.

Roulette-wheel selection associates each individual with a probability depending on its function value, similarly to stochastic tournament selection. More specifically, if we denote with f_i the function value of the i-th individual, then it is assigned a probability:

$$p_i = f_i \bigg/ \sum_{j=1}^{N} f_j$$

Assuming that, $p_0 = 0$, a random number, q, is generated from a uniform distribution in the range $[0,1]$, and the k-th individual is selected to join the parent pool, where $k \in \{1, 2,..., N\}$ is an index such that:

$$\sum_{i=0}^{k-1} p_i \leq q < \sum_{i=0}^{k} p_i \qquad (3)$$

This procedure resembles a roulette-wheel spin with each individual occupying a portion, p_i, of the wheel. Roulette-wheel selection can be described with the following pseudocode:

```
Assign to each individual a probability, p_i.
Do (i = 1...k)
    Generate a uniformly distributed random number in [0,1].
    Find index k such that relation (3) holds.
    Add the k-th individual to the parent pool.
End Do
```

Obviously, roulette-wheel selection offers a high chance of survival to less fit individuals. A thorough discussion and analysis of selection schemes can be found in Baker (1985), Blickle and Thiele (1995), Goldberg and Klösener (1991), Michalewicz (1999).

Crossover or Recombination

Crossover or *recombination* is the procedure of recombining the information carried by two individuals to produce new offspring. This is in direct analogy to the biological reproduction, where DNA sequences of parents are mixed to produce offspring DNA sequences that combine their genetic information. There are different forms of crossover schemes for different representations of the solutions. Since the binary representation of genetic algorithms closely resembles its natural counterpart, we will focus on it in the rest of this section.

Let, $p = \{p_1, p_2,..., p_n\}$ and $q = \{q_1, q_2,..., q_n\}$, be two n-dimensional binary parent vectors, selected randomly from the parent pool generated by the selection procedure. Then, a crossover point, $k \in \{1, 2,..., n-1\}$, is defined, and each parent is divided in two parts that are recombined to produce two offsprings, $o_1 = \{p_1, p_2,..., p_k, q_{k+1}, q_{k+2},..., q_n\}$ and $o_2 = \{q_1, q_2,..., q_k, p_{k+1}, p_{k+2},..., p_n\}$. If we denote with the symbols "$\otimes$" and "$\oplus$" a bit of information of the two parents, p and q, respectively (thus $\otimes$ and $\oplus$ can be either 0 or 1), then crossover can be represented schematically as follows:

parent p: $\otimes\otimes\otimes\otimes\otimes\otimes \mid \otimes\otimes\otimes\otimes$
parent q: $\oplus\oplus\oplus\oplus\oplus\oplus \mid \oplus\oplus\oplus\oplus$

- -

offspring o_1: $\otimes\otimes\otimes\otimes\otimes\otimes \mid \oplus\oplus\oplus\oplus$
offspring o_2: $\oplus\oplus\oplus\oplus\oplus\oplus \mid \otimes\otimes\otimes\otimes$

This procedure is also called *one-point crossover*, as it uses a single crossover point. Similarly, we can have *2-point crossover*, where two crossover points are used:

parent p: ⊗⊗⊗ | ⊗⊗⊗ | ⊗⊗⊗⊗
parent q: ⊕⊕⊕ | ⊕⊕⊕ | ⊕⊕⊕⊕

- -

offspring o_1: ⊗⊗⊗ | ⊕⊕⊕ | ⊗⊗⊗⊗
offspring o_2: ⊕⊕⊕ | ⊗⊗⊗ | ⊕⊕⊕⊕

In general, we can have an arbitrary number of crossover points, producing *multi-point crossover* schemes.

In all the aforementioned schemes, the dimension of the parents is inherited to their offspring. Thus, recombining *n*-dimensional parents will produce *n*-dimensional offspring, which is desirable in most numerical optimization problems. However, there are applications where dimensionality of candidate solutions may not be necessarily fixed. In such cases, the crossover point may differ between the two parents, producing offspring of different dimensionality:

parent p: ⊗⊗⊗ | ⊗⊗⊗⊗⊗⊗⊗
parent q: ⊕⊕⊕⊕⊕⊕ | ⊕⊕⊕⊕

- -

offspring o_1: ⊗⊗⊗ | ⊕⊕⊕⊕
offspring o_2: ⊕⊕⊕⊕⊕⊕ | ⊗⊗⊗⊗⊗⊗⊗

This approach is called *cut-and-splice crossover*. Another common scheme is *uniform crossover*, where each bit is independently compared between the two parents and switched with a probability equal to 0.5.

The presented crossover schemes can also be applied to real-valued cases by recombining the real components of parent vectors. However, arithmetic recombination schemes are usually preferred in such cases. According to these scheme, a real number, a, is randomly (and uniformly) drawn from the range $(0,1)$, and the two parent vectors, p and q, are recombined through a convex linear combination, producing two offspring:

offspring o_1: $o_1 = a\,p + (1-a)\,q$
offspring o_2: $o_2 = (1-a)\,p + a\,q$

Different weights can also be used for the two parents in the linear combination. Also, all the presented crossover and recombination schemes can use more than two parents and produce more than two offspring. Further information on crossover and recombination can be found in Liepins and Vose (1992), Michalewicz (1999), Rowe *et al.* (2002), Vose (1999).

Mutation

Mutation is a fundamental biological operation. It enables organisms to change one or more biological properties radically, in order to fit an environmental change or continue their evolution by producing offspring with higher chances of survival. In nature, mutation constitutes an abrupt change in the genotype of an organism, and it can be either inherited by parents to children or acquired by an organism itself.

A DNA mutation may result in the modification of small part(s) of the DNA sequence or rather big sections of a chromosome. Its effect on the organism depends heavily on the mutated genes. While mutations to less significant genes have small positive or negative effects, there are mutations that trigger radical changes in the behavior of several genes. These mutations alternate genes that control the activation of other genes. Thus, they have a crucial impact that can possibly affect the whole structure of the organism.

Modeling its biological counterpart, mutation in evolutionary algorithms constitutes a means for retaining diversity in the population, hindering individuals from becoming very similar; an undesired behavior that leads to search stagnation and entrapment to local minima. On the other hand, mutation strength shall be balanced enough to permit convergence. For this purpose, mutation is applied with a probability on each component of an individual. If crossover is used, mutation is usually applied on the generated offspring. Otherwise, it can be applied directly on the actual population.

In binary representations, random bit flips with a prespecified probability, called *mutation rate*, is the most common mutation scheme. Let the population consist of N individuals, $P = \{p_1, p_2, ..., p_N\}$, with $p_i = \{p_{i1}, p_{i2}, ..., p_{in}\}$, and $p_{ij} \in \{0,1\}$, for all $i = 1, 2,..., N$, and $j = 1, 2,..., n$. Also, let, $a \in [0,1]$, be a prespecified mutation rate. Then, mutation can be described with the following pseudocode:

```
Do (i = 1...N)
   Do (j = 1...n)
      Generate a uniformly distributed random number, r∈[0,1].
      If (r < a) Then Mutate the component p_ij.
      End If
   End Do
End Do
```

Although the standard mutation scheme in binary representations consists of the aforementioned simple bit flip, arithmetic mutation has also been used in several applications. According to this, binary arithmetic operations are applied on the component(s) of the mutated vector.

In real-valued representations, mutation can be defined as the replacement of a vector component with a random number distributed over its corresponding range. In numerical optimization problems, probabilistic schemes are frequently used. These schemes alternate an individual by adding a random vector drawn from a probability distribution. The Gaussian and uniform distributions are the most common choices in such mutations. Further information on mutation operators can be found in Bäck (1996), Goldberg (1989), Kjellström (1991), Michalewicz (1999), Mitchell (1996), Motwani and Raghavan (1995), Mühlenbein and Schlierkamp-Voosen (1995).

SWARM INTELLIGENCE

Swarm intelligence is a branch of artificial intelligence that studies the collective behavior and emergent properties of complex, self-organized, decentralized systems with social structure. Such systems consist of simple interacting agents organized in small societies (swarms). Although each agent has a very limited action space and there is no central control, the aggregated behavior of the whole swarm exhibits traits of intelligence, i.e., an ability to react to environmental changes and decision-making capacities.

The main inspiration behind the development of swarm intelligence stems directly from nature. Fish schools, bird flocks, ant colonies and animal herds, with their amazing self-organization capabilities and reactions, produce collective behaviors that cannot be described simply by aggregating the behavior of each team member. This observation has stimulated scientific curiosity regarding the underlying rules that produce these behaviors. The study of rules and procedures that promote intelligent behavior and pattern emergence through collaboration and competition among individuals gave rise to the fields of *collective intelligence* and *emergence*. Human teams have also been shown to share many of these properties, rendering collective intelligence and emergence, inter-disciplinary scientific fields that intersect, among others, with mathematics, sociology, computer science, and biology (Goldstein, 1999; Holland, 1998; Lévy, 1999; Szuba, 2001).

In the global optimization framework, swarm intelligence appeared in 1989 as a set of algorithms for controlling robotic swarms (Beni & Wang, 1989). Then, in six years, the three main swarm intelligence optimization algorithms, namely *ant colony optimization, stochastic diffusion search*, and *particle swarm optimization*, were developed. Although there are philosophical and operational differences between evolutionary and swarm intelligence algorithms, they were all categorized as evolutionary computation approaches in the mid-90's. This binding was made due to their inherent similarities, such as stochasticity, use of populations, types of application fields, as well as the scientific audience that was primarily interested on these approaches. Thus, swarm intelligence papers were mainly hosted in special sessions on evolutionary algorithms and published in international scientific journals under the topic of evolutionary computation.

A few years later, there was an exponential increase in the number of works related to swarm intelligence, which made its autonomous presentation indispensable. Today, there are specialized symposia and a large number of special sessions devoted to the latest developments in swarm intelligence. The increasing number of journal papers, as well as the establishment of new journals, reveal the extensive interest of the scientific community in swarm intelligence, whose fundamental algorithms are briefly described in the following sections.

Ant Colony Optimization

Dorigo (1992) introduced a stochastic optimization algorithm for combinatorial problems. The algorithm was inspired by the behavior of ants in search of food, and was named *ant colony optimization* (Bonabeau *et al.*, 1999; Dorigo & Stützle, 2004). Its workings were heavily based on the concept of *stigmergy*, i.e., the indirect stimulation of an agent's action by traces previously left in the environment by other agents. In natural ant colonies, stigmergy arises due to *pheromones* laid on the ground by the ants during their search for food. Initially, ants perform a random search around their nest for food. As soon as a food source is found, they bring food back to their nest, laying pheromones on the ground. The shortest path between the nest and the food source carries more pheromone than the rest paths, due to the more frequent passage of ants and pheromone evaporation. Thus, an ant starting its route from the nest will be strongly stimulated by higher pheromone levels and follow the shortest path to the food source, also laying its own pheromone on it. This is a simple yet efficient way of nature for solving shortest-path problems.

A very similar procedure is followed by the ant colony optimization algorithm to solve combinatorial optimization problems. Artificial ants start from an initial search point (nest) and build the components of a new potential solution one-by-one. For each component, a probabilistic decision is made among alternatives. Each alternative carries a pheromone level that determines its selection probability. The pheromone of each selected alternative is updated at the end of the tour, based on the quality of the obtained solution, so that components of solutions with low function values (shortest route length) are assigned higher pheromone levels. In order to avoid a rapid biasing on a suboptimal route, pheromone evaporation also takes place. A plethora of different approaches that promote elitism within the swarm or use special pheromone update schemes have been proposed in literature. Comprehensive presentations of the ant colony optimization algorithm can be found in Bonabeau *et al.* (1999), Dorigo and Stützle (2004).

Stochastic Diffusion Search

Stochastic diffusion search was initially introduced by Bishop (1989) as a search heuristic for pattern matching. Similarly to ant colony optimization, it uses populations of communicating individuals to perform stochastic search. However, it uses a special communication scheme, where information is exchanged through one-to-one direct communication between individuals. Candidate solutions are partially evaluated by each individual and the information is diffused to other individuals through the aforementioned communication scheme. Thus, promising solutions are associated with portions of the swarm that carry their (partial) information.

The special partial evaluation scheme of stochastic diffusion search has proved very useful in cases where the objective function is computationally expensive and decomposable, i.e., the complete evaluation can be done in several consecutive independent parts. Obviously, this property provides extensive synchronous and asynchronous parallelization capabilities to the algorithm.

The application field of stochastic diffusion search is wide, including, among others, text and object recognition (Bishop, 1989; Bishop & Torr, 1992), robotics (Beattie & Bishop, 1998), face recognition (Grech-Cini & McKee, 1993), and wireless networks (Whitaker & Hurley, 2002). Also, a sound mathematical framework has been developed for its theoretical analysis (Nasuto, 1999; Nasuto & Bishop, 1999, Myatt *et al.*, 2004).

Particle Swarm Optimization

Particle swarm optimization was developed by Kennedy and Eberhart (1995) as a stochastic optimization algorithm based on social simulation models. The algorithm employs a population of search points that moves stochastically in the search space. Concurrently, the best position ever attained by each individual, also called its *experience*, is retained in memory. This experience is then communicated to part or the whole population, biasing its movement towards the most promising regions detected so far. The communication scheme is determined by a fixed or adaptive social network that plays a crucial role on the convergence properties of the algorithm.

The development of particle swarm optimization was based on concepts and rules that govern socially organized populations in nature, such as bird flocks, fish schools, and animal herds. Unlike the ant colony approach, where stigmergy is the main communication mechanism among individuals through their environment, in such systems communication is rather direct without altering the environment.

The book at hand is completely devoted to the particle swarm optimization algorithm and its applications. In the following chapters, we will present its standard concepts and variants, as well as modifications that enhance its performance in detail, while also, discussing various applications. Thus, we postpone our further consideration of particle swarm optimization until the next chapter, which is completely devoted to its detailed description.

THE NO-FREE-LUNCH THEOREM

Wolpert and Macready (1997) developed one of the strongest theoretical results in optimization, namely the *no-free-lunch theorem*. Its main conclusion can be summarized with the statement:

Any two algorithms are equivalent in terms of their performance when it is averaged over all problem instances and metrics.

Alternatively, we can say that if an algorithm A outperforms algorithm B for some problems, then there is exactly the same number of problems where B outperforms A. The theorem holds for finite search spaces and algorithms that do not re-evaluate sampled points. Under these assumptions, and considering an algorithm as a mapping between an objective function (input) and its sequence of function evaluations on the sampled points (output), there is no free lunch if and only if the distribution of objective functions remains invariant under permutations of the space of sampled solutions (Igel & Toussaint, 2004; Wolpert & Macready, 1997).

There is an ongoing discussion regarding the applicability of the theorem in practice (Droste *et al.*, 2002). Although its main assumptions do not hold precisely in practice, they can be verified approximately. Nevertheless, even if it holds approximately, its main conclusion remains significant, yielding that the special characteristics of each problem are the key elements for selecting the most suitable algorithm. Therefore, effort shall be paid in order to reveal these elements; otherwise the choice of algorithm will most probably be suboptimal.

Under this prism, there is a strong merit in developing new algorithms that take full advantage of special characteristics of a given problem, or in fitting the established approaches to specific classes of problems. An extensive archive of developments on the no-free-lunch theorem is provided at the web site: http://www.no-free-lunch.org/.

CHAPTER SYNOPSIS

This chapter introduces the reader to the basic concepts of optimization and provides a brief presentation of the major developments in fields relevant to its main topic, namely the particle swarm optimization algorithm. Short introductions were provided for global optimization problems and algorithms. The main concepts and operations of evolutionary algorithms were briefly discussed along with the most common approaches, providing a background on nature-inspired computation. Swarm intelligence complemented these ideas by considering natural systems at a higher level (self-organization, communication) rather than their building blocks (genes, biological evolution). The final section of the chapter justifies the interest and necessity for further research on more efficient and specialized algorithms.

The chapter did not aim at covering all presented topics in detail. This task would require several books to even enumerate all results per topic. For this purpose, selected references were provided to stimulate the non-expert reader, providing a thread for further inquiry. The next chapter is devoted to the detailed presentation of the particle swarm optimization algorithm, from early precursors to its contemporary variants.

REFERENCES

Archetti, F., & Schoen, F. (1984). A survey on the global optimization problem: general theory and computational approaches. *Annals of Operations Research, 1*(1), 87–110. doi:10.1007/BF01876141

Arnold, D. V. (2002). *Noisy optimization with evolution strategies*. Berlin: Springer.

Bäck, T. (1996). *Evolutionary algorithms in theory and practice: Evolution strategies, evolutionary programming, genetic algorithms*. UK: Oxford University Press.

Bagley, J. D. (1967). *The behavior of adaptive systems with employ genetic and correlation algorithms*. Ph.D. thesis, University of Michigan, USA.

Baker, J. E. (1985). Adaptive selection methods for genetic algorithms. In *Proceedings of the 1ˢᵗ International Conference on Genetic Algorithms and their Applications, Pittsburgh (PA), USA* (pp. 101–111).

Beattie, P. D., & Bishop, J. M. (1998). Self-localisation in the "Senario" autonomous wheelchair. *Journal of Intelligent & Robotic Systems, 22*, 255–267. doi:10.1023/A:1008033229660

Beni, G., & Wang, J. (1989). Swarm intelligence in cellular robotic systems. In P. Dario, G. Sandini & P. Aebischer (Eds.), *Robotics and biological systems: Towards a new bionics, NATO ASI Series, Series F: Computer and System Science Vol. 102* (pp. 703–712).

Beyer, H.-G. (2001). *The theory of evolution strategies*. Berlin: Springer.

Beyer, H.-G., & Schwefel, H.-P. (2002). Evolution strategies: A comprehensive introduction. *Natural Computing, 1*(1), 3–52. doi:10.1023/A:1015059928466

Bienert, P. (1967). *Aufbau einer Optimierungsautomatik für drei Parameter*. Dipl.-Ing. thesis, Technical University of Berlin, Institute of Measurement and Control Technology, Germany.

Bishop, J. M. (1989). Stochastic searching network. In *Proceedings of the 1ˢᵗ IEE Conference on Artificial Neural Networks, London, UK* (pp. 329-331).

Bishop, J. M., & Torr, P. (1992). The stochastic search network. In R. Linggard, D.J. Myers, C. Nightingale (Eds.), *Neural networks for images, speech and natural language* (pp. 370-387). New York: Chapman & Hall.

Blickle, T., & Thiele, L. (1995). *A comparison of selection schemes used in genetic algorithms* (Tech. Rep. 11). Zürich, Switzerland: Swiss Federal Institute of Technology.

Bonabeau, E., Dorigo, M., & Theraulaz, G. (1999). *Swarm intelligence: From natural to artificial systems*. UK: Oxford University Press.

Boros, E., & Hammer, P. L. (Eds.). (2003). *Discrete optimization: The state of the art*. Amsterdam: Elsevier Science.

Branke, J. (2002). *Evolutionary optimization in dynamic environments*. Dordrecht, The Netherlands: Kluwer Academic Publishers.

Bremermann, H. J. (1962). Optimization through evolution and recombination. In M.C. Yovits, G.T. Jacobi, & G.D. Goldstein (Eds.), *Self-organizing systems 1962 (Proceedings of the conference on self-organizing systems, Chicago, Illinois)*. Washington, DC: Spartan Books.

Cavicchio, D. J. (1970). *Adaptive search using simulated evolution*. Ph.D. thesis, University of Michigan, USA.

Chiang, A. C. (1991). *Elements of dynamic optimization*. Prospect Heights, IL: Waveland Press.

De Jong, K., Fogel, D. B., & Schwefel, H.-P. (1997). A history of evolutionary computation. In T. Bäck, D.B. Fogel, & Z. Michalewicz (Eds.), *Handbook of evolutionary computation* (pp. A2.3:1-12). New York: IOP Press.

De Jong, K. A. (1975). *Analysis of behavior of a class of genetic adaptive systems*. Ph.D. thesis, University of Michigan, USA.

Deb, K. (2001). *Multi-objective optimization using evolutionary algorithms*. Chichester, UK: John Wiley & Sons.

Dixon, L. C. W., & Szegö, G. P. (1978). The global optimization problem: an introduction. In L.C.W Dixon & G.P. Szegö (Eds.), *Towards global optimization 2* (pp. 1-15). Amsterdam: North-Holland.

Dorigo, M. (1992). *Optimization, learning and natural algorithms*. Ph.D. thesis, Politecnico di Milano, Italy.

Dorigo, M., & Stützle, T. (2004). *Ant colony optimization*. Cambridge, MA: MIT Press.

Droste, S., Jansen, T., & Wegener, I. (2002). Optimization with randomized search heuristics: the (A)NFL theorem, realistic scenarios, and difficult functions. *Theoretical Computer Science, 287*(1), 131–144. doi:10.1016/S0304-3975(02)00094-4

Ehrgott, M., & Gandibleux, X. (Eds.). (2002). *Multiple criteria optimization: State of the art annotated bibliographic surveys*. Dordrecht, The Netherlands: Kluwer Academic Publishers.

Eiben, A. E., & Smith, J. E. (2003). *Introduction to evolutionary computing*. Berlin, Heidelberg: Springer.

Falkenauer, E. (1997). *Genetic algorithms and grouping problems*. Chichester, UK: John Wiley & Sons Ltd.

Floudas, C. A. (1995). *Nonlinear and mixed-integer optimization: Fundamentals and applications*. New York: Oxford University Press.

Fogel, D. B. (1988). An evolutionary approach to the travelling salesman problem. *Biological Cybernetics, 60*(2), 139–144. doi:10.1007/BF00202901

Fogel, D. B., & Fogel, L. J. (1988). Route optimization through evolutionary programming. In *Proceedings of the 22nd Asilomar Conference on Signals, Systems and Computers, Pacific Grove (CA), USA* (Vol. 2, pp. 679-680).

Fogel, D. B., Fogel, L. J., & Atmar, J. W. (1991). Meta-evolutionary programming. In R.R. Chen (Ed.), *Proceedings of the 25th asilomar conference on signals, systems and computers, Pacific Grove (CA), USA* (Vol. 1, pp. 540-545).

Fogel, D. B., Fogel, L. J., & Porto, V. W. (1990). Evolving neural networks. *Biological Cybernetics, 63*(6), 487–493. doi:10.1007/BF00199581

Fogel, L. J. (1962). Autonomous automata. *Industrial Research, 4*, 14–19.

Fogel, L. J. (1964). *On the organization of intellect*. Ph.D. thesis, University of California at Los Angeles, USA.

Fogel, L. J., Owens, A. J., & Walsh, M. J. (1964). On the evolution of artificial intelligence. In *Proceedings of the 5th National Symposium on Human Factors and Electronics, San Diego (CA) USA* (pp. 63-76).

Fogel, L. J., Owens, A. J., & Walsh, M. J. (1965). Artificial intelligence through a simulation of evolution. In A. Callahan, M. Maxfield, & L.J. Fogel (Eds.), *Biophysics and cybernetic systems* (pp. 131-156). Washington, DC: Spartan Books.

Fogel, L. J., Owens, A. J., & Walsh, M. J. (1966). *Artificial intelligence through simulated evolution*. New York: Wiley.

Friedberg, R. M. (1958). A learning machine: Part I. *IBM Journal, 2*, 2-13.

Goldberg, D. E. (1989). *Genetic algorithms in search, optimization and machine learning*. Reading, MA: Addison Wesley.

Goldberg, D. E. (2002). *The design of innovation: Lessons from and for competent genetic algorithms*. Reading, MA: Addison-Wesley.

Goldberg, D. E., & Klösener, K. H. (1991). A comparative analysis of selection schemes used in genetic algorithms. In G. Rawlins (Ed.), *Foundation of genetic algorithms* (pp. 69-93). San Mateo, CA: Kaufmann.

Goldstein, J. (1999). Emergence as a construct: History and issues. *Emergence: Complexity and Organization, 1*, 49–72.

Grech-Cini, H. J., & McKee, G. T. (1993). Locating the mouth region in images of human faces. In P.S. Schenker (Ed.), *Proceedings of SPIE - The International Society for Optical Engineering, Sensor Fusion, Boston (MA), USA* (Vol. 2059, pp. 458-465).

Hansen, N., & Ostermeier, A. (2001). Completely derandomized self-adaptation in evolution strategies. *Evolutionary Computation, 9* 1), 159-195.

Holland, J. H. (1992). *Adaptation in natural and artificial systems*. Cambridge, MA: MIT Press.

Holland, J. H. (1998). *Emergence from chaos to order*. New York: Perseus Books.

Hollstien, R. B. (1971). *Artificial genetic adaptation in computer control systems*. Ph.D. thesis, University of Michigan, USA.

Horst, R., & Pardalos, P. M. (Eds.). (1995). *Handbook of global optimization*. Dordrecht, The Netherlands: Kluwer Academic Publishers.

Horst, R., & Tuy, J. (2003). *Global optimization: deterministic approaches*. Berlin: Springer-Verlag.

Igel, C., & Toussaint, M. (2004). A no-free-lunch theorem for non-uniform distributions of target functions. *Journal of Mathematical Modelling and Algorithms, 3*, 313–322. doi:10.1023/B:JMMA.0000049381.24625.f7

Kennedy, J., & Eberhart, R. C. (1995). Particle swarm optimization. In *Proceedings of the IEEE International Conference on Neural Networks, Perth, Australia* (pp. 1942–1948).

Kjellström, G. (1991). On the efficiency of Gaussian adaptation. *Journal of Optimization Theory and Applications, 71*(3), 589–597. doi:10.1007/BF00941405

Koza, J. R. (1992). *Genetic programming: On the programming of computers by means of natural selection*. Cambridge, MA: MIT Press.

Langdon, W. B., & Poli, R. (2002). *Foundations of genetic programming*. Berlin: Springer.

Lévy, P. (1999). *Collective intelligence: Mankind's emerging world in cyberspace*. New York: Perseus Books.

Liepins, G., & Vose, M. (1992). Characterizing crossover in genetic algorithms. *Annals of Mathematics and Artificial Intelligence, 5*, 27–34. doi:10.1007/BF01530778

Luenberger, D. G. (1989). *Linear and nonlinear programming*. Reading, MA: Addison-Wesley.

McDonnell, J. R., Andersen, B. D., Page, W. C., & Pin, F. (1992). Mobile manipulator configuration optimization using evolutionary programming. In D.B. Fogel & W. Atmar (Eds.), *Proceedings of the 1st Annual Conference on Evolutionary Programming, La Jolla (CA), USA* (pp. 52-62).

Michalewicz, Z. (1999). *Genetic algorithms + data structures = evolution programs*. Berlin: Springer.

Mitchell, M. (1996). *An introduction to genetic algorithms*. Cambridge, MA: MIT Press.

Motwani, R., & Raghavan, P. (1995). *Randomized algorithms*. MA: Cambridge University Press.

Mühlenbein, H., & Schlierkamp-Voosen, D. (1995). Analysis of selection, mutation and recombination in genetic algorithms. In W. Banzhaf & F.H. Eeckman (Eds.), *Evolution as a computational process, Lecture Notes in Computer Science, Vol. 899* (pp. 142-168). Berlin: Springer.

Myatt, D. M., Bishop, J. M., & Nasuto, S. J. (2004). Minimum stable convergence criteria for Stochastic Diffusion Search. *Electronics Letters, 40*(2), 112–113. doi:10.1049/el:20040096

Nasuto, S. J. (1999). *Analysis of resource allocation of stochastic diffusion search*. Ph.D. thesis, University of Reading, UK.

Nasuto, S. J., & Bishop, J. M. (1999). Convergence analysis of stochastic diffusion search. *Journal of Parallel Algorithms and Applications, 14*(2), 89–107.

Nocedal, J., & Wright, S. J. (2006). *Numerical optimization.* New York: Springer.

Ostermeier, A., Gawelczyk, A., & Hansen, N. (1994). A derandomized approach to self-adaptation of evolution strategies. *Evolutionary Computation, 2*(4), 369–380. doi:10.1162/evco.1994.2.4.369

Polak, E. (1997). *Optimization: algorithms and consistent approximations.* New York: Springer.

Rechenberg, I. (1965). *Cybernetic solution path of an experimental problem.* Royal Aircraft Establishment Library, Translation 1122.

Rowe, J. E., Vose, M. D., & Wright, A. H. (2002). Group properties of crossover and mutation. *Evolutionary Computation, 10*(2), 151–184. doi:10.1162/106365602320169839

Schwefel, H.-P. (1965). *Kybernetische Evolution als Strategie der experimentellen Forschung in der Strömungstechnik.* Dipl.-Ing. thesis, Technical University of Berlin, Hermann Föttinger Institute for Hydrodynamics, Germany.

Schwefel, H.-P. (1995). *Evolution and optimum seeking.* New York: Wiley & Sons.

Spall, J. C. (2003), *Introduction to Stochastic Search and Optimization.* Hoboken, NJ: John Wiley & Sons Ltd.

Szuba, T. (2001). *Computational collective intelligence.* New York: Wiley & Sons.

Torn, A., & Žilinskas, A. (1989). *Global optimization.* Berlin: Springer.

Vose, M. D. (1999). *The simple genetic algorithm: Foundations and theory.* Cambridge, MA: MIT Press.

Whitaker, R. M., & Hurley, S. (2002). An agent based approach to site selection for wireless networks. In *Proceedings of the 2002 ACM Symposium on Applied Computing, Madrid, Spain* (pp. 574-577).

Wolpert, D. H., & Macready, W. G. (1997). No free lunch theorems for optimization. *IEEE Transactions on Evolutionary Computation, 1*(1), 67–82. doi:10.1109/4235.585893

Zhigljavsky, A., & Žilinskas, A. (2008). *Stochastic global optimization.* New York: Springer.

Chapter 2
Particle Swarm Optimization

This chapter is devoted to particle swarm optimization (PSO), from early precursors to contemporary standard variants. The presentation begins with the main inspiration source behind its development, followed by early variants and discussion on their parameters. Severe deficiencies of early variants are also pointed out and their solutions are reported in a relative historical order, bringing the reader to contemporary developments, considered as the state-of-the-art PSO variants today.

MAIN INSPIRATION SOURCE

Bird flocks, fish schools, and animal herds constitute representative examples of natural systems where aggregated behaviors are met, producing impressive, collision-free, synchronized moves. In such systems, the behavior of each group member is based on simple inherent responses, although their outcome is rather complex from a macroscopic point of view. For example, the flight of a bird flock can be simulated with relative accuracy by simply maintaining a target distance between each bird and its immediate neighbors. This distance may depend on its size and desirable behavior. For instance, fish retain a greater mutual distance when swimming carefree, while they concentrate in very dense groups in the presence of predators. The groups can also react to external threats by rapidly changing their form, breaking in smaller parts and re-uniting, demonstrating a remarkable ability to respond collectively to external stimuli in order to preserve personal integrity.

Similar phenomena are observed in physical systems. A typical example is the particle aggregation caused by direct attraction between particles due to Brownian motion or fluid shear. Humans too are characterized by agnate behaviors, especially at the level of social organization and belief formulation. However, these interactions can become very complex, especially in the belief space, where, in contrast

DOI: 10.4018/978-1-61520-666-7.ch002

Copyright © 2010, IGI Global. Copying or distributing in print or electronic forms without written permission of IGI Global is prohibited.

to the physical space, the same point (a belief or an idea) can be occupied concurrently by large groups of people without collisions. The aforementioned aggregating behaviors, characterized by the simplicity of animal and physical systems or the abstractness of human social behavior, intrigued researchers and motivated their further investigation through extensive experimentation and simulations (Heppner & Grenander, 1990; Reynolds, 1987; Wilson, 1975).

Intense research in systems where collective phenomena are met prepared the ground for the development of swarm intelligence, briefly described in the previous chapter. Notwithstanding their physical or structural differences, such systems share common properties, recognized as the five basic principles of swarm intelligence (Millonas, 1994):

1. **Proximity:** Ability to perform space and time computations.
2. **Quality:** Ability to respond to environmental quality factors.
3. **Diverse response:** Ability to produce a plurality of different responses.
4. **Stability:** Ability to retain robust behaviors under mild environmental changes.
5. **Adaptability:** Ability to change behavior when it is dictated by external factors.

Moreover, the social sharing of information among individuals in a population can provide an evolutionary advantage. This general belief, which was suggested in several studies and supported by numerous examples from nature, constituted the core idea behind the development of PSO.

EARLY VARIANTS OF PSO

The early precursors of PSO were simulators of social behavior for visualizing bird flocks. Nearest-neighbor velocity matching and acceleration by distance were the main rules employed to produce swarming behavior by simple agents in their search for food, in simulation experiments conducted by Russell C. Eberhart (Purdue School of Engineering and Technology, Indiana University Purdue University Indianapolis) and James Kennedy (Bureau of Labor Statistics, Washington, DC). After realizing the potential of these simulation models to perform optimization, Eberhart and Kennedy refined their model and published the first version of PSO in 1995 (Eberhart & Kennedy, 1995; Kennedy & Eberhart, 1995).

Putting it in a mathematical framework, let, $A \subset \mathbf{R}^n$, be the search space, and, $f{:}A{\to}Y \subseteq \mathbf{R}$, be the objective function. In order to keep descriptions as simple as possible, we assume that A is also the feasible space of the problem at hand, i.e., there are no further explicit constraints posed on the candidate solutions. Also, note that no additional assumptions are required regarding the form of the objective function and search space. As mentioned in the previous chapter, PSO is a population-based algorithm, i.e., it exploits a population of potential solutions to probe the search space concurrently. The population is called the *swarm* and its individuals are called the *particles*; a notation retained by nomenclature used for similar models in social sciences and particle physics. The swarm is defined as a set:

$S = \{x_1, x_2, \ldots, x_N\},$

of N particles (candidate solutions), defined as:

$$x_i = (x_{i1}, x_{i2}, \ldots, x_{in})^{\mathrm{T}} \in A, \quad i = 1, 2, \ldots, N.$$

Indices are arbitrarily assigned to particles, while N is a user-defined parameter of the algorithm. The objective function, $f(x)$, is assumed to be available for all points in A. Thus, each particle has a unique function value, $f_i = f(x_i) \in Y$.

The particles are assumed to move within the search space, A, iteratively. This is possible by adjusting their *position* using a proper position shift, called *velocity*, and denoted as:

$$v_i = (v_{i1}, v_{i2}, \ldots, v_{in})^{\mathrm{T}}, \quad i = 1, 2, \ldots, N.$$

Velocity is also adapted iteratively to render particles capable of potentially visiting any region of A. If t denotes the iteration counter, then the current position of the i-th particle and its velocity will be henceforth denoted as $x_i(t)$ and $v_i(t)$, respectively.

Velocity is updated based on information obtained in previous steps of the algorithm. This is implemented in terms of a memory, where each particle can store the *best position* it has ever visited during its search. For this purpose, besides the swarm, S, which contains the current positions of the particles, PSO maintains also a *memory* set:

$$P = \{p_1, p_2, \ldots, p_N\},$$

which contains the best positions:

$$p_i = (p_{i1}, p_{i2}, \ldots, p_{in})^{\mathrm{T}} \in A, \quad i = 1, 2, \ldots, N,$$

ever visited by each particle. These positions are defined as:

$$p_i(t) = \arg\min_t f_i(t),$$

where t stands for the iteration counter.

PSO is based on simulation models of social behavior; thus, an information exchange mechanism shall exist to allow particles to mutually communicate their experience. The algorithm approximates the global minimizer with the best position ever visited by all particles. Therefore, it is a reasonable choice to share this crucial information. Let g be the index of the best position with the lowest function value in P at a given iteration t, i.e.,

$$p_g(t) = \arg\min_i f(p_i(t)).$$

Then, the early version of PSO is defined by the following equations (Eberhart & Kennedy, 1995; Eberhart *et al.*, 1996; Kennedy & Eberhart, 1995):

$$v_{ij}(t+1) = v_{ij}(t) + c_1 R_1 \, (p_{ij}(t) - x_{ij}(t)) + c_2 R_2 \, (p_{gj}(t) - x_{ij}(t)), \tag{1}$$

$$x_{ij}(t+1) = x_{ij}(t) + v_{ij}(t+1), \tag{2}$$

Table 1. Pseudocode of the operation of PSO

Input:	Number of particles, *N*; swarm, *S*; best positions, *P*.
Step 1.	**Set** $t \leftarrow 0$.
Step 2.	**Initialize** *S* and **Set** $P \equiv S$.
Step 3.	**Evaluate** *S* and *P*, and define index *g* of the best position.
Step 4.	**While** (termination criterion not met)
Step 5.	**Update** *S* using equations (1) and (2).
Step 6.	**Evaluate** *S*.
Step 7.	**Update** *P* and redefine index *g*.
Step 8.	**Set** $t \leftarrow t+1$.
Step 9.	**End While**
Step 10.	**Print** best position found.

$$i = 1, 2, \ldots, N, \quad j = 1, 2, \ldots, n,$$

where *t* denotes the iteration counter; R_1 and R_2 are random variables uniformly distributed within [0,1]; and c_1, c_2, are weighting factors, also called the *cognitive* and *social* parameter, respectively. In the first version of PSO, a single weight, $c = c_1 = c_2$, called *acceleration constant*, was used instead of the two distinct weights in equation (1). However, the latter offered better control on the algorithm, leading to its predominance over the first version.

At each iteration, after the update and evaluation of particles, best positions (memory) are also updated. Thus, the new best position of x_i at iteration *t*+1 is defined as follows:

$$p_i(t+1) = \begin{cases} x_i(t+1), & \text{if } f(x_i(t+1)) \leq f(p_i(t)), \\ p_i(t), & \text{otherwise.} \end{cases}$$

The new determination of index *g* for the updated best positions completes an iteration of PSO.

The operation of PSO is provided in pseudocode in Table 1. Particles are usually initialized randomly, following a uniform distribution over the search space, *A*. This choice treats each region of *A* equivalently; therefore it is mostly preferable in cases where there is no information on the form of the search space or the objective function, requiring a different initialization scheme. Additionally, it is implemented fairly easily, as all modern computer systems can be equipped with a uniform random number generator.

The previous velocity term, $v_{ij}(t)$, in the right-hand side of equation (1), offers a means of inertial movement to the particle by taking its previous position shift into consideration. This property can prevent it from becoming biased towards the involved best positions, which could entrap it to local minima if suboptimal information is carried by both (e.g., if they both lie in the vicinity of a local minimizer). Furthermore, the previous velocity term serves as a perturbation for the global best particle, x_g. Indeed, if a particle, x_i, discovers a new position with lower function value than the best one, then it becomes the global best (i.e., $g \leftarrow i$) and its best position, p_i, will coincide with p_g and x_i in the next iteration. Thus, the two stochastic terms in equation (1) will vanish. If there was no previous velocity term in equation (1), then the aforementioned particle would stay at the same position for several iterations, until a new

Figure 1. Candidate new positions of the particle $x_i = (0,0)^T$ (cross) with $p_i = (2,1)^T$ (star) and $p_g = (1,3)^T$ (square), for the cases (A) $c_1 = c_2 = 1.0$, and (B) $c_1 = c_2 = 2.0$

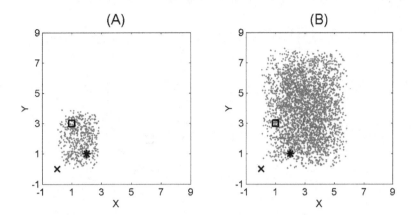

best position is detected by another particle. Contrary to this, the velocity term allows this particle to continue its search, following its previous position shift.

The values of c_1 and c_2 can affect the search ability of PSO by biasing the sampled new positions of a particle, x_i, towards the best positions, p_i and p_g, respectively, as well as by changing the magnitude of search. For example, consider the two cases illustrated in Fig. 1. Let $x_i = (0,0)^T$, denoted with the cross symbol, be the current position of a particle. Also, let, $p_i = (2,1)^T$ and $p_g = (1,3)^T$, be its own best and overall best position, denoted with a star and a square symbol, respectively. Moreover, for simplicity, let its current velocity, v_i, be equal to zero. Then, Fig. 1 represents 1000 possible new positions of x_i for $c_1 = c_2 = 1.0$ (left part) and $c_1 = c_2 = 2.0$ (right part).

Apparently, the magnitude of search differs significantly in the two cases. If a better global exploration is required, then high values of c_1 and c_2 can provide new points in relatively distant regions of the search space. On the other hand, a more refined local search around the best positions achieved so far would require the selection of smaller values for the two parameters. Also, choosing, $c_1 > c_2$, would bias sampling towards the direction of p_i, while in the opposite case, $c_1 < c_2$, sampling towards the direction of p_g would be favored. This effect can be useful in cases where there is special information regarding the form of the objective function. For instance, in convex unimodal objective functions, a choice that promotes sampling closer to p_g is expected to be more efficient, if combined with a proper search magnitude.

We must also notice that PSO operates on each coordinate direction independently. At this point, we shall mention a typical mistake made by several researchers, especially when PSO equations were considered in their vectorial form:

$$v_i(t+1) = v_i(t) + c_1 R_1 \left(p_i(t) - x_i(t)\right) + c_2 R_2 \left(p_g(t) - x_i(t)\right), \tag{3}$$

$$x_i(t+1) = x_i(t) + v_i(t+1), \tag{4}$$

$$i = 1, 2, \ldots, N,$$

Figure 2. Candidate new positions generated by equation (3) for the particle $x_i = (0,0)^T$ (cross), with $p_i = (2,1)^T$ (star), $p_g = (1,3)^T$ (square), and $c_1 = c_2 = 2.0$, when (A) R_1 and R_2 are random n-dimensional vectors, and (B) R_1 and R_2 are random one-dimensional values

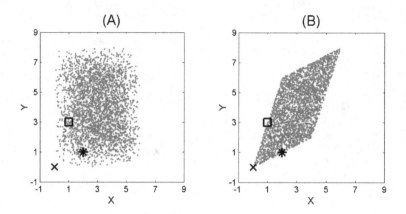

where all vector operations are performed componentwise. In this case, R_1 and R_2 should be considered as random *n*-dimensional vectors with their components uniformly distributed within [0,1]. Instead, R_1 and R_2 were often considered as random one-dimensional values, similarly to equation (1), resulting in a scheme that uses the same random number for all direction components of the corresponding difference vector in equation (3). The effect of this choice is illustrated in Fig. 2, where 1000 possible new positions of the same particle, x_i, as in Fig. 1, are generated with the (correct) configuration of equation (1) (left part) and the (wrong) configuration of equation (3) with random values used instead of random vectors (right part). Obviously, the latter case restricts sampling within a parallelepiped region between the two best positions, p_i and p_g.

In most optimization applications, it is desirable or inevitable to consider only particles lying within the search (feasible) space. For this purpose, bounds are imposed on the position of each particle, x_i, to restrict it within the search space, A. If a particle assumes an undesirable step out of the search space after the application of equation (2), it is immediately clamped at its boundary. In the simplest case, where the search space can be defined as a box:

$$A = [a_1, b_1] \times [a_2, b_2] \times \cdots \times [a_n, b_n], \tag{5}$$

with $a_i, b_i \in \mathbf{R}$, $i = 1, 2, \ldots, n$, the particles are restricted as follows:

$$x_{ij}(t+1) = \begin{cases} a_j, & \text{if } x_{ij}(t+1) < a_j, \\ b_j, & \text{if } x_{ij}(t+1) > b_j, \end{cases}$$

$i = 1, 2, \ldots, N, \quad j = 1, 2, \ldots, n.$

Alternatively, a bouncing movement off the boundary back into the search space has been considered, similarly to a ball bounced off a wall (Kao *et al.*, 2007). The popularity of this approach is limited, since it requires the modeling of particle motion with complex physical equations. In cases where A cannot be defined as a box, special problem-dependent conditions may be necessary to restrict particles.

FURTHER REFINEMENT OF PSO

Early PSO variants performed satisfactorily for simple optimization problems. However, their crucial deficiencies were revealed as soon as they were applied on harder problems with large search spaces and a multitude of local minima. In the following paragraphs, refinements developed to address deficiencies of the original PSO model are reported and discussed.

Swarm Explosion and Velocity Clamping

The first significant issue, verified by several researchers, was the *swarm explosion* effect. It refers to the uncontrolled increase of magnitude of the velocities, resulting in swarm divergence. This deficiency is rooted in the lack of a mechanism for constricting velocities in early PSO variants, and it was straightforwardly addressed by using strict bounds for *velocity clamping* at desirable levels, preventing particles from taking extremely large steps from their current position.

More specifically, a user-defined maximum velocity threshold, $v_{max} > 0$, is considered. After determining the new velocity of each particle with equation (1), the following restrictions are applied prior to the position update with equation (2):

$$|v_{ij}(t+1)| \leq v_{max}, \quad i = 1, 2, \ldots, N, \quad j = 1, 2, \ldots, n. \tag{6}$$

In case of violation, the corresponding velocity component is set directly to the closest velocity bound, i.e.,

$$v_{ij}(t+1) = \begin{cases} v_{max}, & \text{if } v_{ij}(t+1) > v_{max}, \\ -v_{max}, & \text{if } v_{ij}(t+1) < -v_{max}. \end{cases}$$

If necessary, different velocity bounds per direction component can be used. The value of v_{max} is usually taken as a fraction of the search space size per direction. Thus, if the search space is defined as in equation (5), a common maximum velocity for all direction components can be defined as follows:

$$v_{max} = \frac{\min_{i}\{b_i - a_i\}}{k}.$$

Alternatively, separate maximum velocity thresholds per component can be defined as:

$$v_{max,i} = \frac{b_i - a_i}{k}, \quad i = 1, 2, \ldots, n,$$

Figure 3. Swarm diversity during search for k = 2 (dotted line) and k = 10 (solid line). The plots pertain to a swarm of 20 particles with $c_1 = c_2 = 2$, minimizing the 2-dimensional instance of test problem TP_{UO-1}, defined in Appendix A of the book at hand, in the range $[-100,100]^2$ for 500 iterations

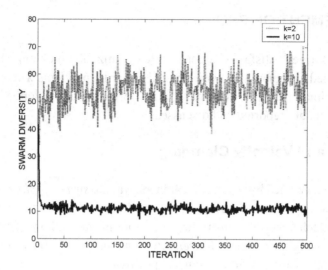

with $k = 2$ being a common choice. Of course, if the problem at hand requires smaller particle steps, then higher values of k shall be used. For example, if the search space has a multitude of minimizers with narrow regions of attraction close to each other, then k shall assume adequately large values to prevent particles from overflying them. On the other hand, k shall not take very small or large values that encumber a satisfactory search progress.

Figure 3 illustrates the impact of a large and a small value of k on swarm diversity, which is defined as the mean of the standard deviations of particles per coordinate direction. The cases for $k = 2$ and $k = 10$ are illustrated for a swarm of 20 particles with $c_1 = c_2 = 2$, minimizing the 2-dimensional objective function, $f(x) = x^\mathrm{T}x$, in the range $[-100,100]^2$ for 500 iterations. Evidently, the value $k = 10$ corresponds to smaller diversity with mild fluctuations, in contrast to $k = 2$, where diversity is almost five times larger, with wide fluctuations, obviously due to the larger position shifts assumed by the particles.

Velocity clamping offered a simple yet efficient solution to the problem of swarm explosion. However, it did not address the problem of convergence. The particles were now able to fluctuate around their best positions, but they were unable either to achieve convergence on a promising position or perform a refined search around it. This problem was addressed by the introduction of a new parameter in the original PSO model, as described in the next section.

The Concept of Inertia Weight

Although the use of a maximum velocity threshold improved the performance of early PSO variants, it was not adequate to render the algorithm efficient in complex optimization problems. Despite the alleviation of swarm explosion, the swarm was not able to concentrate its particles around the most promising solutions in the last phase of the optimization procedure. Thus, even if a promising region of the search space was roughly detected, no further refinement was made, with the particles instead oscillating on wide trajectories around their best positions.

The reason for this deficiency was shown to be a disability to control velocities. Refined search in promising regions, i.e., around the best positions, requires strong attraction of the particles towards them, and small position shifts that prohibit escape from their close vicinity. This is possible by reducing the perturbations that shift particles away from best positions; an effect attributed to the previous velocity term in equation (1). Therefore, the effect of the previous velocity on the current one shall fade for each particle. For this purpose, a new parameter, w, called *inertia weight*, was introduced in equation (1), resulting in a new PSO variant (Eberhart & Shi, 1998; Shi & Eberhart, 1998a; Shi & Eberhart, 1998b):

$$v_{ij}(t+1) = wv_{ij}(t) + c_1R_1\ (p_{ij}(t)\text{-}x_{ij}(t)) + c_2R_2\ (p_{gj}(t)\text{-}x_{ij}(t)), \tag{7}$$

$$x_{ij}(t+1) = x_{ij}(t) + v_{ij}(t+1), \tag{8}$$

$$i = 1, 2,\ldots, N, \quad j = 1, 2,\ldots, n.$$

The rest of the parameters remain the same as for the early PSO variant of equations (1) and (2). The inertia weight shall be selected such that the effect of $v_{ij}(t)$ fades during the execution of the algorithm. Thus, a decreasing value of w with time is preferable. A very common choice is the initialization of w to a value slightly greater than 1.0 (e.g., 1.2) to promote exploration in early optimization stages, and a linear decrease towards zero to eliminate oscillatory behaviors in later stages. Usually, a strictly positive lower bound on w (e.g., 0.1) is used to prevent the previous velocity term from vanishing.

In general, a linearly decreasing scheme for w can be mathematically described as follows:

$$w(t) = w_{\text{up}} - (w_{up} - w_{low})\frac{t}{T_{\text{max}}}, \tag{9}$$

where t stands for the iteration counter; w_{low} and w_{up} are the desirable lower and upper bounds of w; and T_{max} is the total allowed number of iterations. Equation (9) produces a linearly decreasing time-dependent inertia weight with starting value, w_{up}, at iteration, $t = 0$, and final value, w_{low}, at the last iteration, $t = T_{\text{max}}$.

Figure 4 illustrates diversity for a swarm of 20 particles updated with equations (7) and (8), with $c_1 = c_2 = 2$, minimizing the 2-dimensional instance of test problem $\text{TP}_{\text{UO-1}}$, defined in Appendix A of the book at hand, in the range $[-100,100]^2$, with (solid line) and without (dotted line) a decreasing inertia weight. As in Fig. 3, diversity is defined as the mean value of the standard deviations of particles per coordinate direction. Obviously, the use of inertia weight has a tremendous effect on swarm diversity, which almost vanishes after 300 iterations, in contrast to the case of simple velocity clamping, which retains almost the same diversity levels throughout the search.

In addition, we observe that, in the case of inertia weight, there is an increase in swarm diversity for the first almost 100 iterations. This effect can be attributed to the initial value, $w_{\text{up}} = 1.2$, of the inertia weight. Since this value is greater than 1.0, the previous velocity term has a greater impact in equation (7) than in equation (1). This results in a temporary swarm explosion that enhances the exploration capabilities even of a poorly initialized swarm. After almost 90 iterations, the inertia weight assumes values smaller than 1.0 and diversity starts declining towards zero, thereby promoting exploitation. As

Figure 4. Swarm diversity during search, with (solid line) and without (dotted line) inertia weight. The plots pertain to a swarm of 20 particles with $c_1 = c_2 = 2$, minimizing the 2-dimensional instance of test problem TP_{UO-1}, defined in Appendix A of the book at hand, in the range $[-100,100]^2$ for 500 iterations and $v_{max} = 50$. The inertia weight decreases linearly from 1.2 to 0.1

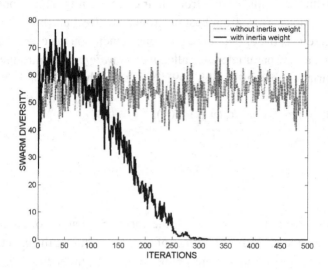

discussed in the next chapter, different schemes for inertia weight adaptation can be used to produce different behaviors of the algorithm.

The amazing performance improvement gained by using the inertia weight PSO variant with velocity clamping, rendered it the most popular PSO approach for a few years. However, although particles were able to avoid swarm explosion and converge around the best positions, they were still getting trapped easily in local minima, especially in complex problems. This deficiency was addressed by introducing a more sophisticated information-sharing scheme among particles, as described in the following section.

The Concept of Neighborhood

The use of inertia weight equipped PSO with convergence capabilities, as described in the previous section. However, it did not suffice to increase its efficiency to the most satisfactory levels in complex, multimodal environments. As depicted in Fig. 4, after a number of iterations, the swarm collapses due to complete diversity loss. This implies that further exploration is not possible and the particles can perform only local search around their convergence point, which most possibly lies in the vicinity of the overall best position. Although the effect of fast convergence can be mild in simple optimization problems, especially in unimodal and convex functions, it becomes detrimental in high-dimensional, complex environments.

This deficiency can be attributed to the global information exchange scheme that allows each particle to know instantly the overall best position at each iteration. Using this scheme, all particles assume new positions in regions related to the same overall best position, reducing the exploration capabilities of the swarm.

The aforementioned problem was addressed by introducing the concept of *neighborhood*. The main idea was the reduction of the global information exchange scheme to a local one, where information is diffused only in small parts of the swarm at each iteration. More precisely, each particle assumes a set of other particles to be its *neighbors* and, at each iteration, it communicates its best position only to these particles, instead of to the whole swarm. Thus, information regarding the overall best position is initially communicated only to the neighborhood of the best particle, and successively to the rest through their neighbors.

To put it formally, let x_i be the *i*-th particle of a swarm, $S = \{x_1, x_2, \ldots, x_N\}$. Then, a neighborhood of x_i is defined as a set:

$$NB_i = \{x_{n_1}, x_{n_2}, \ldots, x_{n_s}\},$$

where, $\{n_1, n_2, \ldots, n_s\} \subseteq \{1, 2, \ldots, N\}$, is the set of indices of its neighbors. The cardinality, $|NB_i|$, of this set is called *neighborhood size*. If g_i denotes the index of the best particle in NB_i, i.e.,

$$p_{g_i} = \underset{j \text{ such that } x_j \in NB_i}{\arg\min} f(p_j),$$

then equations (7) and (8) are modified as follows:

$$v_{ij}(t+1) = wv_{ij}(t) + c_1 R_1 (p_{ij}(t) - x_{ij}(t)) + c_2 R_2 (p_{g_i,j}(t) - x_{ij}(t)), \tag{10}$$

$$x_{ij}(t+1) = x_{ij}(t) + v_{ij}(t+1), \tag{11}$$

$$i = 1, 2, \ldots, N, \quad j = 1, 2, \ldots, n.$$

The only difference between equations (7) and (10) is the index of the best position in the second parenthesis. According to equation (10), the particle will move towards its own best position as well as the best position of its neighborhood, instead of the overall best position in equation (7).

The scheme for determining the neighbors of each particle is called *neighborhood topology*. A common scheme that comes to mind is the formation of neighborhoods based on the actual distances of the particles in the search space. According to this, each particle would be assigned a neighborhood consisting of a number, *s*, of the particles that lie closer to its current position. Although simple, this scheme requires the computation of $N(N+1)/2$ distances between particles at each iteration, given that the search space is equipped with a proper metric. The computational burden for this approach can become prohibitive when a large number of particles is used. Moreover, it exhibits a general trend of forming particle clusters that can be easily trapped in local minima. For these reasons, the specific neighborhood topology was not established as the most promising solution.

The idea of forming neighborhoods based on arbitrary criteria was promoted in order to alleviate the particle clustering effects produced by distance-based neighborhood topologies. The simplest and directly applicable alternative was the formation of neighborhoods based on particle indices. According to this, the *i*-th particle assumes neighbors with neighboring indices. Thus, the neighborhood of x_i can be defined as:

$$NB_i = \{x_{i-r}, x_{i-r+1}, \ldots, x_{i-1}, x_i, x_{i+1}, \ldots, x_{i+r-1}, x_{i+r}\},$$

Figure 5. Common neighborhood topologies of PSO: ring (left) and star (right)

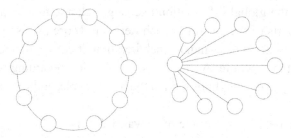

as if the particles were lying on a ring and each one was connected only to its immediate neighbors. This scheme is illustrated in Fig. 5 (left) and it is called *ring topology*, while the parameter r that determines the neighborhood size is called *neighborhood radius*. Obviously, indices are considered to recycle in this topology, i.e., index $i = 1$ follows immediately after the index $i = N$ on the ring.

The PSO variant that uses the overall best position of the swarm can be considered as a special case of the aforementioned ring scheme, where each neighborhood is the whole swarm, i.e., $NB_i \equiv S$, for all $i = 1, 2,..., N$. To distinguish between the two approaches, the variant that employs the overall best position is called the *global PSO variant* (often denoted as *gbest*), while the one with strictly smaller neighborhoods is called the *local PSO variant* (denoted, respectively, as *lbest*). The gbest scheme is also called *star topology*, and is graphically depicted in Fig. 5 (right) where all particles communicate with the best one.

The effect of using lbest instead of gbest on swarm diversity is illustrated in Fig. 6 for a swarm of 20 particles with $c_1 = c_2 = 2$ and decreasing inertia weight, with $T_{max} = 500$, $w_{up} = 1.2$, $w_{low} = 0.1$, and $v_{max} = 5.12$, minimizing the 10-dimensional Rastrigin function, defined as test problem TP_{UO-3} in Appendix A of the book at hand, in the range $[-5.12, 5.12]^{10}$. Solid line represents swarm diversity (as defined in the previous section) for lbest with ring topology and radius $r = 1$, while the dotted line stands for the corresponding gbest case. There is an apparent difference in swarm diversities between gbest and lbest, which becomes more intense when problem dimension increases.

Although ring topology is adequately simple and efficient, different topologies have been proposed in literature (Kennedy, 1999; Mendes *et al.*, 2003). Also, topology can change with time instead of remaining fixed throughout a run. Such dynamic topologies have been used in mutiobjective optimization problems (Hu & Eberhart, 2002). Moreover, each particle can have its own individual (fixed or dynamic) topology, providing high flexibility to the user and the ability to fit any special requirements of the problem at hand. Nevertheless, the vast majority of lbest models in literature are based on ring topology; hence, it can be considered as a standard choice for local PSO variants.

The introduction of neighborhoods enhanced the performance of PSO significantly, offering a boost to research towards the development of more competitive and sophisticated variants that incorporated all presented concepts so far. In the next section, we present the most established contemporary PSO variants, which are widely used in applications and considered as the state-of-the-art nowadays.

Figure 6. Swarm diversity during search using the lbest (solid line) and the gbest (dotted line) PSO variant. The plots pertain to a swarm of 20 particles with $c_1 = c_2 = 2$, minimizing the 10-dimensional Rastrigin function (TP$_{UO-3}$ of Appendix A) in the range $[-5.12, 5.12]^{10}$ for 500 iterations and $v_{max} = 5.12$. The inertia weight decreases linearly from 1.2 to 0.1, while ring topology with radius $r = 1$ is used for the lbest case

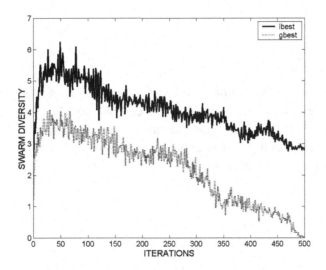

CONTEMPORARY STANDARD PSO

The efficiency of the presented PSO variants attracted the interest of the scientific community. A remarkable number of scientists and engineers were testing PSO against the established evolutionary algorithms in a variety of applications, producing very promising results. The simplicity of PSO allowed scientists from various disciplines, with limited background in computer science and programming skills, to use PSO as an efficient optimization tool in applications where classical optimization methods were inefficient.

The blossoming research prompted researchers to also investigate the theoretical properties of PSO, offering better understanding of its operation and dynamics, as well as the mathematical traits for proper parameter configuration. However, from the very first moment, it became obvious that such a theoretical analysis would be a difficult task due to some particularities: PSO incorporates stochastic elements but the search is not based on probability distributions. Thus, a direct probabilistic analysis based on adaptive distributions was not possible. For this purpose, deterministic approximations of the original PSO model were initially investigated, while stochasticity was introduced in the studied models as a perturbation factor of the considered deterministic systems.

Ozcan and Mohan (1999) published the first theoretical investigation in multi-dimensional spaces, providing closed-form equations for particle trajectories. Their study focused on the early PSO model of equations (1) and (2), and they showed that particles were actually moving on sinusoidal waves per coordinate of the search space, while stochasticity offering a means to manipulate its frequency and amplitude. A few years later, this interesting result was followed by a thorough investigation by Clerc and Kennedy (2002), who considered different generalized PSO models and performed a dynamical system analysis of their convergence.

Clerc and Kennedy's analysis offered a solid theoretical background to the algorithm, and it established one of the investigated models as the *default contemporary PSO variant*. This model is defined by the following equations:

$$v_{ij}(t+1) = \chi\,[v_{ij}(t) + c_1 R_1\,(p_{ij}(t)\text{-}x_{ij}(t)) + c_2 R_2\,(p_{gj}(t)\text{-}x_{ij}(t))], \qquad (12)$$

$$x_{ij}(t+1) = x_{ij}(t) + v_{ij}(t+1), \qquad (13)$$

$$i = 1, 2,\ldots, N, \quad j = 1, 2,\ldots, n.$$

where χ is a parameter called *constriction coefficient* or *constriction factor*, while the rest of the parameters remain the same as for the previously described PSO models. Obviously, this PSO variant is algebraically equivalent with the inertia weight variant defined by equations (7) and (8). However, it is distinguished in literature due to its theoretical properties that imply the following explicit selection of its parameters (Clerc & Kennedy, 2002):

$$\chi = \frac{2}{\left|2 - \varphi - \sqrt{\varphi^2 - 4\varphi}\right|},$$

where $\varphi = c_1 + c_2$, and $\varphi > 4$. Based on this equation, the setting:

$$\chi = 0.729, \quad c_1 = c_2 = 2.05,$$

is currently considered as the default parameter set of the constriction coefficient PSO variant.

The velocity update of equation (12) corresponds to the gbest (global best) PSO model. Naturally, the concept of neighborhood can be alternatively used by replacing the global best component, $p_{gj}(t)$, in the second parenthesis of equation (12), with the corresponding local best component. Thus, the local variant (lbest) of equation (12) becomes:

$$v_{ij}(t+1) = \chi[v_{ij}(t) + c_1 R_1(p_{ij}(t) - x_{ij}(t)) + c_2 R_2(p_{g,j}(t) - x_{ij}(t))], \qquad (14)$$

where g_i denotes the index of the best particle in the neighborhood of x_i. The use of equation (14) instead of equation (12) has the same effect on the exploration/exploitation properties of the constriction coefficient PSO variant, as for the previously described inertia weight variant.

The stochastic parameters, R_1 and R_2, are considered to be uniformly distributed within the range [0,1]. Thus, prospective new positions of a particle are distributed in a rectangle as illustrated in the left part of Fig. 7, similarly to the rectangular areas illustrated in Fig. 1. Alternatively, R_1 and R_2 can be uniformly distributed within a sphere centered at the origin, with radius equal to 1, as illustrated in the right part of Fig. 7. A different approach suggests that R_1 and R_2 are normally distributed with a Gaussian distribution, $N(\mu,\sigma^2)$. In this case, the distribution of new prospective positions can differ significantly from the previous two cases, depending heavily on the values of μ and σ, which, in turn, are usually dependent on the parameters c_1 and c_2. Figure 8 illustrates this dependency for two different cases, namely $\mu = 0$ and $\mu = c/2$, respectively, with $\sigma = c/4$, and $c = c_1 = c_2 = 2.05$.

Figure 7. Candidate new positions of the particle $x_i = (0,0)^T$ (cross) with $p_i = (2,1)^T$ (star) and $p_g = (1,3)^T$ (square), for the cases of (A) rectangular distribution of R_1, R_2, and (B) spherical distribution of R_1, R_2. The velocity of x_i is set to $v_i = (0,0)^T$ for simplicity, and the default parameters, $\chi = 0.729$, $c_1 = c_2 = 2.05$, are used

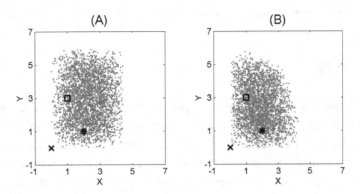

Figure 8. Candidate new positions of the particle $x_i = (0,0)^T$ (cross) with $p_i = (2,1)^T$ (star) and $p_g = (1,3)^T$ (square), for normally distributed values of R_1 and R_2, with mean value (A) $\mu = 0$, and (B) $\mu = c/2$, and standard deviation $\sigma = c/4$. The velocity of x_i is set to $v_i = (0,0)^T$ for simplicity, and the default parameters, $\chi = 0.729$, $c_1 = c_2 = 2.05$, are used

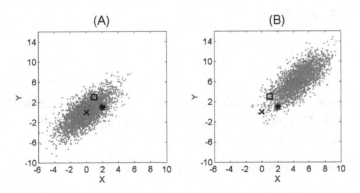

The aforementioned distributions appear in the vast majority of relevant works, with the rectangular one being the most popular. Different distributions are also used, although less frequently. Most of them are reported and thoroughly discussed in Clerc (2006, Chapter 8). Although they have shown potential for enhanced efficiency in specific problems, these approaches cannot be characterized as standard PSO variants. The case of rectangular distributions offers simplicity and efficiency that has established it as the default choice. The case of spherical distributions is quite similar to the rectangular one, although it further restricts the range of possible new particle positions.

On the other hand, the case of Gaussian distributions offers a completely different potential to the algorithm. Depending on their parameters, a particle can move even in a direction opposite to the involved best positions, exhibiting completely different dynamics than the standard PSO models. The efficiency of this model always depends on the problem at hand. If the PSO has detected the most promising regions of the search space, then Gaussian distributions can slow the convergence down. On the other hand, if there is a plethora of local minima with narrow basins of attraction, closely positioned to the global one, then this approach can work beneficially for the algorithm. Nevertheless, distribution selection has an obvious effect on performance; thus, careful choices are needed. In a given problem, experimental evidence in tackling similar problems can be a substantial criterion for such a choice.

CHAPTER SYNOPSIS

This chapter presents a brief history of the most important landmarks in the development of PSO, from its early precursors to contemporary standard variants, describing sources of inspiration behind the development of PSO, as well as its first variants. Major deficiencies, such as the swarm explosion effect and inability to converge, were identified in early variants and addressed either by adding new parameters or by clamping the magnitude of existing ones. Later developments regarding the scope of information exchange among particles offered a better control of the exploration/exploitation properties of PSO. In the last section we presented a variant, which is considered as the standard contemporary PSO variant. This is accompanied by an explicit scheme for determining its parameters, while it incorporates all previous performance-enhancing developments, retaining remarkable simplicity and flexibility.

In the next chapter, we provide a closer inspection of the theoretical properties of PSO, along with several practical issues, such as swarm initialization, parameter selection, termination conditions, and sensitivity analysis of the algorithm.

REFERENCES

Angeline, P. J. (1998). Evolutionary optimization versus particle swarm optimization: philosophy and performance differences. In V.W. Porto, N. Saravanan, D. Waagen, & A.E. Eiben (Eds.), *Evolutionary Programming VII* (pp. 601-610). Berlin: Springer.

Clerc, M. (2006). *Particle swarm optimization*. London: ISTE Ltd.

Clerc, M., & Kennedy, J. (2002). The particle swarm - explosion, stability, and convergence in a multidimensional complex space . *IEEE Transactions on Evolutionary Computation*, 6(1), 58–73. doi:10.1109/4235.985692

Eberhart, R. C., & Kennedy, J. (1995). A new optimizer using particle swarm theory. In *Proceedings of the 6th Symposium on Micro Machine and Human Science, Nagoya, Japan* (pp. 39-43).

Eberhart, R. C., & Shi, Y. (1998). Comparison between genetic algorithms and particle swarm optimization. In: W. Porto, N. Saravanan, D. Waagen, & A.E. Eiben (Eds.), *Evolutionary Programming VII* (pp. 611-616). Berlin: Springer.

Eberhart, R. C., Simpson, P., & Dobbins, R. (1996). *Computational intelligence PC tools*. Boston: Academic Press Professional.

Heppner, F., & Grenander, U. (1990). A stochastic nonlinear model for coordinate bird flocks. In S. Krasner (Ed.), *The Ubiquity of Chaos* (pp. 233-238). Washington, DC: AAAS Publications.

Hu, X., & Eberhart, R. C. (2002). Multiobjective optimization using dynamic neighborhood particle swarm optimization. In *Proceedings of the 2002 IEEE Congress on Evolutionary Computation, Honolulu (HI), USA* (pp. 1677-1681).

Kao, I. W., Tsai, C. Y., & Wang, Y. C. (2007). An effective particle swarm optimization method for data clustering. In *Proceedings of 2007 IEEE Industrial Engineering and Engineering Management, Singapore* (pp. 548-552).

Kennedy, J. (1999). Small worlds and megaminds: effects of neighborhood topology on particle swarm performance. In *Proceedings of 1999 IEEE Congress on Evolutionary Computation, Washington (DC), USA* (pp. 1931-1938).

Kennedy, J., & Eberhart, R. C. (1995). Particle swarm optimization. In *Proceedings of the 1995 IEEE International Conference on Neural Networks, Perth, Australia* (Vol. IV, pp. 1942-1948).

Mendes, R., Kennedy, J., & Neves, J. (2003). Watch thy neighbor or how the swarm can learn from its environment. In *Proceedings of the 2003 IEEE Swarm Intelligence Symposium, Indianapolis (IN), USA* (pp. 88-94).

Millonas, M. M. (1994). Swarms, phase transitions, and collective intelligence. In C.G. Langton (Ed.), *Artificial Life III* (pp. 417-445). Reading, MA: Addison-Welsey.

Ozcan, E., & Mohan, C. K. (1999). Particle swarm optimization: surfing the waves. In *Proceedings of 1999 IEEE Congress on Evolutionary Computation, Washington (DC), USA* (pp. 1939-1944).

Reynolds, C. W. (1987). Flocks, herds, and schools: a distributed behavioral model. *Computer Graphics, 21*(4), 25–34. doi:10.1145/37402.37406

Shi, Y., & Eberhart, R. C. (1998a). A modified particle swarm optimizer. In *Proceedings of the 1998 IEEE International Conference on Evolutionary Computation, Anchorage (AK), USA* (pp. 69-73).

Shi, Y., & Eberhart, R. C. (1998b). Parameter selection in particle swarm optimization. [London: Springer.]. *Lecture Notes in Computer Science, 1447,* 591–600. doi:10.1007/BFb0040810

Wilson, E. O. (1975). *Sociobiology: the new synthesis.* Cambridge, MA: Belknap Press.

Chapter 3
Theoretical Derivations and Application Issues

This chapter deals with fundamental theoretical investigations and application issues of PSO. We are mostly interested in developments that offer new insight in configuring and tuning the parameters of the method. For this purpose, the chapter opens with a discussion on initialization techniques, followed by brief presentations of investigations on particle trajectories and the stability analysis of PSO. A useful technique based on computational statistics is also presented for the optimal tuning of the algorithm on specific problems. The chapter closes with a short discussion on termination conditions.

INITIALIZATION TECHNIQUES

Initialization is perhaps the less studied phase of PSO and other evolutionary algorithms. This may be due to the general demand for developing algorithms that are not very sensitive in the initial conditions. However, it can be experimentally verified that, in various problems, initialization can have a significant impact on performance.

As already mentioned in the previous chapter, uniform random initialization is the most popular scheme in evolutionary computation due to the necessity for equally treating each part of a search space with unrevealed characteristics. However, alternative initialization methodologies that use different probability distributions or employ direct search methods to provide the first steps of the algorithm have proved very useful.

In the following sections, we discuss the most common probabilistic initialization techniques. In addition, we present a scheme based on the nonlinear simplex method of Nelder and Mead, which has been shown to work beneficially for the initialization of PSO (Parsopoulos & Vrahatis, 2002).

DOI: 10.4018/978-1-61520-666-7.ch003

Copyright © 2010, IGI Global. Copying or distributing in print or electronic forms without written permission of IGI Global is prohibited.

Random Probabilistic Initialization

In the framework of PSO, the quantities that need to be initialized prior to application are the particles as well as their velocities and best positions. The best positions consist of the best solutions already detected by each particle, while current particle positions represent candidate new solutions. Since no information on the promising regions of the search space is expected to be available prior to initialization, the initial particles and the corresponding best positions are considered to coincide. Also, in constrained optimization, we are interested in detecting feasible solutions, i.e., solutions that do not violate problem constraints. For this purpose, the initialization of swarm and best positions within the feasible search space, $A \subset \mathbf{R}^n$, is desirable.

The most common technique in evolutionary computation is *random uniform initialization*. According to this, each particle of the initial swarm and, consequently, of the initial best positions is drawn by sampling a uniform distribution over the search space A. The applicability of this approach depends on the form of the search space. If A is given as an n-dimensional bounded box:

$$A = [a_1, b_1] \times [a_2, b_2] \times \cdots \times [a_n, b_n],$$

then any of the available pseudo-random number generators, such as the number-theoretically generated Sobol sequences (Press *et al.*, 1992, Chapter 7), can be directly used to produce uniformly distributed numbers within it.

In practice, it is very common to exploit the one-dimensional pseudo-random generators that accompany all modern computer systems. Thus, each component of a particle is generated as a uniformly distributed pseudo-random value within the interval [0,1] and then scaled in the magnitude of the corresponding direction of A. This procedure is described in the pseudocode of Table 1, where we use the function drand48() provided by the C++ programming language as the pseudo-random generator in [0,1]. The produced value is then scaled in the corresponding direction of the search space, so that the produced particles lie strictly within A. In addition, we scale the produced pseudo-random values of the velocity components, in order to clamp it within its limits, $[-v_{max}, v_{max}]$, as described in the previous chapter. When a large number of subsequent experiments are conducted, re-initialization of the pseudo-random generator with a different seed may be occasionally necessary, in order to obtain unbiased experimental results.

Mathematically speaking, the particles produced by the aforementioned procedure do not exactly follow the multi-dimensional uniform distribution over A. Despite this theoretical deficiency, random uniform initialization became the technique one of choice; a popularity that can be attributed to the following properties:

1. It can be implemented easily in any computer system and programming language.
2. In many applications, except from some complex constrained optimization problems, the feasible search space can be given in the form of a bounded box or approximated by a sequence of such boxes.
3. It is suitable for time-critical applications as the generation of numbers is adequately fast, requiring only minor computational effort.
4. It does not have a partiality for any region of the search space.

Table 1. Pseudocode for random uniform initialization. We employ the drand48() pseudo-random generator of the C++ programming language to produce uniformly distributed pseudo-random values within [0,1]

Input:	Number of particles N, dimension n, velocity bounds $[-v_{max}, v_{max}]$, and search space, $A = [a_1, b_1] \times [a_2, b_2] \times ... \times [a_n, b_n]$
Step 1.	**Do** $(i = 1...N)$
Step 2.	**Do** $(j = 1...n)$
Step 3.	**Set** particle component $x_{ij} = a_j + \text{drand48}() (b_j - a_j)$.
Step 4.	**Set** best position component $p_{ij} = x_{ij}$.
Step 5.	**Set** velocity component $v_{ij} = -v_{max} + 2 \text{ drand48}() v_{max}$.
Step 6.	**End Do**
Step 7.	**End Do**

The last property is very important in PSO. Sutton *et al.* (2006) have empirically shown that initialization affects performance in multi-funnel landscapes. In such landscapes, there is no single global tendency towards the global minimum, while the best local minima are not clustered closely. Preliminary experiments revealed that, if two funnels exist and most (nearly 80%) of the particles are initialized in the one funnel, then PSO has up to five times greater probability to converge into this, rather than the other funnel. This property seems to hold regardless of swarm size, underlining the importance of proper initialization in complex problems.

On the other hand, if there are items of information available regarding the location of the global minimizer in the search space, it makes more sense to initialize the majority of the swarm around it. This can be done by replacing the uniform distribution with a Gaussian one. The mean value and standard deviation of the Gaussian shall depend on the available information. For example, the mean value can be selected to lie somewhere in the conjectured region of the global minimizer, while the standard deviation can be a fraction of the expected distance of the mean value from the global minimizer. It is up to the user to exploit as much available information as possible.

Alternatively to random initialization, a grid that covers the search space with equidistant points can be considered. In this case, particles are initialized on grid nodes, while their initial velocities are taken randomly. This approach is characterized by explicit fairness regarding the covering of different regions of the search space; however, it suffers from a series of deficiencies that reduces its popularity. More specifically, its deterministic nature does not adhere to the philosophy of PSO and, more generally, of stochastic optimization algorithms. In addition, the grid construction can become laborious even in search spaces with simple but not rectangular shape. Finally, experimentation shows that, in most cases, it is not accompanied by a statistically significant improvement in performance.

Regarding velocity, although the most common practice is random initialization, it can be alternatively initialized to zero. In this case, the particles are let to obtain acceleration in the first iteration based solely on their distance from the best positions. Under this assumption, the overall best particle will remain in its initial position until another particle finds a better one. Also, particles initialized close to the overall best position will obtain small velocities, becoming more prone to get stuck in local minima. For these reasons, random initialization of velocities in non-zero values is preferable.

Experimental studies revealed that, in some benchmark problems, initialization that biases the swarm towards more promising regions of the search space can offer a significant reduction in the required computational burden (Parsopoulos & Vrahatis, 2002). In unexplored search spaces, this biasing can be the outcome of a simple direct search algorithm, which has been applied for a few iterations prior to PSO. Such an approach that uses the nonlinear simplex method is presented in the next section.

Initialization Using the Nonlinear Simplex Method

The *nonlinear simplex method* (NSM) was developed by Nelder and Mead (1965) for function minimization tasks. NSM requires only function evaluations, and it is considered a good starting procedure when the figure of merit is to "get something to work quickly", especially in noisy problems. In addition, it is characterized by a geometrical naturalness that makes it attractive to work through (Press *et al.*, 1992, Chapter 10).

NSM is based on the mathematical structure of *simplex*. An *n*-dimensional simplex, also called *n-simplex*, is the convex hull of a set of (*n*+1) affinely independent points in $\mathbf{R}^n$, i.e., (*n*+1) points in general positions in the sense that no *m*-dimensional hyperplane contains more than (*m*+1) of them. Thus, a 2-dimensional simplex is a triangle, while a 3-dimensional simplex is a tetrahedron. In general, we consider only *non-degenerated* simplices, i.e., simplices that enclose a finite inner *n*-dimensional volume. In such cases, if any vertex of the simplex coincides with the origin, the rest *n* vertices define directions that span the *n*-dimensional vector space.

Figure 1. Possible moves of a simplex (the worst and best vertex are denoted): (A) Reflection of the worst vertex against the face of the best vertex; (B) Reflection and expansion; (C) Contraction towards the face of the best vertex; and (D) Multiple contraction towards the best vertex

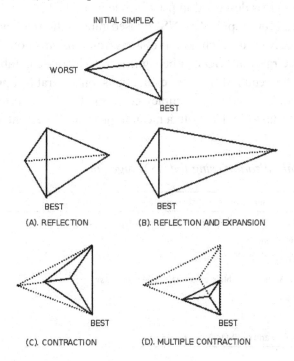

The operation of the NSM starts with an initial simplex and takes a series of steps where its worst vertex, i.e., the one with the highest function value, is mostly moved through its opposite face of the simplex, hopefully to a new point with a lower function value. If possible, the simplex expands in one or another direction to take larger steps. When the method reaches a "valley floor", the simplex is contracted in the transverse direction to ooze down the valley or it can be contracted in all directions, pulling itself around its lowest point. The possible moves of a simplex are illustrated in Fig. 1. An implementation of NSM is provided in (Press *et al.*, 1992, Chapter 10).

In practice, NSM is usually applied until a minimizer is detected. Then, it is restarted at the detected minimizer to make sure that stopping criteria have not been fulfilled by a possible anomalous step. Thus, after restarting, the detected minimizer constitutes a vertex of the new initial simplex, while the rest n vertices are taken randomly. This restarting scheme is not expected to be computationally expensive, since the algorithm has already converged to one of its initial simplex vertices.

The convergence properties of NSM are in general poor, but, in many applications, they have been shown to be very useful, especially in cases of noisy functions and problems with imprecise data. A more efficient variant of NSM was proposed and analyzed by Torczon (1991) along with its convergence properties. The reader is referred to the original paper for more details (Torczon, 1991).

The motives for using NSM for the initialization of PSO lie in its ability to take quick downhill steps, combined with the performance improvement achieved when PSO is seeded with a good initial swarm (Parsopoulos & Vrahatis, 2002). The technique works as follows: suppose that we start NSM with an initial simplex in the n-dimensional search space, A. Then, the $(n+1)$ vertices of the initial simplex will constitute the first $(n+1)$ particles of the swarm. Next, NSM is applied for N-$(n+1)$ iterations, where N is the required swarm size. At each NSM iteration, the new simplex vertex produced by NSM is accepted as a new initial particle into the swarm. Thus, the initial swarm is equipped with information gained in the initial iterations of NSM. As soon as the initial swarm is filled with N particles that served as simplex vertices, NSM stops and PSO starts its operation with the produced initial swarm (Parsopoulos & Vrahatis, 2002). The technique is described in pseudocode in Table 2.

Parsopoulos and Vrahatis (2002) applied the NSM-based initialization technique with the inertia weight PSO variant on several widely used benchmark problems from optimization literature, and their results suggested that convergence rates of PSO can be significantly increased. Table 3 reports the achieved improvement, in terms of the required function evaluations, on several test problems described in Appendix A of the book at hand, for solution accuracy equal to 10^{-3}. The improvement in the considered problems ranged from 2.9% up to 35.1%, with a mean improvement percentage of 11.02%, suggesting

Table 2. Pseudocode for initialization using the NSM algorithm

Input:	Number of particles, N; dimension, n; initial swarm, $S \equiv \varnothing$.
Step 1.	**Generate** the initial n-dimensional simplex with vertices $sv_1, sv_2, \ldots, sv_{n+1}$.
Step 2.	**Set** $S = S \cup \{sv_1, sv_2, \ldots, sv_{n+1}\}$.
Step 3.	**Do** ($i = 1 \ldots N$-$(n+1)$)
Step 4.	**Perform** an NSM step and obtain a new acceptable vertex, sv_{n+1+i}.
Step 5.	**Set** $S = S \cup \{sv_{n+1+i}\}$.
Step 6.	**End Do**
Step 7.	**Set** S to be the initial swarm of PSO.

that NSM-based initialization can be beneficial for PSO. For further details, the reader is referred to the original paper (Parsopoulos & Vrahatis, 2002).

In the following sections, we concisely present the most important theoretical works that contributed towards the better understanding of the basic operation mechanisms of PSO and provided fundamental rules for its parameter setting.

THEORETICAL INVESTIGATIONS AND PARAMETER SELECTION

The theoretical analysis of PSO has proved to be harder than expected. This can be ascribed mostly to the nature of the algorithm. First, it is stochastic, thus it cannot be studied through deterministic-oriented approaches. Secondly, it is not a pure probabilistic algorithm; therefore stochastic investigations using adaptive probability densities are not valid. Moreover, its stochasticity can be regarded as a mutation, similarly to evolutionary algorithms; however, this mutation depends on non-statistical information (best positions) that potentially changes at each iteration.

The lack of a mathematical tool that could be used directly for the theoretical investigation of PSO enforced researchers to approximate it with simplified deterministic models. These models were studied with classical mathematical methodologies, such as dynamical systems theory and stochastic analysis, and the derived conclusions were generalized to the actual PSO case by infusing stochasticity and analyzing its implications. Nevertheless, the practical value of the obtained results varies, since stochasticity can often annul deterministic theoretical results.

In the following sections, we describe the most influential theoretical investigations of PSO, and report their most crucial derivations regarding its dynamics and parameter settings. Our primary aim is to expose the fundamental assumptions on which the theoretical studies were built, as well as underline their impact on the form and parameter configuration of PSO. The insight gained can be very useful in applying PSO to new, challenging problems.

Table 3. Performance improvement of PSO, in terms of the required function evaluations, using the NSM initialization technique. Percentages are derived from results reported in (Parsopoulos & Vrahatis, 2002)

Problem	Dim.	Improvement (%)	Problem	Dim.	Improvement (%)
TP_{UO-1}	2	22.3%	TP_{UO-13}	2	6.2%
TP_{UO-2}	2	7.2%	TP_{UO-14}	2	9.5%
TP_{UO-3}	6	15.0%	TP_{UO-15}	2	4.7%
TP_{UO-4}	2	35.1%	TP_{UO-16}	3	2.9%
TP_{UO-10}	2	19.0%	TP_{UO-17}	6	7.4%
TP_{UO-11}	2	7.9%	TP_{UO-17}	9	5.7%
TP_{UO-12}	2	7.7%	TP_{UO-18}	6	3.7%

One-Dimensional Particle Trajectories

The first study on particle trajectories was conducted by Kennedy (1998). The study considered the early PSO version without inertia weight or constriction coefficient. This version was described by equations (1) and (2) of Chapter Two, which are reproduced below for presentation completeness:

$$v_{ij}(t+1) = v_{ij}(t) + c_1 R_1 (p_{ij}(t)\text{-}x_{ij}(t)) + c_2 R_2 (p_{gj}(t)\text{-}x_{ij}(t)),$$

$$x_{ij}(t+1) = x_{ij}(t) + v_{ij}(t+1),$$

$$i = 1, 2,\ldots, \quad N, j = 1, 2,\ldots, n.$$

Kennedy (1998) assumed a simplified one-dimensional PSO model; therefore, index j can be omitted from the equations. In addition, if we let:

$$\varphi_1 = c_1 R_1 \text{ and } \varphi_2 = c_2 R_2,$$

the equations become:

$$v_i(t+1) = v_i(t) + \varphi_1 (p_i(t)\text{-}x_i(t)) + \varphi_2 (p_g(t)\text{-}x_i(t)),$$

$$x_i(t+1) = x_i(t) + v_i(t+1),$$

$$i = 1, 2,\ldots, N.$$

As discussed in previous chapters, the use of the previous velocity term, $v_i(t)$, for updating the current one, equips particles with the ability to make an oscillatory move around the best positions, rather than moving aggressively towards them. In fact, each particle follows a trajectory around the mean of the best positions (Kennedy, 1998):

$$\frac{\varphi_1 p_i + \varphi_2 p_g}{\varphi_1 + \varphi_2}. \tag{1}$$

On the other hand, the best particle, for which p_i and p_g coincide by definition, performs the following velocity update:

$$v_i(t+1) = v_i(t) + (\varphi_1 + \varphi_2) (p_i(t)\text{-}x_i(t)).$$

The effective parameter that controls velocity for all particles is:

$$\varphi = \varphi_1 + \varphi_2.$$

The weighted mean of best positions in equation (1) is unpredictable when φ_1 and φ_2 assume random values. However, in the long run, the sequence of particle positions, x_i, approximates an average of p_i

Figure 2. Particle trajectories for fixed values of φ in (0,1]

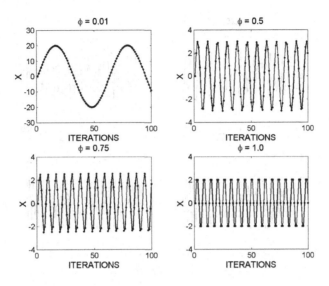

Figure 3. Particle trajectories for fixed values of φ in (1,2]

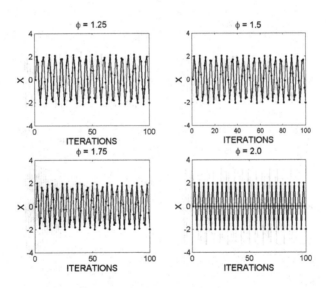

and p_g. This effect is imputed to the uniform distribution of φ_1 and φ_2, which is expected to equalize them on average. If we further simplify the system by considering a solitary particle and a constant best position, we obtain the following PSO model (Kennedy, 1998):

$$v(t+1) = v(t) + \varphi\,(p-x(t)), \tag{2}$$

$$x(t+1) = x(t) + v(t+1), \tag{3}$$

Figure 4. Particle trajectories for fixed values of φ in (2,3]

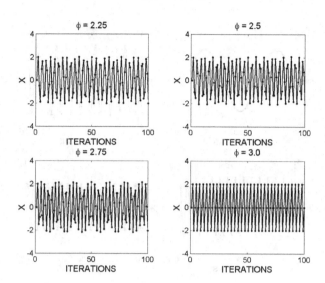

Figure 5. Particle trajectories for fixed values of φ in (3,4]

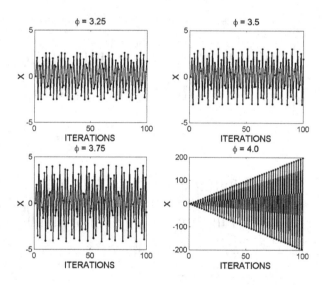

where p is the best position of the (single) particle. The shape of the trajectory traversed by x depends heavily on φ, while v and p have an impact on its amplitude. The assumption of a constant p is not inconsistent with reality, since best positions are expected to vary less frequently and finally stabilize at the last few iterations of a run.

Kennedy (1998) performed an experimental study of the model defined by equations (2) and (3), for varying values of the control parameter, φ. The obtained trajectories for different values of φ in the intervals, (0,1], (1,2], (2,3], (3,4], and for φ > 4, are illustrated in Figs. 2, 3, 4, and 5, respectively. In all figures, the initial values, $x(0) = 0$, $v(0) = 2$, and a fixed, $p = 0$, were used.

There is an apparent influence of φ on the produced particle trajectories. Different values of φ produce oscillations of different amplitude and frequency around the best position, *p*. When φ becomes equal to 4, the particle diverges to infinity, as illustrated in Fig. 5. In all other cases, the particle oscillates without diverging or converging towards *p*, although the period of its cycles depends on φ. Kennedy offers a comprehensive discussion per case, which reveals that there is a critical upper limit, φ = 4, after which divergence occurs (Kennedy, 1998).

An interesting question arises regarding the validity of these observations in the original PSO case, i.e., when φ is stochastic. If the fixed φ assumes a random weight within [0,1], then the system becomes unpredictable, since particles take steps from all possible sequences of trajectories produced by the corresponding fixed values of φ. In addition, the system inherits the undesirable property of swarm explosion (Kennedy, 1998) discussed in the previous chapter. Swarm explosion occurs even when the random values of φ are constrained in ranges that exclude the critical upper limit, φ = 4. This is illustrated in Fig. 6, for uniformly sampled values of φ within the ranges (0,1), (0,2), (0,3), and (0,4). We also notice a remarkable scaling difference in Fig. 6 as the range increases towards φ = 4.

As mentioned in the previous chapter, the explosion effect can be addressed by imposing an upper limit on the absolute values of the velocities. The effect of using a maximum velocity, v_{max} = 5, in the experiments of Fig. 6, is illustrated in Fig. 7. Clearly, the behavior of the particle changes radically, with the oscillation amplitude being significantly reduced. Although in our example the value of v_{max} is arbitrarily selected, estimates of scale and correlation dimension for the problem at hand can be very useful in selecting more appropriate velocity limits (Kennedy, 1998).

The work of Kennedy constituted a milestone in the theoretical investigation of PSO. It identified that a crucial value on φ exists, and, in combination with v_{max}, it has a crucial impact on the behavior of the considered simplified models. The next step was taken one year later by Ozcan and Mohan (1999). They generalized Kennedy's study to multi-dimensional search spaces and derived closed form equations of particle trajectories, as described in the following section.

Figure 6. Particle trajectories for random values of φ, uniformly distributed within the ranges (0,1), (0,2), (0,3), and (0,4)

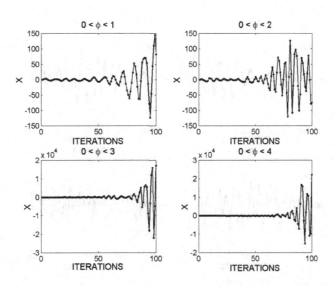

Multi-Dimensional Particle Trajectories

Similarly to Kennedy (1998), the early PSO version without inertia weight or constriction coefficient was studied by Ozcan and Mohan (1998, 1999). The only difference between the two studies was the use of the local PSO variant (lbest) with neighborhoods, instead of the global (gbest) variant used by Kennedy. The original local PSO model:

$$v_{ij}(t+1) = v_{ij}(t) + \varphi_1 \ (p_{ij}(t)\text{-}x_{ij}(t)) + \varphi_2 \ (p_{lj}(t)\text{-}x_{ij}(t)), \tag{4}$$

$$x_{ij}(t+1) = x_{ij}(t) + v_{ij}(t+1), \tag{5}$$

$$i = 1, 2, \ldots, N, \quad j = 1, 2, \ldots, n,$$

where l denotes the best particle in the neighborhood of x_i, and φ_1, φ_2, are samples of a uniform distribution in $[0, c_1]$ and $[0, c_2]$, respectively, was simplified by assuming constant values for p_{ij}, p_{gj}, φ_1, and φ_2. No further simplification regarding the dimensionality or the swarm size was made. Substituting equation (4) to equation (5), a recursive formula for the trajectory of the particle is obtained (Ozcan & Mohan, 1999):

$$x_{ij}(t) - (2 - \varphi_1 - \varphi_2) \, x_{ij}(t\text{-}1) + x_{ij}(t\text{-}2) = \varphi_1 p_{ij}(t) + \varphi_2 p_{lj}(t). \tag{6}$$

The initial position and velocity, $x_{ij}(0)$ and $v_{ij}(0)$, respectively, serve also as initial conditions for the recursion. Thus, the first step is defined as:

$$x_{ij}(1) = x_{ij}(0) \ (1 - \varphi_1 - \varphi_2) + v_{ij}(0) + \varphi_1 p_{ij}(t) + \varphi_2 p_{lj}(t).$$

Figure 7. Particle trajectories for random values of φ, uniformly distributed within the ranges (0,1), (0,2), (0,3), and (0,4), using a maximum velocity, $v_{max} = 5$

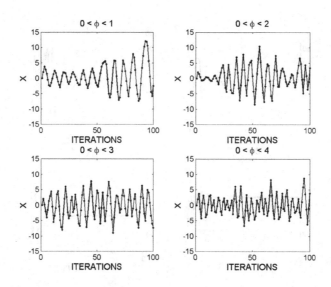

Closed form solutions for equation (6) were obtained using methodologies for solving non-homogeneous linear recurrence equations for the displacement (Ozcan & Mohan, 1999):

$$x_{ij}(t) = \eta_{ij}\,\alpha_{ij}^{t} + \iota_{ij}\,\beta_{ij}^{t} + \kappa_{ij}, \tag{7}$$

where,

$$\delta_{ij} = \sqrt{(2 - \varphi_1 - \varphi_1)^2 - 4}, \tag{8}$$

$$\alpha_{ij} = (2 - \varphi_1 - \varphi_2 + \delta_{ij})/2, \tag{9}$$

$$\beta_{ij} = (2 - \varphi_1 - \varphi_2 - \delta_{ij})/2, \tag{10}$$

$$\eta_{ij} = (0.5 - (\varphi_1 + \varphi_2)/2\delta_{ij})\,x_{ij}(0) + v_{ij}(0)/\delta_{ij} + (\varphi_1 p_{ij} + \varphi_2 p_{lj})/(2\delta_{ij}) - (\varphi_1 p_{ij} + \varphi_2 p_{lj})/(2\varphi_1 + 2\varphi_2), \tag{11}$$

$$\iota_{ij} = (0.5 + (\varphi_1 + \varphi_2)/2\delta_{ij})\,x_{ij}(0) - v_{ij}(0)/\delta_{ij} - (\varphi_1 p_{ij} + \varphi_2 p_{lj})/(2\delta_{ij}) + (\varphi_1 p_{ij} + \varphi_2 p_{lj})/(2\varphi_1 + 2\varphi_2), \tag{12}$$

$$\kappa_{ij} = (\varphi_1 p_{ij} + \varphi_2 p_{lj})/(\varphi_1 + \varphi_2), \tag{13}$$

with special cases arising for $\varphi_1 + \varphi_2 = 2$.

Two major cases, namely real and complex values, are distinguished for δ_{ij} in the analysis of Ozcan and Mohan (1999). In the first, the trajectory is governed by equations (7)-(13), while in the latter, which corresponds to the values:

$$0 < \varphi_1 + \varphi_2 < 4,$$

α_{ij} and β_{ij} become complex numbers. Assuming that:

$$\theta_{ij} = \mathrm{atan}(|\delta_{ij}| / |2 - \varphi_1 - \varphi_2|),$$

and with proper mathematical manipulations, equation (7) becomes (Ozcan & Mohan, 1999):

$$x_{ij}(t) = K_{ij}\,\sin(\theta_{ij}t) + \Lambda_{ij}\,\cos(\theta_{ij}t) + \kappa_{ij}, \tag{14}$$

with,

$$K_{ij} = (2v_{ij}(0) - (\varphi_1 + \varphi_2)\,x_{ij}(0) + \varphi_1 p_{ij} + \varphi_2 p_{lj}) / |\delta_{ij}|,$$

$$\Lambda_{ij} = x_{ij}(0) - \kappa_{ij}.$$

For $\varphi_1 + \varphi_2 > 2$, the value of θ_{ij} shall be increased by π.

Several regions of interest can be identified based on the value $\varphi = \varphi_1 + \varphi_2$ within the range $(0,4)$, while special cases arise for $\varphi = 0$, 2, and 4 (Ozcan & Mohan, 1999). If,

$\varphi \in (0, 2-\sqrt{3}] \cup (2+\sqrt{3}, 4]$,

then it holds that $|\delta_{ij}| \leq 1$. For increasing values of φ, K_{ij} becomes larger, resulting in larger steps of the particle. On the other hand, if,

$\varphi \in (2-\sqrt{3}, 2) \cup (2, 2+\sqrt{3})$,

then it holds that $|\delta_{ij}| > 1$, and the particle moves randomly around a weighted average of the best positions, p_i and p_j, with step sizes taken randomly from a sinusoidal form (Ozcan & Mohan, 1999). For $\varphi > 4$, the amplitude of the movement grows exponentially, resulting in divergence. This verifies the observations of Kennedy (1998) reported in the previous section. For a thorough analysis of the special cases, $\varphi = 0$, 2, and 4, the reader is referred to the original paper by Ozcan and Mohan (1999).

Besides the particles, Ozcan and Mohan (1999) also analyzed the velocities. Equations (4) and (5) produce, after substitution, the following recursive formula:

$$v_{ij}(t) - (2 - \varphi_1 - \varphi_2)\, v_{ij}(t-1) + v_{ij}(t-2) = 0, \tag{15}$$

with initial conditions, $x_{ij}(0)$, $v_{ij}(0)$, and,

$$v_{ij}(1) = v_{ij}(0) - (\varphi_1 + \varphi_2)\, x_{ij}(0) + v_{ij}(0) + \varphi_1 p_{ij}(t) + \varphi_2 p_{lj}(t).$$

Similarly to equation (6), closed form solutions can be given for equation (15):

$$v_{ij}(t) = \rho_{ij}\, \alpha_{ij}^t + \tau_{ij}\, \beta_{ij}^t, \tag{16}$$

where,

$$\rho_{ij} = v_{ij}(0)/2 + ((\varphi_1 + \varphi_2)/2\delta_{ij})\, v_{ij}(0) + ((\varphi_1 + \varphi_2)/\delta_{ij})\, x_{ij}(0) + (\varphi_1 p_{ij} + \varphi_2 p_{lj})/\delta_{ij}, \tag{17}$$

$$\tau_{ij} = v_{ij}(0)/2 - ((\varphi_1 + \varphi_2)/2\delta_{ij})\, v_{ij}(0) - ((\varphi_1 + \varphi_2)/\delta_{ij})\, x_{ij}(0) - (\varphi_1 p_{ij} + \varphi_2 p_{lj})/\delta_{ij}. \tag{18}$$

Analysis akin to particle trajectories can be made also for the velocity regions of interest. Indeed, for values:

$\varphi = \varphi_1 + \varphi_2 \in (0,4)$,

velocity can be given in the form (Ozcan & Mohan, 1999):

$$v_{ij}(t) = Y_{ij}\sin(\theta_{ij}t) + \Phi_{ij}\cos(\theta_{ij}t), \tag{19}$$

with,

$$Y_{ij} = ((\varphi_1 + \varphi_2)\, 2\, x_{ij}(0) + 2(\varphi_1 p_{ij} + \varphi_2 p_{lj}) + \varphi_1 + \varphi_2) / |\delta_{ij}|,$$

$\Phi_{ij} = v_{ij}(0),$

and

$\theta_{ij} = \text{atan}(|\delta_{ij}| \, / \, |2 - \varphi_1 - \varphi_2|),$

while, for $\varphi > 4$, the value of θ_{ij} shall be increased by π. The reader is referred to the original paper for further details (Ozcan and Mohan, 1999).

In the last part of their study, Ozcan and Mohan (1999) considered equipping their PSO model with inertia weight. The corresponding analysis showed that a constant inertia weight can alter the boundaries between areas of interest, and, consequently, modify the convergence properties of the model. Thus, a careless selection of the inertia weight can have a detrimental effect on efficiency. The case of decreasing inertia weight required more thorough investigation, and was put aside as a topic of future research.

The works of Kennedy (1998) and Ozcan and Mohan (1999) offered a good starting point for unraveling the Ariadne's thread of PSO dynamics, attracting the interest of many researchers. Although a small step was taken towards the theoretical analysis of the algorithm by providing rough parameter bounds, complete fine-tuning was still an open problem. It was not but three years later that a very important - perhaps the most influential - theoretical work on PSO, namely the stability analysis due to Clerc and Kennedy (2002) was published. In this work, new variants of the algorithm were proposed and mathematical formulae for determining its parameters were reported, providing the state-of-the-art PSO variants until nowadays.

Stability Analysis of Particle Swarm Optimization

Clerc and Kennedy (2002) considered the original local PSO (lbest) model:

$v_{ij}(t+1) = v_{ij}(t) + \varphi_1 \, (p_{lj}(t)\text{-}x_{ij}(t)) + \varphi_2 \, (p_{lj}(t)\text{-}x_{ij}(t)),$

$x_{ij}(t+1) = x_{ij}(t) + v_{ij}(t+1),$

$i = 1, 2,\ldots, N, \quad j = 1, 2,\ldots, n,$

where l denotes the best particle in the neighborhood of x_i; and φ_1, φ_2, are samples of a uniform distribution bounded by an upper limit, φ_{max}. The velocity is also considered to be clamped by a parameter, v_{max}, as follows:

$|v_{ij}| \le v_{max}$, for all i and j.

Velocity update can be simplified to the algebraically equivalent form:

$v_{ij}(t+1) = v_{ij}(t) + \varphi \, (p_{ij}(t)\text{-}x_{ij}(t)),$

where p_{ij} is no longer the best position of the particle but the aggregated position:

$$p_{ij} = (\varphi_1 p_{ij} + \varphi_2 p_{lj}) / (\varphi_1 + \varphi_2),$$

with $\varphi = \varphi_1 + \varphi_2$. Stripping further the system down, p_{ij} is assumed to be constant and henceforth denoted simply as p. The parameter φ is also considered to be constant, similarly to the analyses described in previous sections.

The first model studied by Clerc and Kennedy (2002) was one-dimensional and assumed a reduced population size, i.e., it consisted of a deterministic one-dimensional particle that moves according to the following equations:

$$v(t+1) = v(t) + \varphi (p-x(t)), \tag{20}$$

$$x(t+1) = x(t) + v(t+1). \tag{21}$$

Notice that all subscripts denoting particle index and direction are dropped. We can now distinguish two time cases for the particle movement: discrete and continuous. We describe each case separately in the following sections.

The Discrete-Time Case

The first part of the study of Clerc and Kennedy (2002) was purely algebraic, analyzing the movement of a particle in discrete time. The system of equations (20) and (21) was written as a dynamical system:

$$\begin{cases} v_{t+1} = v_t + \varphi y_t, \\ y_{t+1} = -v_t + (1-\varphi)y_t, \end{cases} \tag{22}$$

with $t = 0, 1, 2,\ldots$, and $y_t = p - x_t$, where p is the aggregated best position defined in the previous section. In matrix form, the system reads:

$$P_{t+1} = M P_t = M^t P_0, \tag{23}$$

where M^t denotes the t-th power of matrix M, and,

$$P_t = \begin{bmatrix} v_t \\ y_t \end{bmatrix}, \quad M = \begin{bmatrix} 1 & \varphi \\ -1 & 1-\varphi \end{bmatrix}.$$

Its behavior depends on the eigenvalues of M, which are defined as follows:

$$\begin{cases} \lambda_1 = 1 - \dfrac{\varphi}{2} + \dfrac{\sqrt{\varphi^2 - 4\varphi}}{2}, \\ \lambda_2 = 1 - \dfrac{\varphi}{2} - \dfrac{\sqrt{\varphi^2 - 4\varphi}}{2}. \end{cases} \tag{24}$$

The critical value, $\varphi = 4$, previously identified by Kennedy (1998) and Ozcan and Mohan (1999), appears again as a special case, since it constitutes the limit between the cases of two different real eigenvalues, one eigenvalue of multiplicity 2, and two complex conjugate eigenvalues.

Since the matrix M^t defines completely the system in equation (23), we can use linear algebra to make it eager to investigation through further simplification. Thus, if $\lambda_1 \neq \lambda_2$, or equivalently, $\varphi \neq 4$, there exists a matrix, A, which produces the following similarity transformation of M:

$$L = A \, M A^{-1} = \begin{bmatrix} \lambda_1 & 0 \\ 0 & \lambda_2 \end{bmatrix}.$$

Assuming that A has the canonical form:

$$A = \begin{bmatrix} a & 1 \\ c & 1 \end{bmatrix},$$

and by doing the proper mathematical manipulations, it is derived that (Clerc & Kennedy, 2002):

$$A = \begin{bmatrix} \varphi + \sqrt{\varphi^2 - 4\varphi} & 2\varphi \\ \varphi - \sqrt{\varphi^2 - 4\varphi} & 2\varphi \end{bmatrix}.$$

Since it holds that:

$$L = A \, M A^{-1} \Leftrightarrow A^{-1} L A = M,$$

we can go back to equation (23) and substitute M with its equivalent matrix, $A^{-1} L A$:

$$P_{t+1} = M P_t \Leftrightarrow P_{t+1} = A^{-1} L A P_t \Leftrightarrow A P_{t+1} = L A P_t.$$

If we define, $Q_t = A P_t$, then the system is described by:

$$Q_{t+1} = L Q_t = L^t Q_0, \qquad (25)$$

where L is a diagonal matrix, with its main diagonal consisting of the eigenvalues of M. Thus, its t-th power is defined as:

$$L^t = \begin{bmatrix} \lambda_1^t & 0 \\ 0 & \lambda_2^t \end{bmatrix}.$$

The study becomes easier now because cyclic behavior, determined by $Q_t = Q_0$, can be simply achieved by requiring that L^t is the unitary matrix in equation (25), or equivalently:

$$\lambda_1^t = \lambda_2^t = 1.$$

In the case of $\varphi \in (0,4)$, the eigenvalues defined in equation (24) become complex. Thus, we have:

$$\begin{cases} \lambda_1^t = \cos(t\theta) + i\sin(t\theta), \\ \lambda_2^t = \cos(t\theta) - i\sin(t\theta), \end{cases}$$

and the system cycles for $\theta = (2k\pi)/t$. The corresponding solutions of φ are given as follows:

$$\varphi = 2\,(1-\cos(2k\pi/t)), \quad k = 1, 2, \dots, t\text{-}1.$$

On the other hand, if $\varphi > 4$, there is no cyclic behavior of the system, and it is proved that the distance of P_t from the origin is monotonically increasing (Clerc & Kennedy, 2002). In the limit case, $\varphi = 4$, the system will have either an oscillatory behavior, i.e.,

$$P_{t+1} = -P_t,$$

if the initial vector P_0 is an eigenvector of M, or it will have an almost linear increase or decrease of the norm $\|P_t\|$, for $y_0 > 0$ or $y_0 < 0$, respectively (Clerc & Kennedy, 2002).

These derivations provide useful intuition regarding the behavior of the system in the discrete-time case under different values of φ. Let us now study the same system from the continuous-time viewpoint.

The Continuous-Time Case

Clerc and Kennedy (2002) also conducted a continuous-time analysis based on differential equations. More specifically, equations (20) and (21) were merged to produce the following recurrent velocity equation:

$$v(t+2) + (\varphi\text{-}2)\, v(t+1) + v(t) = 0.$$

This becomes a second-order differential equation in continuous-time (Clerc & Kennedy, 2002):

$$\frac{\partial^2 v}{\partial t^2} + \ln(\lambda_1 \lambda_2)\frac{\partial v}{\partial t} + \ln(\lambda_1)\ln(\lambda_2)v = 0,$$

where λ_1 and λ_2 are the solutions of the equation:

$$\lambda^2 + (\varphi\text{-}2)\,\lambda + 1 = 0,$$

resulting in the forms of equation (24). Thus, the general form of the velocity becomes (Clerc & Kennedy, 2002):

$$v(t) = c_1 \lambda_1^t + c_2 \lambda_2^t,$$

while $y(t)$ has the form:

$$y(t) = (c_1 \lambda_1^t (\lambda_1 - 1) + c_2 \lambda_2^t (\lambda_2 - 1)) / \varphi,$$

where c_1 and c_2 depend on $v(0)$ and $y(0)$. If $\lambda_1 \neq \lambda_2$, then c_1 and c_2 have the following expressions (Clerc & Kennedy, 2002):

$$\begin{cases} c_1 = \dfrac{-\varphi y(0) - (1 - \lambda_2)v(0)}{\lambda_2 - \lambda_1}, \\ c_2 = \dfrac{\varphi y(0) + (1 - \lambda_1)v(0)}{\lambda_2 - \lambda_1}. \end{cases}$$

Analysis similar to the discrete-time case was made for the regions of interest, in terms of the value of φ. Thus, it was shown that system explosion depends on whether the condition:

$$\max \{|\lambda_1|, |\lambda_2|\} > 1,$$

holds (Clerc & Kennedy, 2002). After these additional confirmations, Clerc and Kennedy (2002) generalized their model to approximate the original PSO form. Their major developments are reported in the next section.

Generalized System Representations

Clerc and Kennedy (2002) extended their study in more generalized models. The system of equation (22), produced by equations (20) and (21), was generalized by adding new coefficients:

$$\begin{cases} v_{t+1} = \alpha v_t + \beta \varphi y_t, \\ y_{t+1} = -\gamma v_t + (\delta - \eta \varphi)y_t, \end{cases} \qquad (26)$$

with φ being a positive real number. The new system matrix is:

$$M' = \begin{bmatrix} \alpha & \beta \varphi \\ -\gamma & \delta - \eta \varphi \end{bmatrix},$$

and let, λ_1', λ_2', be its eigenvalues. Then, the system of equation (26) can be written in its analytic form as follows:

$$\begin{cases} v(t) = c_1 (\lambda_1')^t + c_2 (\lambda_2')^t, \\ y(t) = \dfrac{1}{\beta \varphi}(c_1 (\lambda_1')^t (\lambda_1' - \alpha) + c_2 (\lambda_2')^t (\lambda_2' - \alpha)), \end{cases} \qquad (27)$$

with,

$$\begin{cases} c_1 = \dfrac{-\beta\varphi y(0) - (\alpha - \lambda_2)v(0)}{\lambda_2' - \lambda_1'}, \\[3mm] c_2 = \dfrac{\beta\varphi y(0) + (\alpha - \lambda_1)v(0)}{\lambda_2' - \lambda_1'}. \end{cases}$$

We can now define two constriction coefficients (recall the corresponding PSO variant described in the previous chapter), χ_1 and χ_2, such that:

$$\lambda_1' = \chi_1\lambda_1 \text{ and } \lambda_2' = \chi_2\lambda_2,$$

where λ_1 and λ_2 are defined by equation (24). Then, the following forms of χ_1, χ_2, are obtained (Clerc & Kennedy, 2002):

$$\begin{cases} \chi_1 = \dfrac{\alpha + \delta - \eta\varphi + \sqrt{(\eta\varphi)^2 + 2\varphi(\alpha\eta - \delta\eta - 2\beta\gamma) + (\alpha - \delta)^2}}{2 - \varphi + \sqrt{\varphi^2 - 4\varphi}}, \\[4mm] \chi_2 = \dfrac{\alpha + \delta - \eta\varphi - \sqrt{(\eta\varphi)^2 + 2\varphi(\alpha\eta - \delta\eta - 2\beta\gamma) + (\alpha - \delta)^2}}{2 - \varphi - \sqrt{\varphi^2 - 4\varphi}}. \end{cases} \tag{28}$$

It is worth noting that the system of equation (26) requires an integer time parameter, t, while its analytical equivalent of equation (27) can also admit any positive real value of t. In the latter case, $v(t)$ and $y(t)$ become complex numbers, and the behavior of the system can be investigated in a 5-dimensional search space.

Clerc and Kennedy (2002) distinguished the following four interesting model classes, characterized by different relations among the five parameters, α, β, γ, δ, and η, so that real constriction coefficients are ensured:

1. *Class 1*: This class of models is obtained for,

    $\alpha = \delta$ and $\beta\gamma = \eta^2$.

    In order to ensure real coefficients, one can simply take the additional condition:

    $\chi_1 = \chi_2 = \chi \in \mathbf{R}$.

    Then, a class of solutions is given by assuming that:

    $\alpha = \beta = \gamma = \delta = \eta = \chi$.

2. *Class 1'*: This class of models is obtained for,

    $\alpha = \beta$ and $\gamma = \delta = \eta$.

    Taking again, $\chi_1 = \chi_2 = \chi$, we obtain:

    $\alpha = (2-\varphi)\chi+\varphi-1$ and $\gamma = \chi$ or $\gamma = \chi/(\varphi-1)$.

3. *Class 1''*: This class of models is obtained for,

    $\alpha = \beta = \gamma = \eta$,

and,

$$\alpha = \frac{2\delta + (\chi_1 + \chi_2)(\varphi - 2) - (\chi_1 - \chi_2)\sqrt{\varphi^2 - 4\varphi}}{2(\varphi - 1)}.$$

The case of $\delta = 1$ has been studied extensively for historical reasons (Clerc & Kennedy, 2002).

4. *Class 2*: This class is obtained for,
    $\alpha = \beta = 2\delta$ and $\eta = 2\gamma$,
    and, for $2\gamma\varphi > \delta$, we have:

$$\begin{cases} \delta = \chi\lambda_1 = \chi \dfrac{2 - \varphi + \sqrt{\varphi^2 - 4\varphi}}{2}, \\ \gamma = \chi \dfrac{2 - \varphi + 3\sqrt{\varphi^2 - 4\varphi}}{4\varphi}. \end{cases}$$

Also, Clerc and Kennedy (2002) provided a set of conditions that involve the maximum value, φ_{max}, of φ such that the system remains continuous and real. The reader is referred to the original paper for more details.

From the analysis above, the following models that use a single constriction coefficient were distinguished and studied further (Clerc and Kennedy, 2002):

1. *Model Type 1*: This model is defined as:

$$\begin{cases} v(t+1) = \chi(v(t) + \varphi y(t)), \\ y(t+1) = -\chi(v(t) + (1 - \varphi)y(t)), \end{cases}$$

with its converge criterion being satisfied for $\chi < \min(|\lambda_1|^{-1}, |\lambda_2|^{-1})$, resulting in the coefficient (Clerc & Kennedy, 2002):

$$\chi = \kappa / |\lambda_2|, \kappa \in (0,1).$$

2. *Model Type 1´*: This model is defined as:

$$\begin{cases} v(t+1) = \chi(v(t) + \varphi y(t)), \\ y(t+1) = -\chi(v(t) + (1 - \varphi)y(t)), \end{cases}$$

with coefficient (Clerc & Kennedy, 2002):

$$\chi = \kappa / |\lambda_2|, \kappa \in (0,1), \varphi \in (0,2).$$

Further investigation revealed that convergence is achieved for:

$$\chi = \frac{2 + \varphi - 2\sqrt{\varphi}}{2}, \quad \varphi \in (0,4),$$

but not for higher values of φ (Clerc & Kennedy, 2002).

3. *Model Type 1''*: This model is defined as,

$$\begin{cases} v(t+1) = \chi(v(t) + \varphi y(t)), \\ y(t+1) = -\chi v(t) + (1 - \chi\varphi)y(t), \end{cases} \quad (29)$$

and, with proper substitutions, it results in the following form:

$$\begin{cases} v(t+1) = \chi[v(t) + \varphi(p - x(t))], \\ x(t+1) = x(t) + v(t+1), \end{cases}$$

which is the constriction coefficient PSO variant, described in the previous chapter as one of the state-of-the-art variants nowadays. Special analysis for this approach resulted in the following relation for the determination of the constriction coefficient (Clerc & Kennedy, 2002):

$$\chi = \begin{cases} \dfrac{2\kappa}{\left| \varphi - 2 + \sqrt{\varphi^2 - 4\varphi} \right|}, & \text{for } \varphi > 4, \\ \kappa, & \text{otherwise}, \end{cases} \quad (30)$$

for $\kappa \in (0,1)$.

So far, the analysis was restricted to one-dimensional particles, with constant φ and a single fixed best position, p. Clerc and Kennedy (2002) generalized their theoretical derivations by using random values of φ and two vector terms for velocity update, as in the original PSO model. Thus, they showed that system explosion can be controlled simply by using proper values of the constriction coefficient.

Different PSO variants can be defined by using either of the aforementioned model types. Then, convergence rates depend on the selected system parameters. For example, the value $\kappa = 1$, which is a limit case for the models above, results in slow convergence, thereby promoting better exploration of the search space. This justifies the default parameter setting:

$$\chi = 0.729, \quad c_1 = c_2 = 2.05,$$

of the constriction coefficient PSO variant reported in the previous chapter, which corresponds to equation (30) evaluated for $\kappa = 1$ and $\varphi_1 = \varphi_2 = 2.05$ (hence, $\varphi = \varphi_1 + \varphi_2 = 4.1$). The nice performance properties of the Type 1'' model were also verified experimentally by Clerc and Kennedy (2002), and it was finally established as the state-of-the-art variant of PSO.

Clerc and Kennedy's analysis (2002) was perhaps the most important theoretical work on PSO. However, it was very complicated for readers with a poor mathematical background. One year after its publication, their work was reconsidered by Trelea (2003). He provided a simplified view of the previous complex analysis and performed a series of experiments on typical benchmark problems. Besides the typical parameter setting, $\chi = 0.729$ and $c_1 = c_2 = 2.05$, he also investigated a promising alternative, namely $\chi = 0.6$ and $c_1 = c_2 = 2.83$. His results verified that the latter can also be a beneficial setting, retaining nice convergence properties. Thus, it can now be considered as an alternative parameter setting for the constriction coefficient version of PSO.

Different PSO variants also became the subject of further theoretical investigations. Zheng *et al.* (2003) presented an analysis for the inertia weight variant. They considered a PSO model with increasing inertia weight, $w \in (0.4, 0.9)$, and $\varphi_1, \varphi_2 \in (0.5, 2.0)$, claiming that it outperforms standard PSO, at least in a small set of widely used benchmark problems. Van den Bergh and Engelbrecht (2006) reviewed the existing theoretical works a few years later, also providing a study on the inertia weight variant. Their results agreed with previous studies, further improving our understanding on the most popular PSO variants. A few more papers by different authors appeared in the past two years (Jiang *et al.*, 2007; Xiao *et al.*, 2007); however, they mostly reproduced or reviewed existing results, thus we refer the interested reader directly to these papers for further details.

The presented trajectory and stability analyses determined a set of efficient PSO variants, and provided instructions on parameter setting to achieve satisfactory convergence properties. However, we must note that all studies consider convergence as the ability to reach a state of equilibrium rather than achieve a true local or global optimum. The latter would require additional information on the mathematical properties of the objective function. At the same time, the obtained guidelines for parameter setting referred to rather general environments, without taking into consideration any peculiarities of the problem at hand. However, if only a specific problem is of interest, then one can design algorithms specifically suited to it. Of course, this does not adhere to the requirement to "provide satisfactory solutions as soon as possible", however it can save much computational effort if instances of the same complex problem are considered repeatedly. Sensitivity analysis using computational statistics has proved to be a very useful tool in such cases (Bartz-Beielstein *et al.*, 2004). An excellent contribution to experimental research in evolutionary computation is provided in Bartz-Beielstein (2006).

In the next section, we expose the design of PSO algorithms through computational statistics methodologies, and illustrate its application on a benchmark and a real world problem.

DESIGN OF PSO ALGORITHMS USING COMPUTATIONAL STATISTICS

Bartz-Beielstein *et al.* (2004) proposed an approach for determining the parameters of PSO, tailored to the optimization problem at hand. The approach employs techniques from computational statistics and statistical experimental design, and is applicable with any optimization algorithm (Bartz-Beielstein, 2006). Its operation was illustrated on a well-known benchmark problem, as well as on a simplified model of a real world application, which involves the optimization of an elevator group controller. The following sections present fundamental concepts of the technique along with results for the aforementioned applications.

Fundamental Concepts of Computational Statistics

Computational statistics is a scientific field that embraces computationally intensive methods (Gentle *et al.*, 2004), such as *experimental design* and *regression analysis*, which can be used to analyze the experimental setting of algorithms on specific test problems. A fundamental issue in these approaches is the determination of variables with significant impact on performance, which can be quantitatively defined as the averaged best function value obtained over a number of independent experiments. Such measures have been used for the empirical analysis of PSO (Shi & Eberhart, 1999), attempting to answer fundamental questions such as:

a.   How do variations of swarm size influence performance?
b.   Are there interactions between swarm size and the inertia weight value?

Addressing such issues can enhance our understanding on the operation of PSO, and help towards the design of more efficient variants.

The approach of Bartz-Beielstein *et al.* (2004) combines three types of statistical techniques, namely *design of experiments* (DOE), *classification and regression trees* (CART), and *design and analysis of computer experiments* (DACE). Thorough descriptions of these methodologies can be found in Breiman *et al.*, (1984), Montgomery (2001), Santner *et al.* (2003), while Bartz-Beielstein and Markon (2004) offer a comparison of the three methodologies for direct search algorithms. Below, we provide a rough description of DACE, which constitutes the core methodology in the study of Bartz-Beielstein *et al.* (2004) on designing PSO algorithms.

DACE: Design and Analysis of Computer Experiments

Let a_d denote a vector, called the *algorithm design*, which contains specific settings of an algorithm. In PSO, a_d can contain parameters such as swarm size, social and cognitive parameters, inertia weight etc. A design can be represented only with one vector, and the optimal one is denoted as a_d^*. Also, let p_d represent a *problem design*, i.e., a structure that contains problem-related information, such as its dimension, available computational resources (number of function evaluations) etc. Then, a run of an algorithm can be treated as a mapping of the two designs, a_d and p_d, to a stochastic output, $Y(a_d, p_d)$.

The main goal in DACE is the determination of the design, a_d^*, which optimizes performance in terms of the required number of function evaluations. DACE was introduced for deterministic computer experiments; therefore, its use in stochastic cases requires its repeated application. The specification of a process model in DACE is similar to the selection of a linear or quadratic model in classical regression, and is analyzed in following paragraphs.

DACE can be very useful for interpolating observations from computationally expensive simulations. For this purpose, a deterministic function shall be evaluated at m different design points. Sacks *et al.* (1989) expressed a dynamic response, $y(x)$, for a d-dimensional input vector, x, as the realization of a regression model, F, and a stochastic process, Z, as follows:

$$Y(x) = F(\beta, x) + Z(x). \tag{31}$$

This model generalizes the classical regression model, $Y(x) = \beta x + \varepsilon$. The stochastic process $Z(x)$ is assumed to have a zero mean, and covariance equal to:

$$V(\omega, x) = \sigma^2 R(\theta, \omega, x),$$

between $Z(\omega)$ and $Z(x)$, where σ^2 is the process variance and $R(\theta, \omega, x)$ is a correlation model. The correlation function should be chosen with respect to the underlying process (Isaaks & Srivastava, 1989). Lophaven *et al.* (2002) provide a useful discussion on seven such models. Bartz-Beielstein *et al.* (2004) used correlations of the form:

$$R(\theta, \omega, x) = \prod_{j=1}^{d} R_j(\theta, \omega_j - x_j),$$

with Gaussian correlation functions:

$$R_j(\theta, h_j) = \exp(-\theta_j h_j^2),$$

where $h_j = \omega_j - x_j$ and $\theta_j > 0$. Then, the regression model can be defined by using ρ functions:

$$f_j: \mathbf{R}^d \rightarrow \mathbf{R}, \quad j = 1, 2, \ldots, \rho,$$

as follows (Sacks *et al.*, 1989):

$$F(\beta, x) = \sum_{j=1}^{\rho} \beta_j f_j(x) = \langle f(x), \beta \rangle,$$

where,

$$f(x) = (f_1(x), f_2(x), \ldots, f_\rho(x))^{\mathrm{T}} \text{ and } \beta = (\beta_1, \beta_2, \ldots, \beta_\rho)^{\mathrm{T}}.$$

DACE also provides the mean square-error of the predictor or an estimation of the prediction error at an untried point. In the next section, we expose the basic steps of its application for producing sequential designs of an algorithm.

Application of DACE on Sequential Designs

The algorithm design, a_d, must be specified prior to experimentation with an algorithm. Designs that use sequential sampling are often more efficient than designs with fixed sample sizes. In this case, an initial design $a_d^{(0)}$ is specified, and information gained in the first runs is exploited to define the next design, $a_d^{(1)}$, and so on. Thus, new design points are chosen in a more sophisticated manner that enhances performance in many practical situations.

Several sequential sampling approaches with adaptation have been proposed for DACE (Sacks *et al.*, 1989). Bartz-Beielstein *et al.* (2004) adopted a sequential sampling based on the expected improvement of the algorithm. The concept of improvement arose in the analysis of Santner *et al.* (2003, p. 178) and is defined as follows: let, $y_{\min}^k$, denote the smallest detected function value after k runs of a heuristic global optimization algorithm; $x \in a_d$ be a component of the design; and $y(x)$ be the response of the algorithm, which is a realization of $Y(x)$ defined in equation (31). Then, the improvement of the algorithm is defined as:

$$\Delta = \begin{cases} y_{\min}^k - y(x), & \text{if } y_{\min}^k - y(x) > 0, \\ 0, & \text{otherwise.} \end{cases}$$

(32)

The discussion in Santner *et al.* (2003) leads to the conclusion that new design points are promising if there is either a high probability that their predicted output is below the current observed minimum and/ or there is great uncertainty in the predicted output. This result is in line with the general intention to avoid sites that guarantee worse results.

The complete sequential design approach consists of the twelve steps reported in Table 4. Step 1 consists of pre-experimental planning. At this stage, the practitioner defines the object of study as well as sources to collect data exactly. Although this step seems trivial, formulating a generally accepted goal is not a simple task in practice. *Discovery, confirmation,* and *robustness,* are the only three possible scientific goals of an experiment. Discovery asks what happens if new operators are implemented. Confirmation analyzes how the algorithm behaves on different problems, and robustness asks for conditions that decrease performance.

Statistical methods like *run-length distributions* (RLD) provide suitable means for measuring performance and describe qualitatively the behavior of optimization algorithms. The construction of an RLD plot requires k runs of the algorithm on a given problem instance, using different random number generator seeds. For each run, the maximum number of function evaluations, $t_{\max}$, is set to a relatively high value. For each successful run, the number of required function evaluations, t_{run}, is recorded. If the run fails, t_{run}, is set to infinity. These results are then represented by an empirical *cumulative distribu-*

Table 4. The twelve steps of the sequential approach for tuning the performance of direct search algorithms

Step	Action
Step 1.	Pre–experimental planning.
Step 2.	Scientific hypothesis.
Step 3.	Statistical hypothesis.
Step 4.	Specifications: a. Optimization problem. b. Constraints. c. Initialization method. d. Termination method. e. Algorithm (important factors). f. Initial experimental design. g. Performance measure.
Step 5.	Experimentation.
Step 6.	Statistical modeling of data and prediction.
Step 7.	Evaluation and visualization.
Step 8.	Optimization.
Step 9.	Termination: if the obtained solution is good enough or the maximum number of iterations has been reached, then go to Step 11.
Step 10.	Update design and go to Step 5.
Step 11.	Rejection/acceptance of the statistical hypothesis.
Step 12.	Objective interpretation of the results from Step 11.

tion function (CDF). Let $t_{run}(j)$ be the run-length for the *j*-th successful run. Then, the empirical CDF is defined as (Hoos, 1998):

$$\Pr(t_{run}(j) \le t) = \frac{T_{run}(j)}{k},$$

where $T_{run}(j)$ denotes the number of indices *j*, such that $t_{run}(j) \le t$.

Step 2 consists of formulating a scientific hypothesis, based on the determined experimental goal. For example, a hypothesis can be a statement such as "the employed scheme improves the performance of the algorithm." Then, in Step 3 a statistical hypothesis, e.g., "there is no difference in means when comparing performances of two competing schemes," needs to be determined. Step 4 requires the specification of an optimization problem, along with its essential elements: possible constraints (e.g., maximum number of function evaluations); the initialization method; termination conditions; the algorithm and its important factors; an initial experimental design; and a performance measure.

Recalling the discussion in the beginning of the current chapter, initialization can be either deterministic or random, with the latter being the most popular for population-based stochastic algorithms. In general, we can have the following types of initialization for the repeated application of an algorithm:

(I-1) **Deterministic initialization with constant seed:** According to this scheme, the algorithm is repeatedly initialized to the same point, $x^{(0)}$, which is defined explicitly by the user.

(I-2) **Deterministic initialization with different seeds:** The algorithm is initialized to a different point for each run. The initial points are user-defined and they all lie within an interval, $[x_{low}, x_{up}]$.

(I-3) **Random initialization with constant seed:** This is the random counterpart of (I-1), where the algorithm assumes the same, randomly selected initial point for each run.

(I-4) **Random initialization with different seeds:** This is the random counterpart of (I-2), where the algorithm is initialized at each run to a different point, randomly selected within an interval, $[x_{low}, x_{up}]$.

The scheme (I-4) is typically used with evolutionary algorithms. On the other hand, (I-1) is a common choice for local gradient-based algorithms such as quasi-Newton methods, where the user must provide an initial point sufficiently close to the global minimum in order to guarantee convergence.

Termination of the algorithm occurs when one or more user-defined conditions are met. The most common termination conditions are:

(T-1) **Domain convergence:** The sequence of candidate solutions produced by the algorithm converges to a minimizer.

(T-2) **Function value convergence:** The function values of the produced candidate solutions converge to a minimum.

(T-3) **Algorithm stalls:** The algorithm is not able to produce any new candidate solutions. Such cases occur when, for example, an evolutionary algorithm looses completely its diversity or velocities in PSO become practically equal to zero.

(T-4) **Exhausted resources:** This case refers to the inevitable termination of the algorithm when the maximum available computational budget (e.g., number of function evaluations or CPU time) is exhausted.

A discussion on the termination conditions commonly used with PSO is provided in the final section of the present chapter.

After determining the algorithm and the initial design, performance measures must be defined. In evolutionary computation, performance is usually measured by statistical moments of the number of function evaluations required to find a solution with a desirable accuracy. The mean value, standard deviation, minimum, and maximum number of function evaluations are such measures, used in the vast majority of experimental works.

In Step 5, the experiment is finally conducted. Preliminary (pilot) runs can give rough estimates regarding the experimental error, run times, and consistency of experimental design. At this stage, RLDs can be very useful. For probabilistic search algorithms, functions may be evaluated several times (Santner *et al.*, 2003). The experimental results provide the base for modeling and prediction in Step 6, where the model is fitted and a predictor is obtained per response.

The model is evaluated in Step 7, where visualization techniques can be also applied. For example, simple graphical methods from exploratory data analysis as well as histograms and scatter plots can be used for the detection of outliers. If inappropriate initial ranges were chosen for the designs (e.g., very wide initial ranges), the visualization of the predictor can provide more suitable (narrower) ranges in the next stage. Several techniques for assessing the validity of the model have been proposed. Cross validation predictions versus actual values, as well as standardized cross validation residuals versus cross validation predictions are widely used. Sensitivity analysis can be used to ascertain the dependence of the statistical model on its factors. Thorough presentations of variance-based methods for sensitivity analysis can be found in Chan *et al.* (1997, 2000), Saltelli (2002), Saltelli *et al.* (2000), while Santner *et al.* (2003, p. 193) provide a description of ANOVA-type decompositions.

Computation of sensitivity indices can be done by decomposing response into average, main effects per input, two-input interactions, and higher-order interactions (Sacks *et al.*, 1989, p. 417). Additionally, graphical methods can be used to visualize the effects of different factors and their interactions on predictors. Predicted values can be plotted to support the analysis, while MSE is used to asses prediction accuracy. At this point, we must underline that statistical models provide only guidelines for further experiments, not proofs that connect factors with particular effects. If the predicted values are inaccurate, experimental setup has to be reconsidered. This especially concerns the specification of scientific goals and ranges of design variables. Otherwise, if further experiments are necessary, new promising design points can be determined in Step 8.

Finally, a termination criterion is checked in Step 9. If it is not fulfilled, based on the expected improvement defined by equation (32), new candidate design points can be generated in Step 10. A new design point is selected, if there is a high probability that the predicted output is below the currently observed minimum and/or it is characterized by large uncertainty. Otherwise, if the termination criterion is fulfilled and the obtained solution is good enough, the final statistical evaluation takes place in Step 11, summarizing the results. A comparison between the first and the improved configuration shall be made. For this purpose, techniques from exploratory data analysis can complement the analysis at this stage. Besides that, graphical representations such as boxplots, histograms, and RLDs, can be used to support the final statistical decision.

Finally, the scientific importance of results remains to be decided in Step 12, since any difference, although statistically significant, can be scientifically meaningless. Thus, it is up to the practitioner to assess the importance of results based on personal experience. The experimental setup should be considered again at this stage, and questions like "have suitable test functions or performance measures been chosen?", or "did floor or ceiling effects occur?" must be answered. Simple test problems may cause such ceiling effects. If two algorithms, A and B, achieve their maximum level of performance (or close to it), then the hypothesis "performance of A is better than performance of B" should not be confirmed (Cohen, 1995). Floor effects describe the same phenomenon on the opposite side of the performance scale, i.e., the test problem is so hard that nearly no algorithm can solve it correctly. Such effects can occur when the number of function evaluations is very small. In these cases, performance profiles can help the practitioner to decide whether ceiling effects have occurred or not.

In the next section, we put forward the sequential design technique for tuning PSO on a benchmark function as well as on a simulated real world problem.

Applications of PSO with the Sequential Design Approach

Bartz-Beielstein *et al.* (2004) applied the procedure described in the previous sections on both the inertia weight PSO variant, defined as:

$$v_{ij}(t+1) = w\, v_{ij}(t) + c_1 R_1\, (p_{ij}(t)\text{-}x_{ij}(t)) + c_2 R_2\, (p_{gj}(t)\text{-}x_{ij}(t)),$$

$$x_{ij}(t+1) = x_{ij}(t) + v_{ij}(t+1),$$

$$i = 1, 2,\ldots, N, \quad j = 1, 2,\ldots, n,$$

and the constriction coefficient variant, defined as:

$$v_{ij}(t+1) = \chi\, [v_{ij}(t) + c_1 R_1\, (p_{ij}(t)\text{-}x_{ij}(t)) + c_2 R_2\, (p_{gj}(t)\text{-}x_{ij}(t))],$$

$$x_{ij}(t+1) = x_{ij}(t) + v_{ij}(t+1),$$

$$i = 1, 2,\ldots, N, \quad j = 1, 2,\ldots, n.$$

We will henceforth denote the two variants as PSO[in] and PSO[con], respectively. In PSO[in], the parameters:

$$w_{scale},\, w_{iterScale} \in [0,1],$$

were used to parameterize the linear decrease of inertia weight, w, from a maximum value, w_{max}, to a scaled value, $w_{max} \times w_{scale}$, over $t_{max} \times w_{iterScale}$ iterations of the algorithm, were t_{max} stands for the maximum allowed number of iterations. For the remaining $(1\text{-}w_{iterScale}) \times t_{max}$ iterations, the inertia weight remains constant to the value $w_{max} \times w_{scale}$. In parallel, all particles were considered to lie within a range, $[x_{min}, x_{max}]^n$, while velocity was clamped within $[-v_{max}, v_{max}]$, as follows:

$$x_{min} \le x_{ij} \le x_{max} \text{ and } -v_{max} \le v_{ij} \le v_{max},$$

$$i = 1, 2,\ldots, N, \quad j = 1, 2,\ldots, n.$$

In the experiments of Bartz-Beielstein *et al.* (2004), the global (gbest) PSO variant was used, i.e., p_g denotes the best particle of the whole swarm. Table 5 reports all exogenous parameters of PSO[in] and PSO[con], with the latter given in terms of its equivalent inertia weight representation. In fact, PSO[con] is equivalent to PSO[in] for the parameter values, $w_{max} = 0.729$, $c_1 = c_2 = 1.4944$, $w_{scale} = 1.0$, and $w_{iterScale} = 0.0$.

The problem design, p_d, which summarizes all information of Step 4 in the sequential design of Table 4, is reported in Table 6. More specifically, the number of experiment; number of runs, k; maximum number of function evaluations, t_{max}; problem dimension, n; initialization and termination method; lower, x_{min}, and upper, x_{max}, bounds for the initialization; as well as the corresponding optimization problems are reported. The considered problems are discussed in the following sections.

The algorithm design, a_d, is described in Table 7, summarizing tasks (e) to (f) of Step 4. An experimental design, e_d, consists of both (problem and algorithm) designs. The results reported in Table 7 refer to PSO[in] of experiment 1 in Table 6, which optimizes the 10-dimensional Rosenbrock function. *Latin hypercube designs* (LHD) were adopted in the experiments, as reported by Bartz-Beielstein *et al.* (2004). We denote with $a_d^{(l)}$ and $a_d^{(u)}$ the lower and upper bounds, respectively, for the generation of the LHD, while a_d^* denotes the parameter settings of the improved design found by the sequential approach.

PSO[con] requires the determination of four exogenous strategy parameters, namely the swarm size, N; the constriction factor, χ; the parameter $\varphi = c_1 + c_2$; and the maximum velocity, v_{max}. Aslett *et al.* (1998) reported that, according to their experience with the stochastic process model, ten times the expected number of important factors constitutes an adequate number of runs for the initial LHD. Thus, an LHD with at least $m = 15$ design points was chosen. This is the minimum number of design points to fit a

Table 5. The exogenous parameters of both PSO variants, with the constriction coefficient variant represented in an equivalent inertia weight form

Symbol	Range	PSO[in]	PSO[con]
N	**N**	40	40
c_1	$\mathbf{R_+}$	2	1.4944
c_2	$\mathbf{R_+}$	2	1.4944
w_{max}	$\mathbf{R_+}$	0.9	0.729
w_{scale}	$\mathbf{R_+}$	0.4	1.0
$w_{iterScale}$	$\mathbf{R_+}$	1.0	0.0
v_{max}	$\mathbf{R_+}$	100	100

Table 6. Problem design, p_d for the PSO runs. The number of experiment; number of runs, k; maximum number of function evaluations, t_{max}; dimension of the problem, n; initialization and termination methods; lower, x_{min}, and upper, x_{max}, initialization bounds; as well as the optimization problem under consideration are reported

Exp.	k	t_{max}	n	Init.	Term.	x_{min}	x_{max}	Problem
1	50	2500	10	(I-4)	(T-4)	15	30	Rosenbrock
2	50	1000	12	(I-3)	(T-4)	-10	10	S-Ring

Table 7. Algorithm design, a_d, for the PSO[in] variant that corresponds to Experiment 1 of Table 6, which optimizes the 10–dimensional Rosenbrock function. We denote with $a_d^{(l)}$ and $a_d^{(u)}$ the lower and upper bounds, respectively, for the generation of the LHD, while a_d* denotes the parameter settings of the improved design found by the sequential approach

Design	N	c_1	c_2	w_{max}	w_{scale}	$w_{iterScale}$	v_{max}
$a_d^{(l)}$	5	1.0	1.0	0.7	0.2	0.5	10
$a_d^{(u)}$	100	2.5	2.5	0.99	0.5	1	750
a_d*	21	2.25413	1.74587	0.788797	0.282645	0.937293	11.0496

DACE model that consists of a second order polynomial regression model and a Gaussian correlation function, since the former requires:

$$1 + \sum_{i=1}^{4} i = 11$$

design points, while the latter requires 4 design points. Note that, for $m = 15$, there are no degrees of freedom left to estimate the mean square-error of the predictor (Santner *et al.*, 2003). Let us now analyze the application of PSO on each considered problem, individually.

Application on the Rosenbrock Function

The first test case considered by Bartz-Beielstein *et al.* (2004) was the optimization of the Rosenbrock function. This is a well-known benchmark problem, defined as TP_{UO-2} in Appendix A of the book at hand, and reproduced below for presentation consistency:

$$f(x) = \sum_{i=1}^{n-1} [(1 - x_i)^2 + 100(x_{i+1} - x_i^2)^2],$$ (33)

with $x \in \mathbf{R}^n$. The selection of this simple problem was based on the intention of Bartz-Beielstein *et al.* (2004) to demonstrate the workings of DACE in a simple manner, as well as reveal its potential to improve performance in terms of the required number of function evaluations.

Let us now analyze each step of Table 4 for the case of PSO[in] on the Rosenbrock function, as a comprehensive example of the complete application of DACE (Bartz-Beielstein *et al.*, 2004):

(Step 1) **Pre-experimental planning:** Pre-experimental tests to explore the optimization potential supported the assumption that tuning might improve the performance of PSO on the Rosenbrock function. RLDs revealed the existence of a configuration able to complete the run successfully using less than 8000 iterations for nearly 80% of the cases. This was less than half the number of function evaluations used in the reference study of Shi and Eberhart (1999), justifying the usefulness of further analysis.

(Step 2) **Scientific hypothesis:** There exists a parameterization, a_d*, of PSO[in] that improves efficiency significantly.

(Step 3) **Statistical hypothesis:** PSO[in] with the parameterization $a_d{}^*$ outperforms PSO[in] with the default parameterization, $a_d{}^{(0)}$, used in (Shi & Eberhart, 1999).

(Step 4) **Specification:** The 10-dimensional instance of the Rosenbrock function defined by equation (33) is the optimization problem under consideration. Its global minimizer is $x^* = (1, 1,\ldots, 1)^{\mathrm{T}}$, with global minimum, $f^* = 0$. In accordance to the experimental design of Shi and Eberhart (1999), the mean fitness value of the best particle of the swarm was recorded in 50 runs, and it is denoted as $f_{\mathrm{B}}{}^{(50)}$. For the production of RLD plots, a threshold value to distinguish successful from unsuccessful runs was specified. Thus, a run configuration was classified as successful, if $f_{\mathrm{B}}{}^{(50)} < f_{\mathrm{Shi}}$, where $f_{\mathrm{Shi}} = 96.1715$ is the corresponding value reported in (Shi & Eberhart, 1999). The initial problem design of Table 6 was used, together with the corresponding algorithm design of Table 7, which covers a wide range of interesting parameter settings (regions of interest). No problem-specific knowledge for the Rosenbrock function was used, expecting that the sequential approach would guide the search towards promising regions.

(Step 5) **Experimentation:** Table 8 reports the optimization process. Each line in Table 8 corresponds to one optimization step in the sequential approach. At each step, two new designs are generated and the best one is re-evaluated. The number, k, of runs of the algorithm designs is increased (doubled) if a design has performed best twice or more, with starting value, $k = 2$. For example, design 14 performs best at iteration 1 and iteration 3. It has been evaluated 4 times; therefore, the number of evaluations is set to 4 for every newly generated design. This provides a fair comparison and reduces the risk of incorrectly selecting a worse design.

(Step 6) **Statistical modeling and prediction:** The response is modeled as the realization of a regression model and a random process, as described in equation (31). For this purpose, a Gaussian correlation function and a regression model with a polynomial of order 2, were used. Hence, the model is defined by:

$$Y(x) = \sum_{j=1}^{\rho} \beta_j f_j(x) + Z(x),$$

where $Z(x)$ is a random process with zero mean, and covariance defined as:

$V(\omega, x) = \sigma^2 R(\theta, \omega, x),$

with correlation function:

$$R(\theta, \omega, x) = \prod_{j=1}^{n} \exp(-\theta_j (\omega_j - x_j)^2).$$

Additionally, at certain stages, a tree-based regression model was constructed to determine parameter settings that produce outliers.

(Step 7) **Evaluation and visualization:** The MSE and predicted values can be plotted to support the numerical analysis. For this purpose, the DACE toolbox of Lophaven *et al.* (2002) can be used. For example, the interaction between c_1 and c_2 is shown in Fig. 8. Values of c_1 and c_2 with $c_1 + c_2 > 4$ generate outliers that might disturb analysis. To alleviate these outliers, a design correction was applied by requiring $c_1 = c_2 - 4$ if $c_1 + c_2 > 4$. The right part of Fig. 8 illustrates the estimated MSE. Since no design point has been placed for $1 < c_1 < 1.25$ and $2.25 < c_2 < 2.5$, the MSE is relatively high. This might be an interesting region where a new design point will be placed during the next

Table 8. Optimizing the 10-dimensional Rosenbrock function with PSO[in]. Each row represents the best algorithm design at the corresponding tuning stage. Note that function values (reported in the first column) can worsen (increase) although the design is improved. This happens due to noise in the results. The probability that a seemingly good function value is in fact worse exists. However, it decreases during the sequential procedure, because the number of re–evaluations is increased. The number of repeats is doubled if a configuration performs best twice. These configurations are marked with an asterisk

y	N	c_1	c_2	w_{max}	w_{scale}	$w_{iterScale}$	v_{max}	Conf.
6.61557	26	1.45747	1.98825	0.712714	0.481718	0.683856	477.874	14
18.0596	39	1.30243	1.84294	0.871251	0.273433	0.830638	289.922	19
71.4024	26	1.45747	1.98825	0.712714	0.481718	0.683856	477.874	14*
78.0477	30	2.21960	1.26311	0.944276	0.289710	0.893788	237.343	3
75.6154	30	2.21960	1.26311	0.944276	0.289710	0.893788	237.343	3*
91.0935	18	1.84229	1.69903	0.958500	0.256979	0.849372	95.1392	35
91.5438	21	1.05527	1.25077	0.937259	0.498268	0.592607	681.092	43
93.7541	11	1.58098	2.41902	0.728502	0.469607	0.545451	98.9274	52
93.9967	93	1.71206	1.02081	0.966302	0.378612	0.972556	11.7651	20
99.4085	39	1.30243	1.84294	0.871251	0.273433	0.830638	289.922	19*
117.595	11	1.13995	2.31611	0.785223	0.236658	0.962161	56.9096	57
146.047	12	1.51468	2.48532	0.876156	0.392995	0.991074	261.561	1
147.410	22	1.72657	2.27343	0.710925	0.235521	0.574491	50.5121	54
98.3663	22	1.72657	2.27343	0.710925	0.235521	0.574491	50.5121	54*
41.3997	21	2.25413	1.74587	0.788797	0.282645	0.937293	11.0496	67*
43.2249	21	2.25413	1.74587	0.788797	0.282645	0.937293	11.0496	67*
53.3545	21	2.25413	1.74587	0.788797	0.282645	0.937293	11.0496	67*

iteration. Figure 9 depicts the same situation with Fig. 8, after the design correction. In this case, a high MSE is associated with the region $c_1 + c_2 > 4$, but no design point is placed there.

(Step 8) **Optimization:** Based on the expected improvement defined in equation (32), two new design points, $a_d^{(1)}$ and $a_d^{(2)}$, are generated. These designs are evaluated and their performance is compared to that of the current best design. Then, the best design found so far is re-evaluated. The iteration terminates if a design was evaluated for $k = 50$ times, and the solution is obtained. The parameter values in the final model, as reported in Table 8, are, $N = 21$, $c_1 = 2.25413$, $c_2 = 1.74587$, $w_{max} = 0.788797$, $w_{scale} = 0.282645$, $w_{iterScale} = 0.937293$, and $v_{max} = 11.0496$ (Bartz-Beielstein *et al.*, 2004). This step has also covered the termination procedure (Step 9) and design update (Step 10) of the sequential design reported in Table 4.

(Step 11) **Rejection/acceptance of statistical hypothesis:** Finally, the configuration of Shi and Eberhart (1999) is compared to the optimized one. The final (tuned) and the first configuration are applied 50 times each. Histograms and boxplots are illustrated in Fig. 10 for both PSO variants. Clearly, the tuned design of PSO[in] exhibits significant performance improvement. The corresponding statistical analysis is reported in Table 9 (Bartz-Beielstein *et al.*, 2004). Performing a classical *t*-test indicates that the null hypothesis "there is no difference in the mean performance of the two algorithms" can be rejected at the 5% level.

Figure 8. Predicted values (left) and MSE (right). As we can see in the left figure, $c_1+c_2 > 4$ produces outliers that complicate the analysis

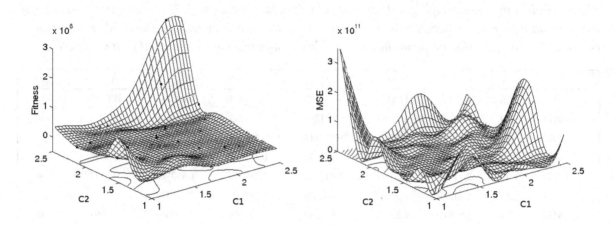

Figure 9. Predicted values (left) and MSE (right). The design correction avoids settings with $c_1+c_2 > 4$ that produce outliers (left). Therefore, a high mean squared error exists in the excluded region (right)

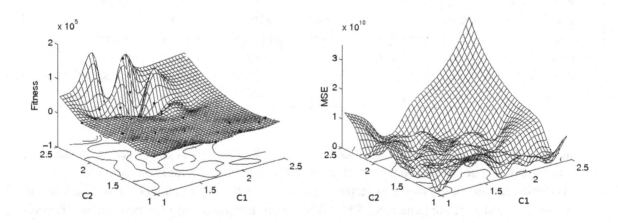

(Step 12) **Objective interpretation:** The statistical results from Step 11 suggest that PSO with the tuned design performs better (on average) than the default design. Comparing the parameters of the improved design, a_d^*, reported in Table 7, with the default setting of PSO, no significant differences are observed except for v_{max}, which appears to be relatively small. A swarm size of 20 particles was shown to be a good value for this problem.

The analysis and tuning procedure described above were based solely on the average function value in 50 runs. However, in a different optimization context, this measure may be irrelevant, and the best function value (minimum) or the median could be used alternatively. In this case, a similar optimization procedure with that presented can be conducted, although the resulting optimal designs may differ from the ones reported above.

Figure 10. Histograms and boxplots for both PSO variants. Left: Solid lines and light bars represent the improved design. Right: The default configuration is denoted with the index 1, whereas index 2 denotes the improved variant. Top: Both plots indicate that the tuned inertia weight PSO version performs better than the default version. Bottom: No clear difference can be detected when comparing the default with the improved constriction factor PSO variant

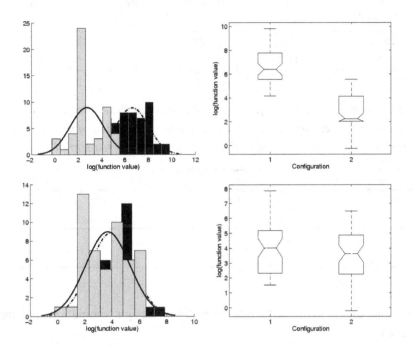

The aforementioned procedure focused on the case of PSO[in] to provide a comprehensive example of the sequential design. PSO[con] was tuned in a similar manner (Bartz-Beielstein *et al.*, 2004). The initial LHD for PSO[con] is reported in Table 10, where $a_d^{(l)}$ and $a_d^{(u)}$ denote the lower and upper bound of the region of interest, respectively, a_d^* is the improved design found by the sequential procedure, and $a_d^{(\mathrm{Clerc})}$ is the default design recommended in (Clerc & Kennedy, 2002). From the numerical results reported in Table 9 and the corresponding graphical representations (histograms and boxplots) in Fig. 10, we can derive that there is no significant difference between the performance of a_d^* and $a_d^{(\mathrm{Clerc})}$. This constitutes an experimental verification of the analysis of Clerc and Kennedy (2002) presented previously in the current chapter.

Application on a Real World Problem

Bartz-Beielstein *et al.* (2004) also applied the sequential design approach on a real-world problem, namely the optimization of an elevator group controller. The construction of elevators for high-rise buildings is a challenging task. The elevator group controller is a central part of an elevator system, with the duty of assigning elevator cars to service calls in real-time, while optimizing the overall service quality, traffic throughput, and/or energy consumption. The *elevator supervisory group control* (ESGC) problem can be classified as a combinatorial optimization problem (Barney, 1986; Markon & Nishikawa, 2002;

Table 9. Results for the Rosenbrock function. Default designs, $a_d^{[Shi]}$ and $a_d^{[Clerc]}$, from (Shi & Eberhart, 1999) and (Clerc & Kennedy, 2002), respectively, as well as the improved design obtained for k = 50 runs, are reported

Design	Algorithm	Mean	Median	StD	Min	Max
$a_d^{(Shi)}$	PSO[in]	1.8383×10^3	592.1260	3.0965×10^3	64.6365	18519
a_d^*	PSO[in]	39.7029	9.4443	55.3830	0.7866	254.1910
$a_d^{(Clerc)}$	PSO[con]	162.02	58.51	378.08	4.55	2.62×10^3
a_d^*	PSO[con]	116.91	37.65	165.90	0.83	647.91

Table 10. Algorithm design, a_d of PSO[con] for the 10-dimensional Rosenbrock function. We denote with $a_d^{(l)}$ and $a_d^{(u)}$ the lower and upper bounds, respectively, for the generation of the LHD, while a_d^* denotes the parameter settings of the improved design that was found by the sequential approach

Design	N	χ	φ	v_{max}
$a_d^{(l)}$	5	0.68	3.0	10
$a_d^{(u)}$	100	0.8	4.5	750
a_d^*	17	0.759	3.205	324.438
$a_d^{(Clerc)}$	20	0.729	4.1	100

So & Chan, 1999), and, due to many difficulties in analysis, design, simulation, and control, has been studied for a long time.

The elevator group controller determines the floors where the elevator cars should go. Since it is responsible for the allocation of elevators to hall calls, a *control strategy*, π, also called *policy*, is needed. One important goal in designing an efficient controller is the minimization of the passenger *waiting time*, which is defined as the time needed until passenger can enter an elevator after having requested service. *Service time* includes the waiting time, and, additionally, the time that a passenger spends in the elevator car. During a day, different traffic patterns can be observed. For example, in office buildings, "up-peak" traffic is observed in the morning, when people start working, and "down-peak" traffic is observed in the evening. Most of the day, traffic is balanced with much lower intensity than at peak times. Lunchtime traffic consists of two (often overlapping) phases, where people first leave the building for lunch or head for a restaurant floor, and then get back to work (Markon, 1995).

Fujitec, one of the world's leading elevator manufacturers, developed a controller that uses a neural network and a set of fuzzy controllers. The weights on the output layer of the neural network can be modified and optimized, thereby resulting in a more efficient controller. The associated optimization problem is complex, since the distribution of local optima in the search space is unstructured, and there are flat plateaus of equal function values. Additionally, the objective function is contaminated by noise and changes dynamically, as it is influenced by the stochastic behavior of customers. Experiments have shown that gradient-based optimization techniques cannot be applied successfully on such problems, suggesting the application of stochastic population-based algorithms (Beielstein *et al.*, 2003).

Elevator group controllers and their related policies are usually incomparable. To enable comparability for benchmark tests, a simplified elevator group control model, called *S-ring* (stands for "sequential

Figure 11. Fujitec's simulator for the visualization of the elevator system dynamics in the case of a building with 15 floors and 6 elevators

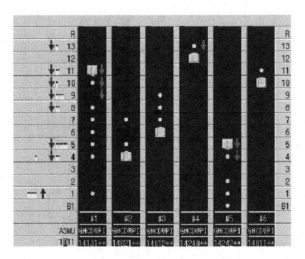

ring"), was developed. This model enables fast and reproducible simulations, and it is applicable to different buildings and traffic patterns. In addition, it is characterized by scalability, and it can be easily extended. Thus, it is appropriate as a test problem generator.

The approach presented in (Bartz-Beielstein *et al.*, 2004) uses the simulator of Fujitec, which is depicted in Fig. 11. This simulator is characterized by high accuracy but heavy computational cost. The corresponding coarse (surrogate) S-ring model is depicted in Fig. 12. This is fast to solve, albeit at the cost of lower accuracy. Their approach also incorporates *space mapping* (SM) techniques, which are used to iteratively update and optimize surrogate models (Bandler *et al.*, 2004), while the main goal is the computation of an improved solution with a minimal number of function evaluations.

To put it more formally, let *i* denote a site in an elevator system. A 2-bit state, (s_i, c_i), is associated with it, where s_i is set to 1 if a server is present on the *i*-th floor, or 0 otherwise, while c_i is set to 1 if there is at least one waiting passenger at site *i*, or 0 otherwise. Figure 12 depicts a typical S-ring configuration. The state of a system at time *t* can be described by a vector:

$$x(t) = (s_0(t), c_0(t),\ldots, s_{d-1}(t), c_{d-1}(t)) \in \{0,1\}^{2d}. \tag{34}$$

Thus, the state of the system depicted in Fig. 12 is given by a vector of the form, $x(t)=(0,1,0,0,\ldots,0,1,0,0)^\mathrm{T}$, where there is a customer waiting on the first ($c_0 = 1$) but no elevator present ($s_0 = 0$), and so on. The dynamics of the system are modeled through a state transition table. The state evolution is sequential, scanning the sites from *d*-1 down to 0, and then again around from *d*-1.

The ascending and descending elevator movements can be considered as a loop. This motivates the ring structure. Each time step considers one of the floor queues, where passengers may arrive with a specific probability. Consider the situation at the third site (the upwards direction on the third floor) in Fig. 12. Since a customer is waiting and a server is present, the controller has to make a decision. The elevator car can either serve ("take" decision) or ignore the customer ("pass" decision). The former decision would change the values of the corresponding bits of $x(t)$ from (1,1) to (0,1), while the latter would change it to (1,0).

Figure 12. The S-ring elevator model. The case of a building with 6 floors and 3 elevators is illustrated

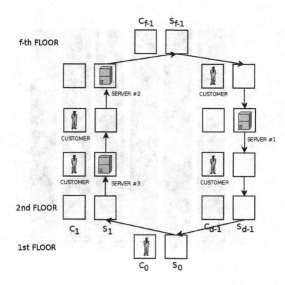

The operational rules of this model are very simple. Thus, it is easily reproducible and suitable for benchmark testing. Despite its simplicity, it is hard to find the optimal policy, π^*, even for a small S-ring. The actual π^* is not obvious, and its difference from heuristic suboptimal policies is non-trivial (Bartz-Beielstein *et al.*, 2004). The transition of the S-ring model from simulation to the corresponding optimization problem requires the introduction of an objective function. The function that counts the sites with waiting customers at time t is defined as:

$$Q(t) = Q(x,t) = \sum_{i=0}^{d-1} c_i(t).$$

The steady-state, time-average number of sites with waiting customers in queue is given by:

$$Q_S = \lim_{T \to \infty} \frac{1}{T} \int_0^T Q(t)dt,$$

with probability 1.

For a given S-ring configuration, the basic optimal control problem is the detection of a policy, π^*, such that the expected number, Q_S, of sites with waiting passengers that is the steady-state, time-average in the system is minimized, i.e.,

$$\pi^* = \arg\min_\pi Q(\pi).$$

The policy can be represented by a $2d$-dimensional vector, $y \in \mathbf{R}^{2d}$. Let, $\theta : \mathbf{R} \to \{0,1\}$, be the Heaviside function, defined as:

$$\theta(z) = \begin{cases} 0, & z < 0, \\ 1, & z \geq 0, \end{cases}$$

$x = x(t)$ be the state at time t, as defined by equation (34), and $y \in \mathbf{R}^{2d}$ be a weight vector. Then, a *linear discriminator* or *perceptron*,

$$\pi(x, y) = \theta(\langle x, y \rangle), \tag{35}$$

where,

$$\langle x, y \rangle = \sum_i x_i y_i,$$

can be used to model the decision process in a compact manner. For a given vector, y, which represents a policy, and a given vector, x, which represents the state of the system, a "take" decision occurs if $\pi(x,y) \geq 0$. Otherwise the elevator ignores the customer.

The most obvious heuristic policy is the greedy one, i.e., to always serve the customer if possible, which is represented by the $2d$-dimensional vector $y_0 = (1, 1, \ldots, 1)^{\mathrm{T}}$. This vector guarantees that the result in equation (35) always equals 1, which is interpreted as a "take" decision. However, this policy is not optimal, except in the case of heavy traffic. This means that a good policy must occasionally bypass some customers to protect the system from a *bunching effect*, which occurs when nearly all elevator cars are positioned in close proximity to each other.

The perceptron S-ring can serve as a benchmark problem for different optimization algorithms, since it relies on a fitness function that maps $\mathbf{R}^{2d}$ to $\mathbf{R}$. In general, a policy, π, can be realized as a look-up table of the system state, x. Then, the optimal policy, π^*, can be found by enumerating all possible policies and selecting the one with the lowest value of $Q(\pi)$. Since this count grows exponentially with d, the enumerative approach would not work for any but the smallest cases.

Bart-Beielstein *et al.* (2004) used the S-ring simulator to define a 12-dimensional optimization problem with noisy function values, and applied the sequential design technique with PSO on it. The number of function evaluations was limited to 1000 for each optimization run, which appears to be a realistic choice for real world applications. The related problem design was reported as Experiment 2 in Table 6 in the previous section, along with the design of the Rosenbrock function.

Similarly to the analysis for the Rosenbrock function, the constriction coefficient PSO variant, PSO[con], as well as the inertia weight variant, PSO[in], were analyzed. The former requires only 4 exogenous strategy parameters, while 7 parameters have to be specified for the latter. Table 11 contains the results in terms of the obtained solution values. Optimizing PSO[in] improved its robustness as observed in Table 11. The average function value decreased from 2.61 to 2.51, which is a significant difference. However, it is very important to note that the minimum function value could not be improved, but increased slightly from 2.4083 to 2.4127, i.e., the tuning procedure was able to find an algorithm design that prevents outliers and produces robust solutions at the cost of an aggressive exploratory behavior. Nevertheless, if the requirement of finding a solution with a minimum function value was specified as the optimization goal, then different optimal designs would have been detected (Bartz-Beielstein *et al.*, 2004).

Although function values look slightly better, the tuning process produced no significant improvement for PSO[con]. It seems that PSO[con] was unable to escape plateaus of equal fitness. This is an already identified property of the employed gbest PSO version (Bartz-Beielstein *et al.*, 2004), and it occurred independently from the parameterization of exogenous strategy parameters. Besides PSO[con], Bartz-Beielstein *et al.* (2004) also used the NSM method, described in the beginning of this chapter, as well as a quasi-Newton gradient-based approach against PSO[in]. However, all algorithms were outperformed

Table 11. Results for the S-ring model. Default designs, $a_d^{[Shi]}$ and $a_d^{[Clerc]}$, from (Shi & Eberhart, 1999) and (Clerc & Kennedy, 2002), respectively, as well as the improved design for k = 50 runs, are reported

Design	Algorithm	Mean	Median	StD	Min	Max
$a_d^{(Shi)}$	PSO[in]	2.6152	2.5726	0.4946	2.4083	5.9988
a_d^*	PSO[in]	2.5171	2.5112	0.0754	2.4127	2.6454
$a_d^{(Clerc)}$	PSO[con]	4.1743	2.6252	1.7021	2.5130	5.9999
a_d^*	PSO[con]	4.1707	2.6253	1.7055	2.5164	5.9999

by the inertia weight PSO variant. Whether this improved result was caused by the scaling property of the inertia weight is subject to further investigation (Bartz-Beielstein *et al.*, 2004).

Summarizing, the sequential approach provided effective and efficient means to improve the performance of PSO. Thus, it can be considered as a useful tool to support the practitioner in selecting a suitable algorithm configuration. Bartz-Beielstein *et al.* (2004) considered only the mean fitness values as performance measure; different results may be received if different goals were determined. A drawback of the sequential design approach, which is common to all statistical methods in this field, is the determination of a good initial design. This may be a very interesting direction for future research.

TERMINATION CONDITIONS

In the previous sections, we mentioned some of the most common termination conditions used in practice. This is perhaps the most user-dependent phase of the optimization procedure for any optimization algorithm. The decision for stopping the algorithm can depend on several criteria, related to the available problem information, resources, or the ability of the algorithm to attain further solutions.

Let $\{x_i\}_{i=1,2,...}$ denote the sequence of solutions produced by an algorithm, with $x_i = (x_{i1}, x_{i2}, ..., x_{in})^T$, and $\{f_i\}_{i=1,2,...}$ be the corresponding sequence of function values, i.e., $f_i = f(x_i)$, for all i. For example, in PSO this sequence may consist of the overall best positions and their function values. Let, also, x^* be a (local or global) minimizer of the objective function, and $f^* = f(x^*)$ be the corresponding (local or global) minimum. Subsequently, we can roughly distinguish four categories of stopping conditions, which are described in the following sections. All cases refer to the unconstrained optimization problem, while slight modifications in formulations may be required under the presence of constraints.

Convergence in Search Space

If $\|\cdot\|$ denotes a distance measure in the search space, A, then convergence in search space is defined as:

$$\lim_{i \to \infty} \| x_i - x^* \| = 0.$$

Since the available number of iterations is always finite, the convergence criterion can be relaxed as follows: for any desirable accuracy, $\varepsilon > 0$, there is an integer, $k > 0$, such that:

$$\|x_i - x^*\| \leq \varepsilon, \text{ for all } i \geq k. \tag{36}$$

Thus, the algorithm is terminated as soon as a solution adequately close to the minimizer is detected. In practice, the minimizer, x^*, is unknown; thus, we can identify convergence by monitoring the gradient at the approximating solutions. However, in order to extract sound conclusions through gradients, the existence of strong mathematical properties, such as continuous differentiability, are required for the objective function. If $f(x)$ is twice differentiable in A, and its gradient and Hessian matrix at x_i are defined as:

$$\nabla f(x_i) = \left(\frac{\partial f(x_i)}{\partial x_1}, \frac{\partial f(x_i)}{\partial x_2}, \dots, \frac{\partial f(x_i)}{\partial x_n} \right)^{\mathrm{T}}, \quad \nabla^2 f(x_i) = \begin{pmatrix} \dfrac{\partial^2 f(x_i)}{\partial x_1^2} & \cdots & \dfrac{\partial^2 f(x_i)}{\partial x_1 \partial x_n} \\ \vdots & \ddots & \vdots \\ \dfrac{\partial^2 f(x_i)}{\partial x_n \partial x_1} & \cdots & \dfrac{\partial^2 f(x_i)}{\partial x_n^2} \end{pmatrix}$$

respectively, then x_i is identified as an approximation of the minimizer, x^*, with accuracy, $\varepsilon > 0$, if it holds that:

$$\|\nabla f(x_i)\| < \varepsilon \text{ and } z^{\mathrm{T}} \nabla^2 f(x_i) z > 0, \tag{37}$$

for all non-zero vectors z (i.e., the Hessian matrix is positive definite).

The aforementioned termination conditions for unconstrained optimization can be applied reliably only in noiseless functions with nice mathematical properties. Moreover, they require the computation of first-order and second-order derivatives, which are not always available in complex problems. In addition, there is no way to distinguish whether the obtained minimizer is the global one, unless additional restrictions (e.g., convexity) are considered in the form of the objective function. PSO and, in general, evolutionary algorithms have been designed to solve problems where the aforementioned required mathematical properties are not necessarily met. Thus, this type of termination condition is of limited practical interest.

Convergence in Function Values

There are several optimization problems where the global minimum, f^*, is *a priori* known due to the form of the objective function. For example, neural network training is equivalent to the minimization of a function that is usually defined as the summed square-error of the network's output. This function, by construction, has the global minimum equal to zero. Similarly, fixed points of nonlinear mappings can be detected by minimizing a sum of square or absolute errors. Again, this objective function has the global minimum equal to zero.

In such cases, the following condition of convergence in function values can be used as the termination criterion:

$$|f_i - f^*| \leq \varepsilon, \tag{38}$$

for user-defined accuracy, $\varepsilon > 0$. Although this condition has milder mathematical requirements than that of equation (37), its applicability is questionable in general cases, since f^* is known or can be bounded below only for specific types of functions. However, it has been recognized as the most popular termination condition for performance studies on benchmark problems. In these cases, the global minimum and minimizers are known, and the practitioner is interested in the required number of function evaluations for the detection of a global minimizer with a prespecified accuracy. Thus, the algorithm is usually stopped as soon as it finds a solution with function value adequately close to the known global minimum. Due to its popularity, we will refer to this termination criterion repeatedly in the rest of the book at hand.

Computational Budget Limitations

In modern applications, the available time for computation is usually limited. Time critical decisions and on-line control of systems require algorithms that provide satisfactory solutions within very restrictive time frames. An example of such a time-critical problem is the elevator controller described previously in the current chapter. In addition, concurrent application servers and computer clusters usually have a multitude of processors available to a large number of users, who require the fastest execution of their programs with the shortest waiting time in the scheduler queue. Thus, queued jobs have explicit time constraints, translated in months, days, minutes, or seconds, based on their priority. Therefore, limitations are usually posed on the available computational time (CPU time) for the execution of an algorithm.

Limitations are also imposed for reasons of comparison. In order to have fair comparisons among algorithms on a specific problem, they must assume the same computational budget. However, a significant issue arises at this point. The time needed for the execution of a program depends heavily on the implementation, programming language, programmer skills, and machine load at the time of execution. Thus, any comparison between two algorithms, in terms of the required CPU time, without taking these factors into consideration is condemned to be biased.

For this reason, researchers have made a compromise. The most computationally expensive part in solving a complex problem is expected to be the evaluation of the objective function, which may be computed through complex mathematical procedures (e.g., finite element simulations or integrations of dynamical systems) or become available directly from experimental devices. Thus, the time required for all function evaluations during the execution of an algorithm is expected to constitute the largest fraction of the overall computation burden. For this purpose, the required number of function evaluations serves very often as a performance measure for optimization algorithms.

Based on the aforementioned discussion, we can define two termination conditions. If t is the CPU time (e.g., in seconds) required by the algorithm from the beginning of its execution, and q denotes the corresponding number of function evaluations required so far, then the following termination conditions are defined:

$$t \geq t_{max}, \tag{39}$$

and,

$$q \geq q_{max}, \tag{40}$$

where t_{max} is the maximum available CPU time, and q_{max} is the maximum allowed number of function evaluations. Thus, the algorithm will stop as soon as its CPU time exceeds the available time frame or the number of function evaluations required so far exceeds an upper limit. For reasons explained above, the condition of equation (40) is preferred against that of equation (39).

However, there are algorithms that, although requiring only a low number of function evaluations, operate based on very complicated and time-consuming procedures. Therefore, their execution time can be comparable with that of algorithms that need more function evaluations to provide results of the same quality. Techniques in evolutionary multi-objective optimization with sophisticated archiving procedures are typical examples of such algorithms. In these cases, it would be unfair to use a termination condition that monitors only the number of function evaluations. Hence, a combination of the two termination conditions described above would be more reasonable.

Finally, we must note that, in evolutionary algorithms, it is very common to use the number of generations (iterations) instead of function evaluations as a stopping criterion. This is equivalent to the condition of equation (40), assuming that the population size and number of function evaluations per population member are fixed at each iteration of the algorithm. However, if there is a variable number of function evaluations per iteration, then this stopping criterion is not valid.

The presented termination conditions have been widely used in PSO literature, especially in cases where no information regarding the objective function is available or a new and unexplored problem is considered. Naturally, if required, they can be used in conjunction with other termination conditions, such as the ones presented in previous sections.

Search Stagnation

The final category of termination criteria consists of performance-related conditions. Monitoring the progress of an algorithm during the optimization procedure provides insight regarding its efficiency and potential for further improvement of the obtained solutions. The lack of such potential is called *search stagnation*, and it can be attributed to several factors.

In evolutionary algorithms, search stagnation can be identified by monitoring changes of the overall best solution within a specific number of iterations. An algorithm is considered to suffer stagnation, if its best solution has not been improved for a number, t_{frame}, of consecutive iterations, which is defined as a fraction of the maximum number of iterations, t_{max}:

$$t_{frame} = h\, t_{max}, \quad h \in (0,1).$$

Alternatively, one can identify search stagnation by monitoring diversity of the population, which is usually defined as its spread in the search space. The standard deviation of the population is a commonly used diversity measure. If it falls under a prespecified (usually problem-dependent) threshold, then the population is considered to be collapsed on a single point, having limited potential for further improvement. Moreover, special features of the algorithm can be used to define diversity. For example, in PSO, if velocities of all particles become smaller than a threshold, then the swarm can be regarded as immobilized. Thus, its ability for further improvement is limited, and it is questionable whether its further execution can offer any gain.

Search stagnation has been widely used as a stopping criterion in evolutionary computation literature. However, the user must take special care regarding the employed stagnation measures. Some of them depend on the scaling of the problem and need a preprocessing stage to identify their thresholds, while some others can be misleading especially in flat areas of the objective function, where diversity may be satisfactory but search is inefficient.

On the Proper Selection of the Termination Condition

Unfortunately, there is no general rule for selecting a proper termination condition applicable to all algorithms and problems. The number of cases, where only one of the presented termination criteria is adequate, is very limited. Thus, the user has to combine several criteria to ensure that the algorithm is not stalled and worth continuing its execution.

The most common termination condition in PSO literature is the combination of equations (38) and (40), along with a measure of search stagnation. Thus, PSO stops as soon as it exceeds the maximum allowed number of function evaluations or has found a solution within the desired accuracy or has not improved its performance for a number of consecutive iterations. These are also the termination criteria that will be used in most applications presented in the rest of the book.

Less frequently, the termination condition defined by equation (40) stands alone, i.e., the algorithm is left to perform a prespecified number of function evaluations and then stops. This is very useful in producing performance plots (e.g., best function value against iterations) for graphical comparisons. Nevertheless, proper termination conditions must be carefully selected, taking into consideration any possible special features of the algorithms or problem-related peculiarities, in order to derive sound conclusions and perform fair comparisons.

CHAPTER SYNOPSIS

This chapter was devoted to a series of critical issues in the theory and practice of PSO. The initialization procedure was discussed and, besides the typical random initialization, a sophisticated technique based on a direct search algorithm was presented. We also described most influential theoretical developments of PSO. Particle trajectory studies, as well as the stability analysis of PSO, were briefly presented, and rules on parameter selection and tuning were derived. The contemporary state-of-the-art variants of PSO stemmed from these studies. Moreover, we analyzed a useful sequential design technique based on computational statistics for the optimal tuning of PSO on specific tasks. Its application on a benchmark problem, as well as on a real world problem, revealed its potential for designing better PSO algorithms. The chapter concluded with the most common termination conditions, underlining the pitfalls that accompany each choice.

REFERENCES

Aslett, R., Buck, R. J., Duvall, S. G., Sacks, J., & Welch, W. J. (1998). Circuit optimization via sequential computer experiments: design of an output buffer. *Journal of the Royal Statistical Society. Series C, Applied Statistics, 47*(1), 31–48. doi:10.1111/1467-9876.00096

Bandler, J. W., Cheng, Q. S., Dakroury, S. A., Mohamed, A. S., Bakr, M. H., Madsen, K., & Søndergaard, J. (2004). Space mapping: the state of the art. *IEEE Transactions on Microwave Theory and Techniques, 52*(1), 337–361. doi:10.1109/TMTT.2003.820904

Barney, G. (1986). *Elevator traffic analysis, design and control.* MA: Cambridge University Press.

Bartz-Beielstein, T. (2006). *Experimental research in evolutionary computation.* Heidelberg, Germany: Springer.

Bartz-Beielstein, T., & Markon, S. (2004). Tuning search algorithms for real-world applications: a regression tree based approach. In *Proceedings of the 2004 IEEE Congress on Evolutionary Computation, Portland (OR), USA* (pp. 1111–1118).

Bartz-Beielstein, T., Parsopoulos, K. E., & Vrahatis, M. N. (2004). Design and analysis of optimization algorithms using computational statistics. *Applied Numerical Analysis & Computational Mathematics, 1*(3), 413–433. doi:10.1002/anac.200410007

Beielstein, T., Ewald, C.-P., & Markon, S. (2003). Optimal elevator group control by evolution strategies. In E. Cantú-Paz *et al.* (Eds.), *Lecture Notes in Computer Science, Vol. 2724* (pp. 1963–1974). Berlin: Springer.

Breiman, L., Friedman, J. H., Olshen, R. A., & Stone, C. J. (1984). *Classification and regression trees.* Florence, KY: Wadsworth.

Chan, K., Saltelli, A., & Tarantola, S. (1997). Sensitivity analysis of model output: variance-based methods make the difference. In *Proceedings of the 29th conference on Winter simulation (WSC'97), Atlanta (GA), USA* (pp. 261–268).

Chan, K. S., Tarantola, S., Saltelli, A., & Sobol', I. M. (2000). Variance based methods. In A. Saltelli, K. Chan & E.M. Scott (Eds.), *Sensitivity Analysis (Probability and Statistics Series)* (pp. 167–197). New York: John Wiley & Sons.

Clerc, M., & Kennedy, J. (2002). The particle swarm - explosion, stability, and convergence in a multidimensional complex space. *IEEE Transactions on Evolutionary Computation, 6*(1), 58–73. doi:10.1109/4235.985692

Cohen, P. R. (1995). *Empirical methods for artificial intelligence.* Cambridge, MA: MIT Press.

Gentle, J. E., Härdle, W., & Mori, Y. (2004). *Handbook of computational statistics.* Berlin: Springer.

Hoos, H. H. (1998). *Stochastic local search – methods, models, applications.* PhD thesis, Technische Universität Darmstadt, Germany.

Isaaks, E. H., & Srivastava, R. M. (1989). *An introduction to applied geostatistics.* UK: Oxford University Press.

Jiang, M., Luo, Y. P., & Yang, S. Y. (2007). Stochastic convergence analysis and parameter selection of the standard particle swarm optimization algorithm. *Information Processing Letters, 102*(1), 8–16. doi:10.1016/j.ipl.2006.10.005

Kennedy, J. (1998). The behavior of the particles. In V.W. Porto, N. Saravanan, D. Waagen & A.E. Eiben (Eds.), *Evolutionary programming VII, Lecture Notes in Computer Science, Vol. 1447* (pp. 581-589). Berlin: Springer-Verlag.

Lophaven, S. N., Nielsen, H. B., & Søndergaard, J. (2002). *Aspects of the Matlab Toolbox DACE* (Tech. Rep. IMM-REP-2002-13). Informatics and Mathematical Modelling, Technical University of Denmark.

Markon, S. (1995). *Studies on applications of neural networks in the elevator system*. PhD thesis, Kyoto University, Japan.

Markon, S., & Nishikawa, Y. (2002). On the analysis and optimization of dynamic cellular automata with application to elevator control. In *Proceedings of the 10th Japanese-German Seminar on Nonlinear Problems in Dynamical Systems, Theory and Applications, Ishikawa, Japan.*

Montgomery, D. C. (2001). *Design and analysis of experiments*. New York: John Wiley & Sons.

Nelder, J. A., & Mead, R. (1965). A simplex method for function minimization. *The Computer Journal, 7*, 308–313.

Ozcan, E., & Mohan, C. K. (1998). Analysis of a simple particle swarm optimization problem. In C. Dagli *et al.* (Eds.), In *Proceedings of the Conference on Artificial Neural Networks in Engineering (ANNIE'98), St. Louis (MO), USA* (pp. 253-258).

Ozcan, E., & Mohan, C. K. (1999). Particle swarm optimization: surfing the waves. In *Proceedings of 1999 IEEE Congress on Evolutionary Computation, Washington (DC), USA* (pp. 1939-1944).

Parsopoulos, K. E., & Vrahatis, M. N. (2002). Initializing the particle swarm optimizer using the nonlinear simplex method. In A. Grmela & N.E. Mastorakis (Eds.), *Advances in Intelligent Systems, Fuzzy Systems, Evolutionary Computation* (pp. 216-221). WSEAS Press.

Press, W. H., Teukolsky, S. A., Vetterling, W. T., & Flannery, B. P. (1992). *Numerical recipes in fortran 77*. MA: Cambridge University Press.

Sacks, J., Welch, W. J., Mitchell, T. J., & Wynn, H. P. (1989). Design and analysis of computer experiments. *Statistical Science, 4*(4), 409–435. doi:10.1214/ss/1177012413

Saltelli, A. (2002). Making best use of model evaluations to compute sensitivity indices. *Computer Physics Communications, 145*(2), 280–297. doi:10.1016/S0010-4655(02)00280-1

Saltelli, A., Tarantola, S., & Campolongo, F. (2000). Sensitivity anaysis as an ingredient of modeling. *Statistical Science, 15*(4), 377–395. doi:10.1214/ss/1009213004

Santner, T. J., Williams, B. J., & Notz, W. I. (2003). *The design and analysis of computer experiments*. Berlin: Springer.

Shi, Y., & Eberhart, R. C. (1999). Empirical study of particle swarm optimization. In *Proceedings of the 1999 IEEE Congress of Evolutionary Computation, Washington (DC), USA* (pp. 1945–1950).

So, A. T., & Chan, W. L. (1999). *Intelligent suilding systems*. Dordrecht, The Netherlands: Kluwer Academic Publishers.

Sutton, A. M., Whitley, D., Lunacek, M., & Howe, A. (2006). PSO and multi-funnel landscapes: how cooperation might limit exploration. In *Proceedings of the 8ᵗʰ Annual Conference on Genetic and Evolutionary Computation (GECCO'06), Seattle (WA), USA* (pp. 75-82).

Torczon, V. (1991). On the convergence of the multidirectional search algorithm. *SIAM Journal on Optimization, 1*, 123–145. doi:10.1137/0801010

Trelea, I. C. (2003). The particle swarm optimization algorithm: convergence analysis and parameter selection. *Information Processing Letters, 85*(6), 317–325. doi:10.1016/S0020-0190(02)00447-7

Van den Bergh, F., & Engelbrecht, A. P. (2006). A study of particle swarm optimization particle trajectories. *Information Sciences, 176*(8), 937–971. doi:10.1016/j.ins.2005.02.003

Xiao, R.-Y., Li, B., & He, X.-P. (2007). The particle swarm: parameter selection and convergence. In D.-S. Huang, L. Heutte & M. Loog (Eds.), *Advanced Intelligent Computing Theories and Applications* (Vol. 2, pp. 396-402). Berlin: Springer.

Zheng, Y.-L., Ma, L.-H., Zhang, L.-Y., & Qian, J.-X. (2003). On the convergence analysis and parameter selection in particle swarm optimization. In *Proceedings of the 2ⁿᵈ International Conference on Machine Learning and Cybernetics (ICMLC2003), Xi'an, China* (pp. 1802-1807).f

Chapter 4
Established and Recently Proposed Variants of Particle Swarm Optimization

In this chapter, we describe established and recently proposed variants of PSO. Due to the rich PSO literature, the choice among different variants proved to be very difficult. Thus, we were compelled to set some criteria and select those variants that best suit them. For this purpose, we considered the following criteria:

1. Sophisticated inspiration source.
2. Close relationship to the standard PSO.
3. Wide applicability in problems of different types.
4. Performance and theoretical properties.
5. Number of reported applications.
6. Potential for further development and improvements.

Thus, we excluded variants based on complicated hybrid schemes that combine other algorithms, where it is not evident which algorithm triggers which effect, as well as over-specialized schemes that refer only to one problem type or instance.

Under this prism, we selected the following methods: *unified PSO, memetic PSO, composite PSO, vector evaluated PSO, guaranteed convergence PSO, cooperative PSO, niching PSO, TRIBES*, and *quantum PSO*. Albeit possibly omitting an interesting approach, the aforementioned variants sketch a rough picture of the current status in PSO literature, exposing the main ideas and features that constitute the core of research nowadays.

DOI: 10.4018/978-1-61520-666-7.ch004

Copyright © 2010, IGI Global. Copying or distributing in print or electronic forms without written permission of IGI Global is prohibited.

UNIFIED PARTICLE SWARM OPTIMIZATION

In the previous chapters, we emphasized the importance of the two main phases, *exploration* and *exploitation*, of the search procedure in a population-based algorithm such as PSO. The former phase is responsible for the detection of the most promising regions of the search space, while the latter promotes convergence of particles towards the best solutions. These two phases can take place either once or successively during the execution of the algorithm. Their impact on its performance necessitates the special handling of inherent features that cause transition from one phase to another, altering its search dynamics.

In the second chapter of the book at hand, we discussed the concept of neighborhood and underlined the influence of its size on the convergence properties of the algorithm. Two PSO variants were distinguished, namely the global (gbest) where the whole swarm is considered as the neighborhood of each particle, and the local (lbest) where neighborhoods are strictly smaller. Although gbest is a special case of lbest, we distinguish between them due to their different outcome in the exploration/exploitation properties of PSO.

More specifically, the global variant converges faster towards the overall best position than the local one, in the most common neighborhood topologies; therefore, it stands out for its exploitation ability. On the other hand, the local variant has better exploration properties, since information regarding the best position of each particle is gradually communicated to the rest through their neighbors. Thus, the overall best position attracts the particles gradually, providing the opportunity of avoiding suboptimal solutions. Apparently, the choice of neighborhood topology and size significantly affects the trade-off between exploration and exploitation, albeit there is no formal procedure to determine it optimally.

In practice, the most common neighborhood configuration of PSO consists of a ring topology applied either to gbest or lbest with radius equal to one. Under such configurations, the algorithm is biased towards exploitation or exploration, depending on the complexity of the problem at hand. Neighborhoods with larger radii that implicitly interpolate between the two extremes are used less frequently. The development of *unified particle swarm optimization* (UPSO) was motivated by the intention to combine the two extremal variants (in terms of their exploration/exploitation properties) in a generalized manner, aiming to produce new schemes that combine their properties.

UPSO was developed by Parsopoulos and Vrahatis (2004) as an algorithm that harnesses the global and local PSO variant in a unified scheme, without imposing additional computational burden in terms of function evaluations. The constriction coefficient velocity update of PSO was used, although the unified scheme can be defined for different schemes as well. Let N be the swarm size, and n denote the dimension of the problem at hand. Also, let $G_i(t+1)$ denote the velocity update of the i-th particle, x_i, for the global PSO variant with constriction coefficient, which is defined as:

$$G_{ij}(t+1) = \chi \left[v_{ij}(t) + c_1 R_1 \left(p_{ij}(t) - x_{ij}(t) \right) + c_2 R_2 \left(p_{gj}(t) - x_{ij}(t) \right) \right], \tag{1}$$

and $L_i(t+1)$ denote the corresponding velocity update for the local PSO variant:

$$L_{ij}(t+1) = \chi \left[v_{ij}(t) + c_1 R_1 \left(p_{ij}(t) - x_{ij}(t) \right) + c_2 R_2 \left(p_{lj}(t) - x_{ij}(t) \right) \right], \tag{2}$$

where g denotes the index of the overall best position, while l denotes the best position in the neighborhood of x_i (to simplify our notation we use the index l instead of the index g_i used in Chapter Two). Then, the main UPSO scheme is defined as (Parsopoulos & Vrahatis, 2004):

$$v_{ij}(t+1) = u\, G_{ij}(t+1) + (1-u)\, L_{ij}(t+1), \tag{3}$$

$$x_{ij}(t+1) = x_{ij}(t) + v_{ij}(t+1), \tag{4}$$

$$i = 1, 2, \ldots, N, \quad j = 1, 2, \ldots, n,$$

where $u \in [0,1]$ is a new parameter, called *unification factor*, which controls the influence of the global and local velocity update. The rest of the parameters are the same as for the standard PSO.

According to equations (3) and (4), the new position shift of a particle in UPSO consists of a weighted combination of the gbest and lbest position shifts. The unification factor balances the influence of these two search directions. The original global and local PSO variants constitute special cases of UPSO for $u = 1$ and $u = 0$, respectively. All intermediate values, $u \in (0,1)$, define UPSO variants that combine the exploration/exploitation properties of gbest and lbest.

The position update of a particle for different values of the unification factor is illustrated in Fig. 1. More specifically, let $x_i = (0,0)^T$ be the current position of a particle, denoted with the cross symbol, and $p_i = (1,2)^T$ be its own best position, denoted with a square. Also, let $p_g = (2,5)^T$ be the overall best position, denoted with a circle, and $p_l = (5,-3)^T$ be the best position in its neighborhood, denoted with a triangle. Then, Fig. 1 illustrates the distribution of 3000 possible new positions of x_i, for varying values of u ranging from 0.0 (local PSO) up to 1.0 (global PSO) with increments of 0.1. For simplicity, the current velocity, v_i, of the particle is considered to be equal to zero in all cases.

Evidently, lower values of u correspond to distributions biased towards the local best position (denoted with a triangle), since the lbest position shift dominates in equation (3). Increasing u towards 1.0 results in a contraction of the distribution shape and expansion towards the overall best position (denoted with a circle). Thus, u can control the distribution of new positions for each particle, controlling its exploration/exploitation properties.

Besides the main UPSO scheme, Parsopoulos and Vrahatis (2004) added further stochasticity to enhance its exploration properties by introducing a new stochastic parameter in equation (3). This addition produces the following two different schemes:

$$v_{ij}(t+1) = r_3 u\, G_{ij}(t+1) + (1-u)\, L_{ij}(t+1), \tag{5}$$

which is mostly based on the local variant, and, alternatively:

$$v_{ij}(t+1) = u\, G_{ij}(t+1) + r_3\, (1-u)\, L_{ij}(t+1), \tag{6}$$

which is mostly based on the global variant (Parsopoulos & Vrahatis, 2004). The stochastic parameter, r_3, follows a Gaussian distribution:

$$r_3 \sim N(\mu, \sigma^2),$$

Figure 1. The distribution of 3000 possible new positions (light grey points) of a particle using the update scheme of UPSO defined by equations (3) and (4), for unification factor values ranging from 0.0 (local PSO) up to 1.0 (global PSO) with increments of 0.1. The cross symbol denotes the current position of the particle, $x_i = (0,0)^T$; the square denotes its own best position, $p_i = (1,2)^T$; the circle denotes the overall best position, $p_g = (2,5)^T$; and the triangle denotes the best position, $p_l = (5,-3)^T$, in the neighborhood of x_i. For simplicity, the current velocity, v_i is considered to be equal to zero in all cases

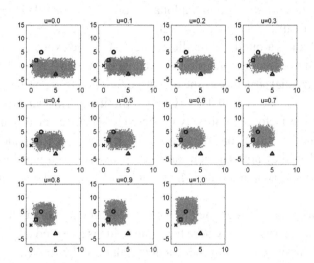

with mean value μ and standard deviation σ. The use of r_3 imitates the mutation operation in evolutionary algorithms. However, in the case of UPSO, mutation is biased towards directions consistent with PSO dynamics, in contrast to the pure random mutation used in evolutionary algorithms. Following the assumptions of Matyas (1965), a proof of convergence in probability was derived for the UPSO scheme of equations (5) and (6) (Parsopoulos & Vrahatis, 2004).

The effect of mutation is illustrated in Fig. 2, using the same data as in Fig. 1. The distributions of new positions are illustrated for the standard UPSO scheme with $u = 0.2$ (right part), as well as for its mutated counterpart of equation (5) with $\mu = 0.0$ and $\sigma = 0.1$ (left part). Mutated UPSO is clearly more biased towards the local best position (denoted with a triangle). In addition, the spread of possible positions is wider, promoting exploration as intended.

At this point, we must note that the choice between equations (5) and (6) shall take into consideration the value of the unification factor. In cases where $u < 0.5$, the local search direction, L_i, has the dominant weight in equation (3), hence the algorithm is mostly based on it. In this case, it is better to use the mutation scheme of equation (5) where the non-dominant (i.e., the global) search direction is mutated. Otherwise, the dominant local search direction will probably be degraded, reducing efficiency. The opposite must hold for $u > 0.5$, where the global search direction, G_i, is in charge of equation (3). In this case, it is better to use the mutation scheme of equation (6), where the local search direction is mutated.

The selection of appropriate values for μ and σ depends on the problem at hand. Usually, $\mu = 0.0$ and small values of σ are adequate to enhance the exploration capability of PSO. However, different values may prove to be better, depending on the shape of the objective function. Parsopoulos and Vrahatis

(2007) offered an extensive experimental study on the selection and adaptation of the unification factor. Their analysis is briefly reported in the next section.

Parameter Selection and Adaptation in UPSO

As already mentioned, there is an obvious dependence of UPSO dynamics on the unification factor, since it controls the balance between its exploration/exploitation properties. Small values favor the local position shift component, thereby resulting in better exploration, while large values favor the global component, promoting exploitation. Values around the middle point, $u = 0.5$, are expected to produce more balanced schemes with respect to their exploration/exploitation capabilities. However, such balanced versions fail to take advantage of any special structure of the objective function (e.g., unimodality, convexity etc). In such cases, unification factors closer to 0.0 or 1.0 may exhibit superior performance. Moreover, online adaptation of the unification factor is intuitively expected to enhance performance.

Parsopoulos and Vrahatis (2007) considered a set of different selection and adaptation schemes of the unification factor. Also, they applied them on a set of widely used benchmark functions to reveal the potential of UPSO for self-adaptation in different environments, and distinguish its most promising configurations.

Following the categorization of Angeline (1995) for evolutionary parameters, the unification factor can be considered either at *swarm-level*, where the same value is assigned to all particles, or at *particle-level*, where each particle assumes its own independent value. In the first case, particles have the same bent for exploration/exploitation, resulting in swarms with aggregating behavior. In the latter case, each particle has its own special exploration/exploitation trade-off, resulting in swarms with higher behavior diversity. These adaptation schemes are described in the following sections.

Quantized Unification Factor

This is a swarm-level scheme, where all particles have the same fixed unification factor value. Parsopoulos and Vrahatis (2007) considered the following set:

Figure 2. Distribution of 3000 possible new positions of a particle using UPSO, (A) with mutation, and (B) without mutation. The same data as in Fig. 1 are used. The unification factor is u = 0.2 and the mutated UPSO of equation (5) is used with μ = 0.0 and σ = 0.1

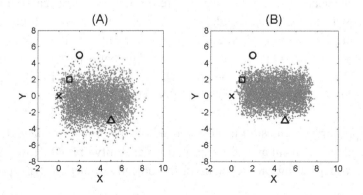

$W = \{0.0, 0.1, 0.2, \ldots, 1.0\},$

of equidistant values of $u \in [0,1]$, and they studied the performance of each value separately to gain intuition regarding the most promising values per problem.

Increasing Unification Factor

This is also a swarm-level scheme with all particles having the same unification factor, which is initialized at 0.0 and increases up to 1.0 during the execution of the algorithm. Consequently, exploration is favored in the first stages where u lies closer to 0.0 (lbest), while exploitation is promoted at final stages where u assumes higher values close to 1.0, approximating gbest.

Let t denote the iteration number; $u(t)$ be the unification factor at iteration t; and T_{max} denote the maximum number of iterations. Three different increasing schedules were considered in the study of Parsopoulos and Vrahatis (2007):

1. **Linear:** Unification factor is linearly increased from 0.0 up to 1.0, according to the scheme:

$$u(t) = \frac{t}{T_{max}},$$

which corresponds to a smooth and relatively slow transition from exploration to exploitation.

2. **Modular:** Unification factor increases repeatedly from 0.0 to 1.0 every q iterations, according to the scheme:

$$u(t) = \frac{t \bmod (q+1)}{q},$$

which repeatedly modifies the dynamics of the algorithm from exploration to exploitation, throughout its execution. The value of q is selected as a reasonable fraction of the maximum number of iterations. In the experiments reported in Parsopoulos and Vrahatis (2007), the value $q = 10^2$ was used for $T_{max} = 10^4$.

3. **Exponential:** Unification factor increases from 0.0 to 1.0 exponentially, according to the scheme:

$$u(t) = \exp\left(\frac{t \log(2.0)}{T_{max}}\right) - 1.0.$$

This scheme results in a mild transition from exploration to exploitation in early iterations, while it is accelerated in later stages of execution.

Figure 3 graphically illustrates the unification factor under the aforementioned increasing schemes for 1000 iterations.

Figure 3. The linear, modular, and exponential increasing scheme of the unification factor for 1000 iterations

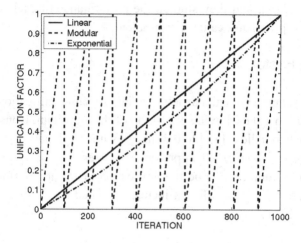

Sigmoid Unification Factor

This is also a swarm-level scheme, similar to the increasing schemes described in the previous section. All particles assume the same unification factor, which is increased according to the relation:

$$u(t) = F_{\mathrm{sig}}\left(t - \frac{T_{\max}}{20}, \lambda\right)$$

where,

$$F_{sig}(x, \lambda) = \frac{1}{1 + \exp(-\lambda x)}.$$

This scheme results in a sigmoid transition from exploration to exploitation. It is considered separately from the other increasing schemes due to the form of the sigmoid, which depends on the value of the parameter λ. Parsopoulos and Vrahatis (2007) investigated the values, $\lambda = 10^{-1}$, 10^{-2}, and 10^{-3}. The corresponding sigmoids are depicted in Fig. 4 for $T_{\max} = 10^3$. Sigmoid has been used as an activation function in artificial neural networks (Parsopoulos & Vrahatis, 2007).

Swarm Partitioning

This is a particle-level scheme, where the swarm is divided in partitions consisting of a prespecified number of particles, called *partition size*. All particles in the same partition share the same unification factor, while each partition has a different value. Parsopoulos and Vrahatis (2007) considered a swarm with 11 partitions, each assuming a value of u from the set $W = \{0.0, 0.1, 0.2, \ldots, 1.0\}$. Depending on the swarm size, it is possible to have partitions with different sizes. For example, a swarm of size 100 can be partitioned in 10 partitions of size 9, and one partition of size 10. The slightly different size of the last partition is not expected to modify the dynamic of the scheme significantly.

Figure 4. The sigmoid scheme for adapting the unification factor, for $\lambda=10^{-1}$, 10^{-2}, and 10^{-3}

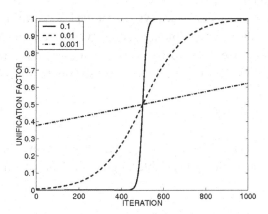

Special care shall be taken in determining the neighborhoods of partitioned swarms. If a neighborhood consists mostly of particles of the same partition, then they will share the same unification factor, thereby biasing its search capabilities. For example, in the ring topology, if the first k particles, $x_1, x_2, \ldots, x_k$, are assigned to partition 1, the next k particles, $x_{k+1}, x_{k+2}, \ldots, x_{2k}$, to partition 2, and so on, then most of the neighborhoods will have the aforementioned undesired property. Better spread of particles of different partitions in different ring neighborhoods can be achieved by assigning particles to partitions in a non-sequential manner, such that the i-th particle is assigned to the $(1 + (i-1) \bmod k)$-th partition (Parsopoulos & Vrahatis, 2007). Put simply, the first k particles of the swarm are assigned to partitions 1 to k, respectively, one particle per partition. Then, it starts over by assigning x_{k+1} to partition 1, x_{k+2} to partition 2, and so on. The procedure is illustrated schematically below for $k = 11$ partitions:

| x_1 | x_{12} | $\cdots$ | $\leftarrow$ Partition 1 |

| x_2 | x_{13} | $\cdots$ | $\leftarrow$ Partition 2 |

$\vdots$ $\vdots$ $\vdots$

| x_{11} | x_{22} | $\cdots$ | $\leftarrow$ Partition 11 |

Using this scheme, particles with different unification factors are allowed to interact by sharing information with their neighbors, while a satisfactory appearance frequency of different unification factor values is retained in the swarm.

Dominance of the Best

This is a particle-level scheme based on swarm partitioning. More specifically, the swarm is divided in partitions, and, after a number of iterations, the partition that contains the best particle of the swarm gains an additional particle from the partition where the worst particle belongs. The minimum size of a partition is set to 1 to prevent its elimination and the consequent loss of behavioral diversity. If the size of the worst partition is equal to 1, then a particle from the immediately next worst partition with size higher than 1 is attributed to the best partition when required. If all but one partitions have sizes equal to 1, then no particle migration takes place.

Dominance of the best incorporates an award system for the best partition to strengthen the influence of the best unification factor in the swarm, although it preserves the existence of all different values of u by imposing a minimum partition size. This mechanism prevents the swarm from being conquered rapidly by the best performing unification factor, which could be detrimental for its overall performance.

Self-Adaptive Unification Factor

In this particle-level scheme, each particle, $x_i = (x_{i1}, x_{i2}, \ldots, x_{in})^{\mathrm{T}}$, has its own unification factor, u_i, which is incorporated in the particle as an additional variable, augmenting the dimension of the problem. Hence, the particle x_i is defined as:

$$x_i = (x_{i1}, x_{i2}, \ldots, x_{in}, u_i)^{\mathrm{T}} \in A \times [0,1], \quad i = 1, 2, \ldots, N,$$

where A is the original search space and n is the problem dimension.

According to the self-adaptive scheme, UPSO is allowed to determine the optimal unification factor for each particle individually, by capturing online possible special structure of the problem at hand.

Empirical Analysis of UPSO

Parsopoulos and Vrahatis (2007) performed an extensive experimental analysis of the aforementioned schemes for selection and adaptation of the unification factor. Their experiments used the test problems TP_{UO-1}-TP_{UO-5}, reported in Appendix A of the book at hand. The corresponding dimension, search space, and desired accuracy per test problem are reported in Table 1.

For each test problem, 100 independent experiments were conducted per scheme. The swarm size was set to 30 in all cases. This choice was based on the promising results reported in Trelea (2003). The swarm was allowed to perform a maximum number, $T_{max} = 10^4$, of iterations. The particles were constrained within the ranges reported in Table 1, and a maximum value, v_{max}, equal to half the corresponding range of the search space was imposed on velocities. The rest of the parameters were set to their default values, $\chi = 0.729$, $c_1 = c_2 = 2.05$ (Clerc & Kennedy, 2002). For the computation of the local PSO component in UPSO, a ring neighborhood topology of radius $r = 1$ was used.

In the sigmoid scheme, the parameters, $\lambda = 10^{-1}$, 10^{-2}, and 10^{-3}, were used for the transition of the unification factor from 0.0 to 1.0 in the first 1000 iterations. The decision to complete the transition in 1000 iterations, instead of the maximum number, was based on the observation that, in most cases, PSO required less than 1000 iterations to converge (Parsopoulos & Vrahatis, 2007).

In swarm partitioning, since the swarm size was 30 and the number of partitions was 11, some partitions of 3 particles and some of 2 particles were considered. More specifically, partitions with u ranging from 0.0 up to 0.7 consisted of 3 particles, while the remaining partitions consisted of 2 particles. The same also held for the initialization of dominance of the best, where additionally a partition update was performed every 20 iterations.

In the self-adaptive scheme, the unification factor of each particle was randomly initialized using a uniform distribution over the range [0.3,0.6]. Initially, the full range [0,1] was used; however, the algorithm was unable to reach the desired goal within the maximum number of iterations, although it was moving close to the global minimizers. This inability is attributed to the increased dimension of the

Table 1. Configuration of the test problems. All problems are defined in Appendix A of the book at hand

Problem	Dimension	Range	Accuracy
TP_{UO-1}	30	$[-100, 100]^{30}$	10^{-2}
TP_{UO-2}	30	$[-30, 30]^{30}$	10^{2}
TP_{UO-3}	30	$[-5.12, 5.12]^{30}$	10^{2}
TP_{UO-4}	30	$[-600, 600]^{30}$	10^{-1}
TP_{UO-5}	2	$[-100,100]^{2}$	10^{-5}

particles after the inclusion of u. Thus, either the maximum number of iterations should be increased or a smaller range for u should be used in order to have a fair comparison of the self-adaptive scheme with the rest. To this end, Parsopoulos and Vrahatis (2007) retained the maximum number of iterations because the same value was used in related works (Trelea, 2003), and they preferred to constrain u within [0.3,0.6]. The selection of the range [0.3,0.6] was based on the promising results obtained through the quantized scheme for this range.

Besides a plethora of tables containing the statistical analyses of their results in terms of the required number of function evaluations to reach the global minimizer with the desired accuracy, Parsopoulos and Vrahatis (2007) also performed t-tests to study the statistical significance of their results. Each different scheme was tested against all other schemes to a significance level of 99%.

Due to spatial limitations, we omit the detailed presentation of all the tables reported in Parsopoulos and Vrahatis (2007); however, we provide a short discussion per problem in the following paragraphs. In addition, in Table 2, we report the overall best approach per test problem, in terms of its success percentage over 100 experiments and the mean required number of iterations. For reasons of comparison, the second best approach, as well as the gbest and lbest PSO variants, are also reported. The adaptation schemes are denoted as follows: "Quan" for quantized; "Sig" for sigmoid; "Self" for self-adaptive; and "Part" for swarm partitioning.

In TP_{UO-1}, all UPSO schemes were 100% successful, outperforming the local and global PSO variants, which correspond to the values $u = 0.0$ and $u = 1.0$ of the quantized scheme, respectively. Most of the algorithms had statistically significant differences in their performance. Only the case of $u = 1.0$ (pure global PSO) in the quantized scheme exhibited a success rate smaller than 100%. At a first glance, this contradicts the claim that $u = 1.0$ promotes exploitation, since the problem is unimodal and convex, thus, it should be solved efficiently by an exploitation-oriented PSO variant. The explanation for this inferior performance lies in the high dimensionality of the problem. The gbest PSO converges rapidly towards the global minimizer, although with a different convergence rate per coordinate direction. Thus, it can be trapped in suboptimal solutions, although close to the actual global minimizer. The quantized scheme with $u \in [0.4,0.7]$, along with the sigmoid with $\lambda = 10^{-3}$, and the self-adaptive scheme, exhibited the best performance with small differences among them, followed by the dominance of the best and swarm partitioning.

In TP_{UO-2}, the quantized scheme with $u = 0.2$ was the most promising. As expected, exploration-oriented UPSO variants with u closer to 0.0 performed better than exploitation-oriented variants with u closer to 1.0. The sigmoid scheme with $\lambda = 10^{-3}$ and the self-adaptive scheme also performed well, without having statistically significant differences. In this problem, dominance of the best exhibited better performance than swarm partitioning.

Table 2. Results for the overall best performing and the second best performing scheme in the experiments of Parsopoulos and Vrahatis (2007), along with those of the standard gbest and lbest PSO variants. The adaptation schemes are denoted as follows: "Quan" for quantized; "Sig" for sigmoid; "Self" for self-adaptive; and "Part" for swarm partitioning. For each case, the success percentage over 100 experiments, as well as the mean required number of iterations, are reported

Problem		Overall Best	Second best	gbest PSO	lbest PSO
$TP_{UO\text{-}1}$	Method	Quan ($u = 0.5$)	Quan ($u = 0.4$)	Quan ($u = 1.0$)	Quan ($u = 0.0$)
	Success	100%	100%	91%	100%
	Mean	1.921×10^2	1.930×10^2	1.231×10^3	5.698×10^2
$TP_{UO\text{-}2}$	Method	Sig ($\lambda = 10^{-3}$)	Quan ($u = 0.2$)	Quan ($u = 1.0$)	Quan ($u = 0.0$)
	Success	100%	100%	68%	100%
	Mean	1.931×10^2	2.401×10^2	3.583×10^3	4.673×10^2
$TP_{UO\text{-}3}$	Method	Self	Quan ($u = 0.5$)	Quan ($u = 1.0$)	Quan ($u = 0.0$)
	Success	100%	100%	52%	95%
	Mean	1.273×10^2	1.313×10^2	4.895×10^3	9.628×10^2
$TP_{UO\text{-}4}$	Method	Quan ($u = 0.5$)	Quan ($u = 0.3$)	Quan ($u = 1.0$)	Quan ($u = 0.0$)
	Success	100%	100%	90%	100%
	Mean	1.794×10^2	1.957×10^2	1.299×10^3	5.317×10^2
$TP_{UO\text{-}5}$	Method	Quan ($u = 0.3$)	Part	Quan ($u = 1.0$)	Quan ($u = 0.0$)
	Success	100%	100%	76%	99%
	Mean	4.074×10^2	4.102×10^2	2.674×10	8.956×10^2

$TP_{UO\text{-}3}$ is highly multimodal. Thus, it was anticipated that balanced UPSO versions would perform better. Indeed, the value $u = 0.5$ proved to be the best, while unification factors higher than 0.7 exhibited poor performance. The self-adaptive scheme had the best overall performance. Also, the modular scheme exhibited superior performance in this problem. This suggests that, in highly multimodal functions, the iterative modification of the unification factor from 0.0 to 1.0 can provide better results than fixed values.

In $TP_{UO\text{-}4}$, the value $u = 0.5$ was again the most promising, with exploration-oriented UPSO variants outperforming the exploitation-oriented ones. Self-adaptive and dominance of the best were also shown to be efficient, with all algorithms having marginal performance differences among them. Finally, in $TP_{UO\text{-}5}$, the quantized scheme with unification factor $u = 0.3$ was the best among all quantized schemes, while swarm partitioning and self–adaptive had the best overall performance among all schemes.

Summarizing the results, $u = 0.2$ and $u = 0.3$ were the only cases of the quantized scheme with success percentages of 100% for all test problems, although, in some cases they were outperformed by more balanced UPSO versions, such as $u = 0.5$. Also, the linearly increasing and self-adaptive schemes were successful in all test problems, with the latter always outperforming both the local and global PSO variants, with respect to the mean number of iterations. Moreover, it was observed that the final unification factor (i.e., that of the solution) in the adaptive schemes came rarely in line with the best unification factor observed in the quantized scheme. This is an indication that, in a given problem the adaptive schemes were able to capture its shape and change their behavior accordingly, based on their performance during execution.

The highest diversities of unification factor values at the solutions were observed for TP_{UO-3} and TP_{UO-5}, while narrower ranges were obtained for the rest of the problems. Overall, there was an UPSO version that outperformed the standard PSO variants in all test problems. The values $u = 0.2$ and $u = 0.3$ were shown to be the most effective of the unification factor, while $u = 0.5$ constitutes a good choice for a more balanced search and faster convergence, although at the risk of slightly reduced success rates. The linearly decreasing, sigmoid, and self-adaptive schemes were robust and reliable, with the latter exhibiting considerably better performance.

The reported results reveal the potential of UPSO to be a very promising approach. This is also reflected in its number of applications in different scientific fields. Such applications will be reported in following chapters of the book at hand. The next section, on memetic PSO, presents a recently proposed and efficient approach that incorporates local search.

MEMETIC PARTICLE SWARM OPTIMIZATION

Memetic PSO (MPSO) is a hybrid algorithm that combines PSO with local search techniques. MPSO consists of two main components: a global one, which is responsible for global search, and a local one, which performs more refined search around roughly detected solutions. In the next section, we briefly describe the fundamental concepts of memetic algorithms (MAs) and provide the necessary background for the presentation of MPSO.

Fundamental Concepts of Memetic Algorithms

MAs comprise a family of population-based heuristic search algorithms, designed for global optimization tasks. The main inspiration behind their development was the concept of the *meme*, as coined by Dawkins (1976), which represents a unit of cultural evolution that admits refinement. Memes can also represent models of adaptation in natural systems that combine evolutionary adaptation with individual learning within a lifetime. MAs include a stage of individual optimization or learning, usually in the form of a local search procedure, as part of their search operation.

MAs were first proposed by Moscato (1989), where simulated annealing was used for local search with a competitive and cooperative game between agents interspersed with the use of a crossover operator, to tackle the traveling salesman problem. This method gained wide acceptance due to its ability to solve difficult problems.

Although MAs bear a similarity with genetic algorithms (GAs) (Goldberg, 1989), they mimic cultural rather than biological evolution. Indeed, GAs employ a combination of selection, recombination, and mutation, similar to that applied to genes in natural organisms. However, genes are usually not modified during a lifetime, whereas memes are. Therefore, most MAs can be interpreted rather as a cooperative/competitive algorithm of optimizing agents.

In general, an evolutionary MA can be described with the procedure reported in Table 3. In particular, at the beginning the population is initialized within the search space. The local search algorithm is initialized on one or more population members and performs local search from each one. Then, the produced solutions are evaluated and evolutionary operators are applied to produce offspring. Local search is applied on these offspring and selection chooses the individuals that will constitute the parent

Table 3. Pseudocode of a generic memetic evolutionary algorithm

Input:	Population, evolutionary operators, local search algorithm.
Step 1.	**Initialize** population.
Step 2.	**Apply** local search.
Step 3.	**Evaluate** population.
Step 4.	**While** (termination condition is false)
Step 5.	**Apply** recombination.
Step 6.	**Apply** mutation.
Step 7.	**Apply** local search.
Step 8.	**Evaluate** population.
Step 9.	**Apply** selection.
Step 10.	**End While**
Step 11.	**Report** best solution.

population in the next generation. The termination condition can include various criteria, such as time expiration and/or maximum generations limit.

The first implementations of MAs were hybrid algorithms combining GAs as the global search components with a local search (GA–LS) (Belew *et al.*, 1991; Hart, 1994; Hinton & Nowlan, 1987; Keesing & Stork, 1991; Muhlenbein *et al.*, 1988). The GA–LS hybrid scheme was interesting due to the interaction between its local and global search components. An important issue of these interactions, also met in natural systems, is the *Baldwin effect* (Belew, 1990; Hinton & Nowlan, 1987), where learning proves to speed up the rate of evolutionary change. Similar effects have been observed in several GA–LS schemes (Belew *et al.*, 1991; Hinton & Nowlan, 1987; Keesing & Stork, 1991). MAs have been successfully applied also in combinatorial optimization, especially for the approximate solution of NP-hard optimization problems, where their success can be attributed to the synergy of the employed global and local search components (Krasnogor, 2002; Land, 1998; Merz, 1998).

Recently Proposed Memetic PSO Schemes

MPSO combines PSO with a local search algorithm. Besides the selection of the most appropriate local search method, three major questions arise spontaneously. Henceforth, we will call these questions the *fundamental memetic questions* (FMQs):

(FMQ 1) When local search has to be applied?
(FMQ 2) Where local search has to be applied?
(FMQ 3) What computational budget shall be accredited to the local search algorithm?

It is a matter of user experience to provide appropriate answers to these questions prior to the application of an MA to a complex problem. For this purpose, a preprocessing phase for the determination of the most promising choices can be valuable in unstudied problems.

Petalas *et al.* (2007b) proposed an MPSO approach that uses the random walk with directional exploitation local search method (Rao, 1992), and studied its performance against the standard PSO on several benchmark problems. In another application framework, related to learning in fuzzy cognitive maps, Petalas *et al.* (2007a) extended their scheme by using the local search algorithms of Hook and Jeeves (Hook & Jeeves, 1961) and Solis and Wets (Solis & Wets, 1981). More recently, an entropy-based MPSO method was proposed and studied for the detection of periodic orbits of nonlinear mappings (Petalas *et al.*, 2007c). Let us take a closer look at each of the aforementioned schemes in the following sections.

Memetic PSO with Random Walk with Directional Exploitation Local Search

In (Petalas *et al.*, 2007b), the following schemes regarding the point of application of the local search (which is related to FMQ 2) were proposed:

(Scheme 1) Local search is applied on the overall best position, p_g, of the swarm, where g is the index of the best particle.

(Scheme 2) For each best position, p_i, $i = 1, 2,..., N$, a random number, $r \in [0,1]$, is generated and, if $r < \varepsilon$ for a prescribed threshold, $\varepsilon > 0$, local search is applied on p_i.

(Scheme 3) Local search is applied on the overall best position, p_g, as well as on some randomly selected best positions, p_i, $i \in \{1, 2,..., N\}$.

(Scheme 4) Local search is applied on the overall best position, p_g, as well as on the best positions, p_i, $i \in \{1, 2,..., N\}$, for which it holds that, $\|p_g - p_i\| > c\Delta(A)$, where $c \in (0,1)$, and $\Delta(A)$ is the diameter of the search space A or an approximation of it.

These schemes can be applied either in every iteration of the algorithm or at some iterations. Of course, different approaches can be also considered, e.g., application of local search to all particles. However, empirical studies suggest that such schemes are costly in terms of function evaluations, while, in practice, only a small percentage about 5% of the particles has to be considered as application point for local search (Petalas *et al.*, 2007b). These conclusions were verified also by Hart (1994) for the case of GA–LS hybrid schemes.

A pseudocode for the MPSO algorithm is reported in Table 4. In addition, a proof of convergence in probability was derived in (Petalas *et al.*, 2007b) for the approach with the random walk with directional exploitation (RWDE) (Rao, 1992) local search algorithm, which is sketched below.

RWDE is an iterative stochastic optimization method. It generates a sequence of approximations of the optimizer by assuming random search direction. RWDE can be applied both on discontinuous and non-differentiable functions. Moreover, it has been shown to be effective in cases where other methods fail due to difficulties posed by the form of the objective function, e.g., sharply varying functions and shallow regions (Rao, 1992). Let $x(t)$ be the approximation generated by RWDE at the t-th iteration, with $x(1)$ denoting the initial value. Let also, λ_{init}, be a starting step length and, T_{max}, be the maximum number of iterations, while $f(x)$ denotes the objective function. Then, RWDE is described by the pseudocode of Table 5.

Alternatively to the rather simple RWDE, more sophisticated stochastic local search algorithms can be used (Hoos & Stützle, 2004). Petalas *et al.* (2007b) justified their choice of RWDE by stating that their main goal was to designate MPSO as an efficient and effective method, rather than conducting a thorough investigation of the employed local search and its convergence properties. Additionally, the simplicity of RWDE, along with its adequate efficiency, was also taken into consideration. Moreover, RWDE does not make any continuity or differentiability assumptions on the objective function; thus, it

Table 4. Pseudocode of the MPSO algorithm

Input:	N, χ, c_1, c_2, x_{min}, x_{max} (lower and upper bounds), f(x) (objective function)
Step 1.	**Set** $t \leftarrow 0$.
Step 2.	**Initialize** $x_i(t)$, $v_i(t)$, $p_i(t)$, $i = 1, 2,..., N$.
Step 3.	**Evaluate** $f(x_i(t))$, $i = 1, 2,..., N$.
Step 4.	**Update** indices, g_i, of best particles.
Step 5.	**While** (stopping condition not met)
Step 6.	**Update** velocities, $v_i(t+1)$, and particles, $x_i(t+1)$, $i = 1, 2,..., N$.
Step 7.	**Constrain** particles within bounds $[x_{min}, x_{max}]$.
Step 8.	**Evaluate** $f(x_i(t+1))$, $i = 1, 2,..., N$.
Step 9.	**Update** best positions, $p_i(t+1)$, and indices, g_i.
Step 10.	**If** (local search is applied) **Then**
Step 11.	**Choose** a position $p_q(t+1)$, $q \in \{1, 2,..., N\}$, according to Schemes 1-4.
Step 12.	**Apply** local search on $p_q(t+1)$ and obtain a solution, y.
Step 13.	**If** $(f(y) < f(p_q(t+1)))$ **Then** $p_q(t+1) \leftarrow y$.
Step 14.	**End If**
Step 15.	**Set** $t \leftarrow t+1$.
Step 16.	**End While**

Table 5. Pseudocode of the RWDE algorithm

Input:	Initial point, x(1); initial step, λ_{init}; maximum iterations, T_{max}.
Step 1.	**Set** $t \leftarrow 0$ and $\lambda \leftarrow \lambda_{init}$.
Step 2.	**Evaluate** $f_1 = f(x(1))$.
Step 3.	**While** $(t < T_{max})$
Step 4.	**Set** $t \leftarrow t+1$.
Step 5.	**Generate** a unit-length random vector, $z(t)$.
Step 6.	**Evaluate** $f' = f(x(t)+\lambda z(t))$.
Step 7.	**If** $(f' < f_t)$ **Then**
Step 8.	**Set** $x(t+1) = x(t)+\lambda z(t)$.
Step 9.	**Set** $\lambda \leftarrow \lambda_{init}$ and $f_{t+1} \leftarrow f'$.
Step 10.	**Else**
Step 11.	**Set** $x(t+1) = x(t)$.
Step 12.	**If** $(f' > f_t)$ **Then Set** $\lambda \leftarrow \lambda/2$.
Step 13.	**End If**
Step 14.	**End While**
Step 15.	**Report** solution.

is consistent with the PSO framework that requires function values solely. For all these reasons, RWDE was preferred instead of gradient-based local search algorithms (Petalas *et al.*, 2007b).

Petalas *et al.* (2007b) performed extensive experiments with the MPSO with RWDE on test problems TP_{UO-1}-TP_{UO-9}, defined in Appendix A of the book at hand. The dimension of each problem, the search

Table 6. Configuration of the test problems. All problems are defined in Appendix A of the book at hand

Problem	Dimension	Range	Accuracy
TP_{UO-1}	30	$[-100, 100]^{30}$	10^{-2}
TP_{UO-2}	30	$[-30, 30]^{30}$	10^{2}
TP_{UO-3}	30	$[-5.12, 5.12]^{30}$	10^{2}
TP_{UO-4}	30	$[-600, 600]^{30}$	10^{-1}
TP_{UO-5}	2	$[-100,100]^{2}$	10^{-5}
TP_{UO-6}	30	$[-32,32]^{30}$	10^{-3}
TP_{UO-7}	4	$[-1000,1000]^{4}$	10^{-6}
TP_{UO-8}	30	$[-50,50]^{30}$	10^{-6}
TP_{UO-9}	30	$[-50,50]^{30}$	10^{-2}

space, as well as the required accuracy in their experiments is reported in Table 6. The maximum number of iterations for every problem was set to $T_{max} = 10^4$. Three different swarm sizes, $N = 15$, 30, and 60, were used to study the corresponding scaling properties of MPSO. Both the global and local PSO variants were equipped with RWDE, resulting in the corresponding memetic schemes, which are henceforth denoted as $MPSO_g^{[RW]}$ and $MPSO_l^{[RW]}$, respectively.

The memetic approaches were compared against their standard PSO counterparts. For this purpose, 50 experiments were performed for each test problem and algorithm. An experiment was considered successful if the global minimizer was detected with the required accuracy within the maximum number of iterations. In an attempt to achieve the best possible performance of the considered memetic approaches per test problem, Petalas *et al.* (2007b) used different configuration of RWDE for each problem, based on observations in preliminary experiments (preprocessing phase). These configurations are reported in Table 7.

The first column of the table denotes the problem, while the second stands for the swarm size, N. The third and fourth column report the number of iterations and initial step size used by RWDE, respectively. The fifth column has the value "Yes" in cases where RWDE was applied only on the best particle of the swarm. On the other hand, if RWDE was applied with a probability on the best position of each particle, then this probability is reported in the sixth column. Finally, the last column shows the frequency of application of the local search. For instance, the value "1" corresponds to application of local search at each iteration, while "20" corresponds to application every 20 iterations.

For the local variants of PSO and MPSO, a ring neighborhood of radius $r = 1$ was used for TP_{UO-1}-TP_{UO-6} and TP_{UO-8}, while a radius equal to $r = 2$ provided better results for TP_{UO-7} and TP_{UO-9} (Petalas *et al.*, 2007b). The results of the best performing MPSO variants (among those reported in Table 7) and those of the corresponding standard PSO variants, are reported in Table 8 (Petalas *et al.*, 2007b). More specifically, for each test problem and algorithm, the swarm size, N, the number of successes (over 50 experiments), as well as the mean number of function evaluations of the best performing variant per algorithm, are reported. The best performing variant was defined as the one with the highest number of successes and the smallest mean number of function evaluations (Petalas *et al.*, 2007b). Instead of iterations, function evaluations were used as the performance measure, since memetic approaches do not perform necessarily the same number of function evaluations per iteration due to the application of

Table 7. Configuration of the RWDE local search in (Petalas et al., 2007b). For each test problem and swarm size, N, the number of iterations (Iter) and initial step size (Step) are reported for each algorithm. The value "Yes" under the column "Best" denotes that RWDE was applied only on the best particle of the swarm. If RWDE was applied with a probability on the best position of each particle, then this probability is reported under the "Prob" column, while column "Freq" shows the frequency of application of RWDE (e.g., "1" corresponds to application at each iteration, while "20" corresponds to application every 20 iterations)

Problem	N	MPSO$_g^{[RW]}$					MPSO$_l^{[RW]}$				
		Iter.	Step	Best	Prob.	Freq.	Iter.	Step	Best	Prob.	Freq.
TP$_{UO-1}$	15	5	1.0	Yes	-	1	10	1.0	Yes	-	1
	30	5	1.0	Yes	-	1	10	1.0	Yes	-	1
	60	5	1.0	Yes	-	1	5	1.0	Yes	-	1
TP$_{UO-2}$	15	10	1.0	Yes	-	1	8	0.5	Yes	-	50
	30	5	1.0	Yes	-	1	5	1.0	Yes	-	30
	60	5	1.0	Yes	-	1	5	1.0	Yes	-	1
TP$_{UO-3}$	15	5	1.0	-	0.2	1	5	1.0	Ye	-	20
	30	5	1.0	-	0.2	1	10	1.0	Yes	-	20
	60	5	1.0	-	0.1	1	5	1.0	Yes	-	1
TP$_{UO-4}$	15	5	4.0	Yes	-	1	10	8.0	Yes	-	1
	30	5	4.0	Yes	-	1	10	8.0	Yes	-	1
	60	5	4.0	Yes	-	1	10	8.0	Yes	-	1
TP$_{UO-5}$	15	8	1.0	-	0.3	1	8	1.0	-	0.3	2
	30	8	1.0	-	0.2	1	8	1.0	-	0.1	1
	60	8	1.0	-	0.1	1	8	1.0	-	0.1	2
TP$_{UO-6}$	15	5	1.0	-	0.5	1	5	1.0	Yes	-	20
	30	5	1.0	-	0.5	1	5	1.0	Yes	-	20
	60	5	1.0	-	0.4	1	5	1.0	Yes	-	20
TP$_{UO-7}$	15	5	1.0	Yes	-	20	5	1.0	Yes	-	20
	30	5	1.0	Yes	-	20	5	1.0	Yes	-	20
	60	5	1.0	Yes	-	20	5	1.0	Yes	-	20
TP$_{UO-8}$	15	5	1.0	-	0.8	1	5	1.0	Yes	0.3	1
	30	5	1.0	-	0.5	1	5	1.0	Yes	-	2
	60	5	1.0	-	0.3	1	5	1.0	Yes	-	2
TP$_{UO-9}$	15	5	1.0	-	0.6	1	3	1.0	-	0.5	1
	30	5	1.0	-	0.5	1	5	1.0	-	0.1	1
	60	5	1.0	-	0.3	1	10	1.0	Yes	-	2

local search. In addition, the reported mean number of function evaluations was computed for all approaches using only information from their successful experiments, in order to provide an estimation of the expected number of function evaluations in a successful run (Petalas *et al.*, 2007b).

Clearly, the memetic variants outperformed their corresponding standard PSO variants. More specifically, for the global variants, $\text{MPSO}_g^{[RW]}$ has higher success rates than PSO_g in all problems. In some cases, this comes at the cost of some extra function evaluations, although in most problems $\text{MPSO}_g^{[RW]}$ was computationally cheaper than PSO_g. The impact of swarm size appears to be similar for both the memetic and the corresponding standard PSO variants, with larger swarms requiring more function evaluations but exhibiting better success rates (Petalas *et al.*, 2007b).

Similar conclusions are also derived for the local variants. The superiority of PSO_l over PSO_g, in terms of success rate, is inherited to its memetic counterpart, $\text{MPSO}_l^{[RW]}$, over $\text{MPSO}_g^{[RW]}$. Moreover, $\text{MPSO}_l^{[RW]}$ performed better than PSO_l in almost all cases, achieving high success rates with significantly smaller number of function evaluations (Petalas *et al.*, 2007b). $\text{MPSO}_l^{[RW]}$ had better success rates than $\text{MPSO}_g^{[RW]}$ in all test problems, although at the cost of slower convergence. This is also an indication that the neighborhood radius in memetic PSO approaches has the same effect on convergence as for the standard PSO methods.

Interestingly, in many cases $\text{MPSO}_g^{[RW]}$ outperformed even PSO_l, which is a promising indication that the use of RWDE with global PSO variants can enhance significantly their exploration capabilities. Petalas *et al.* (2007b) also performed *t*–tests in cases where the superiority of an algorithm over another was not clear. The tests were conducted using the null hypothesis that the mean numbers of required function evaluations between the two algorithms are equal at significance level 99%. Their conclusions suggested that in TP_{UO-1}, where $\text{MPSO}_g^{[RW]}$ was compared with PSO_l (they both achieved 50 successes), the null hypothesis could be rejected, i.e., $\text{MPSO}_g^{[RW]}$ performed better than PSO_l, for swarm size equal to 15. The same holds also for the $\text{MPSO}_l^{[RW]}$ against PSO_l and for $\text{MPSO}_g^{[RW]}$ against $\text{MPSO}_l^{[RW]}$, which seems to be a natural consequence attributed to the simplicity and unimodality of TP_{UO-1}. Similar tests were performed also for TP_{UO-7}, where $\text{MPSO}_g^{[RW]}$ was proved to be statistically superior to PSO_l.

Overall, MPSO outperformed PSO, exhibiting significantly better performance in most test problems. Petalas *et al.* (2007b) presented results also for constrained, minimax, and integer optimization problems to justify further the superiority of memetic PSO approaches over the standard PSO. The reader is referred to the original paper for a thorough presentation.

RWDE was not the only local search algorithm combined with PSO in memetic schemes. Different approaches that employ the Hook and Jeeves (Hook & Jeeves, 1961) and the Solis and Wets (Solis & Wets, 1981) local search were proposed and applied on learning problems in fuzzy cognitive maps (Petalas *et al.*, 2007a). Besides the different local search methods, an entropy-based MPSO (Petalas *et al.*, 2007c) was proposed in an attempt to tackle the issues raised by the FMQs described in the previous section. We present this approach in the next section.

Entropy-Based Memetic PSO

The *entropy-based memetic PSO* (henceforth denoted as E-MPSO) was introduced by Petalas *et al.* (2007c), and it is based on the concept of *Shannon's information entropy* (SIE) (Shannon, 1964). SIE has been used as diversity measure in genetic programming (Burke *et al.*, 2004; Rosca, 1995). Entropy has also been used for developing diversity-preserving techniques in multiobjective evolutionary algorithms (Cui *et al.*, 2001), as well as for determining the initial conditions of local search in parallel memetic schemes (Tang *et al.*, 2006).

Let P be a population divided in k phenotype classes, and q_k be the proportion of P occupied by partition k at a given time. Then, SIE is defined as (Rosca, 1995):

Table 8. Results of the best performing variants (among those reported in Table 7) of MPSO$_g^{[RW]}$, MPSO$_l^{[RW]}$, PSO$_g$ and PSO$_l$. For each problem and algorithm, the swarm size, N, the number of successes (over 50 experiments), as well as the mean number of function evaluations of the best performing variant, are reported

Problem		MPSO$_g^{[RW]}$	MPSO$_l^{[RW]}$	PSO$_g$	PSO$_l$
TP$_{UO-1}$	N	15	15	60	15
	Success	50	50	48	50
	Mean	6009.7	7318.3	16360.0	8467.5
TP$_{UO-2}$	N	15	15	60	15
	Success	50	50	39	50
	Mean	9275.5	7679.0	26420.0	8004.9
TP$_{UO-3}$	N	60	30	60	30
	Success	50	50	40	50
	Mean	19494.8	13815.3	10042.5	16848.0
TP$_{UO-4}$	N	15	15	60	15
	Success	50	50	49	50
	Mean	5956.5	6588.2	14891.0	8006.4
TP$_{UO-5}$	N	30	60	60	60
	Success	50	50	44	50
	Mean	12425.4	18080.2	11877.3	27240.0
TP$_{UO-6}$	N	60	15	60	15
	Success	50	50	20	50
	Mean	79473.6	12978.2	25116.0	12733.5
TP$_{UO-7}$	N	15	15	60	15
	Success	50	50	50	50
	Mean	2094.3	2685.8	5616.0	2686.2
TP$_{UO-8}$	N	30	30	60	60
	Success	49	50	33	50
	Mean	78764.5	38916.5	30172.7	74986.8
TP$_{UO-9}$	N	30	15	30	60
	Success	50	50	38	50
	Mean	34710.7	17579.2	17416.6	25918.8

$$\mathrm{SIE}(P) = -\sum_k q_k \log q_k,$$

representing the amount of chaos in the system. Large values of entropy correspond to small values of q_k, i.e., the partitions are almost equally significant. On the other hand, small values of entropy correspond to larger values of q_k, i.e., a significant number of individuals are concentrated in a few partitions. Therefore, high entropy indicates higher population diversity, in direct analogy to physical systems (Shannon, 1964).

In the context of PSO, let $S = \{x_1, x_2, \ldots, x_N\}$ be a swarm consisting of N particles, and $P = \{p_1, p_2, \ldots, p_N\}$ be the corresponding best positions. Then, at a given iteration, t, SIE is defined as:

$$\mathrm{SIE}_t(P) = -\sum_{i=1}^{N} q_i(t) \log q_i(t), \tag{7}$$

where,

$$q_i(t) = \frac{f(p_i(t))}{\displaystyle\sum_{i=1}^{N} f(p_i(t))},$$

i.e., $q_i(t)$ is the contribution of the function value $f(p_i(t))$ of the best position $p_i(t)$ to the sum of all best position function values. This quantity has been used widely as a *fitness measure* in evolutionary algorithms.

The SIE of equation (7) measures the spread of best position values, providing information regarding the behavior of PSO. High SIE values correspond to widely spread function values of the best positions, while a smaller SIE indicates similar function values. Also, a rapidly changing value of SIE is an indication of rapidly changing best positions, while slight changes of SIE indicate that the relative differences among function values of the population remain almost unchanged, an effect produced also by search stagnation (Petalas *et al.*, 2007c).

Therefore, SIE can be used as a criterion for deciding, at swarm level, regarding the application or not of local search. More specifically, the changes in the value of SIE are monitored in equidistant moments (e.g., every k iterations), and, if the difference of two consequent measurements is smaller than a user-defined threshold, T_{SIE}, i.e.,

$$|\mathrm{SIE}_t(P) - \mathrm{SIE}_{t-k}(P)| \leq T_{\mathrm{SIE}},$$

then the local search component of the algorithm is evoked.

However, only some of the best positions will serve as initial conditions for the local search. For this purpose, a randomized non-elitist selection of best positions is performed (Petalas *et al.*, 2007c). Thus, for each best position, p_i, $i = 1, 2, \ldots, N$, a random value, R_i, uniformly distributed within $[0,1]$, is drawn. If,

$$R_i \leq q_s,$$

where q_s is a user-defined selection probability, then p_i is selected as an initial condition for local search; otherwise it is ignored. The non-elitism can prevent from premature convergence to local minima, while selection pressure imposed by the user-defined threshold prevents from excessively large numbers of function evaluations. A pseudocode describing the operation of E–MPSO is reported in Table 9. The reader is referred to Petalas *et al.* (2007c) for further details.

All PSO variants studied so far in this chapter were essentially using a single swarm. In the next section, we present a multi-swarm approach especially suited for multiobjective problems.

VECTOR EVALUATED PARTICLE SWARM OPTIMIZATION

Vector evaluated PSO (VEPSO) was introduced by Parsopoulos and Vrahatis (2002a, 2002b) as a multi-swarm PSO variant for multiobjective optimization (MO) problems, and it was extended to parallel implementations in (Parsopoulos *et al.*, 2004). In MO problems, a set of K objective functions, $f_1(x),...,f_K(x)$, needs to be minimized concurrently. The concept of optimality must change to fit the MO framework. The main goal in MO problems is the detection of Pareto optimal points, i.e., points were a small perturbation to any direction will cause an immediate increase to at least one of the objective functions. A detailed description of the MO framework, along with relative applications of PSO, is provided in a separate chapter later in the book. Thus, in the following paragraphs, we will focus only on the description of the VEPSO approach, assuming that the reader is familiar with the fundamental concepts and definitions of MO. For the unfamiliar reader, Chapter Eleven offers a nice introduction to these fundamental concepts.

Let $f_k:A{\rightarrow}\mathbf{R}$, $i = 1, 2,..., K$, be a set of n-dimensional objective functions that need to be minimized concurrently. VEPSO utilizes a set of K swarms, $S_1, S_2,..., S_K$, one for each objective function. The i-th particle of the k-th swarm is denoted as $x_i^{[k]}$; the corresponding best position as $p_i^{[k]}$; and its velocity as $v_i^{[k]}$. The swarm S_k is evaluated only with its corresponding objective function, f_k, for all $k = 1, 2,..., K$. Swarms exchange information among them by sharing their individual findings to direct search towards the Pareto optimal set, i.e., the set of all Pareto optimal points. Assuming that the swarm S_k consists of N_k particles, $k = 1, 2,..., K$, then its update equations in VEPSO are defined as (Parsopoulos & Vrahatis, 2002a, 2002b):

Table 9. Pseudocode of the E-MPSO algorithm

Input:	N (swarm size), P (best positions), parameters T_{SIE}, q_s, k.
Step 1.	**Set** $t \leftarrow 0$ and $SIE_{prev} \leftarrow SIE_t(P)$.
Step 2.	**While** (stopping condition not met)
Step 3.	**Set** $t \leftarrow t+1$.
Step 4.	**Update** swarm and best positions, P.
Step 5.	**If** (mod(t, k) = 0) **Then**
Step 6.	**Compute** $SIE_t(P)$.
Step 7.	**If** ($|SIE_t(P) - SIE_{prev}| \leq T_{SIE}$) **Then**
Step 8.	**Do** ($i = 1...N$)
Step 9.	**Draw** a random number, $R_i \in [0,1]$.
Step 10.	**If** ($R_i \leq q_s$) **Then**
Step 11.	**Apply** local search on $p_i(t)$.
Step 12.	**Update** $p_i(t)$ if better solution found.
Step 13.	**End If**
Step 14.	**End Do**
Step 15.	**End If**
Step 16.	**Set** $SIE_{prev} \leftarrow SIE_t(P)$.
Step 17.	**End If**
Step 18.	**End While**

$$v_{ij}^{[k]}(t+1) = \chi^{[k]}[v_{ij}^{[k]}(t) + c_1^{[k]}r_1(p_{ij}^{[k]}(t) - x_{ij}^{[k]}(t)) + c_2^{[k]}r_2(p_{g_s,j}^{[s]}(t) - x_{ij}^{[k]}(t))], \tag{8}$$

$$x_{ij}^{[k]}(t+1) = x_{ij}^{[k]}(t) + v_{ij}^{[k]}(t+1), \tag{9}$$

$$k = 1, 2, \dots, K, \quad i = 1, 2, \dots, N_K, \quad j = 1, 2, \dots, n,$$

where, $\chi^{[k]}$ is the constriction coefficient of the k-th swarm; $c_1^{[k]}$, $c_2^{[k]}$, are its cognitive and social coefficient, respectively; r_1, r_2, are random values uniformly distributed within [0,1]; and g_s is the index of the overall best position of the s-th swarm, with $s \in \{1, 2, \dots, K\}$ and $s \neq k$.

VEPSO permits the parameter configuration of each swarm independently. Thus, the number of particles as well as the values of the PSO parameters per swarm may differ. The inertia weight variant of VEPSO can be defined in direct analogy with the constriction coefficient case of equation (8). A notable characteristic is the insertion of the best position of another swarm in S_k. This information-exchanging scheme among swarms has a prominent position in VEPSO. It can be clearly viewed as a *migration* scheme, where particles migrate from one swarm to another according to a connecting topology. Figure 5 illustrates the *ring migration topology*, which corresponds to the following choice of s in equation (8) (Parsopoulos *et al.*, 2004):

$$s = \begin{cases} K, & i=1, \\ i-1, & i=2,\dots,K. \end{cases}$$

Similarities with the corresponding ring neighborhood topology defined for PSO in Chapter Two, are obvious. An alternative could be a random assignment of s for each swarm. Further constraints can be posed on the selection of s, e.g., allow the best position of a swarm to migrate only to one of the rest (this property holds for the ring topology but not for the random case).

The performance of VEPSO with two swarms was investigated in (Parsopoulos & Vrahatis, 2002a, 2002b). One of its advantages is the potential for direct parallelization of the algorithm by distributing one swarm per machine. In Parsopoulos *et al.* (2004) a parallel VEPSO approach was proposed, where up to 10 swarms were distributed over 10 machines of a small computer cluster and their communication was performed through Ethernet interconnections. Experiments on widely used MO test problems revealed that the number of swarms in combination with the Ethernet speed has a crucial impact on performance, since execution time of the swarms becomes comparable to communication time between machines, thereby raising synchronization issues. The reader is referred to the original paper for further details.

COMPOSITE PARTICLE SWARM OPTIMIZATION: A META-STRATEGY APPROACH

Composite PSO (COMPSO) was introduced by Parsopoulos and Vrahatis (2002a) as a *meta-strategy* that employs an evolutionary algorithm to control the parameters of PSO during optimization. For this purpose, the *differential evolution* (DE) algorithm was used. DE is a probabilistic population-based algorithm that works similarly to PSO, using differences of vectors instead of pure probabilistic sampling, to produce new candidate solutions. In addition, it is a greedier algorithm than PSO, due to a selection

Figure 5. Schematic representation of the ring migration topology for K swarms

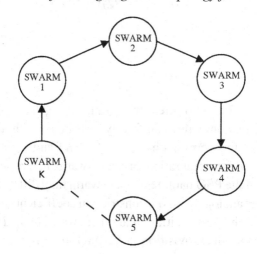

mechanism that retains only the best individuals in a single population. Thus, it can be quite fast but more prone to get stuck in local minima than PSO. The selection of a different algorithm than PSO itself for controlling its parameters was dictated by observations in early experiments. These observations implied that PSO was unable to control its own parameters during optimization, at least in an efficient manner (Parsopoulos & Vrahatis, 2002a).

The meta-strategy approach assumes a standard PSO swarm for the minimization of the objective function in the search space, and a DE population of PSO parameter values that minimizes a performance criterion in the space of parameters. These two minimization procedures take place concurrently by switching repeatedly between the swarm and the parameter population (Parsopoulos & Vrahatis, 2002a). The performance criterion for the latter is usually related to the current performance of the swarm, as it is reflected to the achieved progress in minimizing the objective function.

In the next paragraphs, we briefly describe the DE algorithm and present the COMPSO approach, along with conclusions derived by Parsopoulos and Vrahatis (2002a) for widely used benchmark problems.

The Differential Evolution Algorithm

DE was introduced by Storn and Price (1997). It is a population-based optimization algorithm that utilizes a population, $P = \{x_1, x_2,..., x_N\}$, of N individuals to probe the search space. Each individual is an n-dimensional vector, $x_i = (x_{i1}, x_{i2},..., x_{in})^{\mathrm{T}}$, $i = 1, 2,..., N$, where n is the dimension of the problem. At each iteration, and for each individual, x_i, $i = 1, 2,..., N$, a *mutation* operator is applied on x_i to produce a new candidate solution. There are several established mutation operators:

$$v_i(t+1) = x_{r_1}(t) + F(x_{r_2}(t) - x_{r_3}(t)), \tag{10}$$

$$v_i(t+1) = x_g(t) + F(x_{r_1}(t) - x_{r_2}(t)), \tag{11}$$

$$v_i(t+1) = x_i(t) + F(x_g(t) - x_i(t)) + F(x_{r_1}(t) - x_{r_2}(t)), \tag{12}$$

$$v_i(t+1) = x_g(t) + F(x_{r_1}(t) - x_{r_2}(t)) + F(x_{r_3}(t) - x_{r_4}(t)), \tag{13}$$

$$v_i(t+1) = x_{r_1}(t) + F(x_{r_2}(t) - x_{r_3}(t)) + F(x_{r_4}(t) - x_{r_5}(t)), \tag{14}$$

where r_1, r_2, r_3, r_4, and r_5, are mutually different randomly selected indices that differ also from i, while $F \in [0,2]$ is a user-defined parameter. The index g denotes the best individual of the population, i.e., the one with the lowest function value. Thus, the new candidate solution is produced by combining individuals from the current population. The similarity of some operators, e.g., equation (12), with the velocity update in PSO is apparent, suggesting an intimate relation between the two algorithms.

After mutation, a *crossover* operator is applied to produce a *trial vector*, $u_i = (u_{i1}, u_{i2}, \ldots, u_{in})^T$, as follows:

$$u_{ij}(t+1) = \begin{cases} v_{ij}(t+1), & \text{if } (R_j \leq CR) \text{ or } (j = \text{rnbr}(i)), \\ x_{ij}(t), & \text{if } (R_j > CR) \text{ and } (j \neq \text{rnbr}(i)), \end{cases}$$

$$j = 1, 2, \ldots, n,$$

where R_j is the j-th evaluation of a uniform random number generator in $[0,1]$; CR is a user-defined *crossover constant* in $[0,1]$; and $\text{rnbr}(i)$ is a randomly selected index from $\{1, 2, \ldots, n\}$. If the trial vector improves the function value of x_i, then it replaces it in the population:

$$x_i(t+1) = \begin{cases} u_i(t+1), & \text{if } f(u_i(t+1)) < f(x_i(t)), \\ x_i(t), & \text{otherwise.} \end{cases}$$

Thus, DE always stores the detected best positions in its population and operates directly on them, in contrast to PSO where the best positions are maintained in a separate population. This feature renders DE a greedy algorithm, where convergence is achieved quickly but probably at the cost of low efficiency.

The Composite PSO Approach

Parsopoulos and Vrahatis (2002a) used the inertia weight PSO variant, defined by equations (7) and (8) of Chapter Two for the COMPSO approach. The three parameters of the swarm are the inertia weight, w, and the cognitive and social parameter, c_1 and c_2, respectively. For these parameters, a DE population of 3-dimensional individuals is defined, where the components of each individual correspond to a setting of the three PSO parameters.

Let $S_t = \{x_1(t), x_2(t), \ldots, x_N(t)\}$ be the swarm that operates in the search space of the problem at hand, at the t-th iteration. Then, the particles are updated as follows (Parsopoulos & Vrahatis, 2002a):

$$v_{ij}(t+1) = w\, v_{ij}(t) + c_1 R_1\, (p_{ij}(t) - x_{ij}(t)) + c_2 R_2\, (p_{gj}(t) - x_{ij}(t)), \tag{15}$$

$$x_{ij}(t+1) = x_{ij}(t) + v_{ij}(t+1), \tag{16}$$

$$i = 1, 2, \ldots, N, \quad j = 1, 2, \ldots, n.$$

Let g denote the best particle at each iteration. For a given swarm, S_t, a DE population, $P = \{q_1, q_2, \ldots, q_M\}$, of 3-dimensional individuals, $q_m = \{w_m, c_{1m}, c_{2m}\}$, $m = 1, 2, \ldots, M$, is defined. The swarm S_t is updated and evaluated using individually each parameter set q_m in equation (15). The function value of the best particle for the updated swarm is adopted as the function value of the individual (parameter set) q_m. Then, the DE population is updated using the DE operators, producing a new population of parameters. The procedure continues again with the update of the swarm with each set of parameters, individually, and so on. DE is executed for a specific number of iterations, although different termination criteria can be used. The best parameter set provided by the DE after its termination is used by PSO to produce the next iteration of the swarm. The algorithm is presented in pseudocode in Table 10.

Parsopoulos and Vrahatis (2002a) applied COMPSO to a set of widely used benchmark problems reported in Appendix A of the book at hand. For each problem, they performed 25 independent experiments using the DE operator defined in equation (10). The parameters of PSO were bounded as follows:

$$0.4 \leq w \leq 1.2, \quad 0.1 \leq c_1, c_2 \leq 4.$$

Table 10. Pseudocode of the COMPSO algorithm

Input:	Initial swarm, S_0; velocities, V_0; and t = -1.
Step 1.	**Set** $t \leftarrow t+1$ and $t' \leftarrow 0$.
Step 2.	**Generate** a population, P_t, of vectors $q_m = \{q_{m1}, q_{m2}, q_{m3}\}$, $m = 1, 2, \ldots, M$.
Step 3.	**Do** $(m = 1 \ldots M)$
Step 4.	**Set** $w \leftarrow q_{m1}, c_1 \leftarrow q_{m2}, c_2 \leftarrow q_{m3}$.
Step 5.	**Determine** temporary swarm, S_t', and velocities, V_t', by using the parameters, w, c_1 and c_2, of Step 4 in equations (15) and (16).
Step 6.	**Evaluate** S_t' and find the index g of its best particle, x_g.
Step 7.	**Set** $f(q_m(t')) \leftarrow f(x_g)$.
Step 8.	**End Do**
Step 9.	**Apply** DE mutation, crossover and selection on P_t and generate a new population, $P_{t'+1}$.
Step 10.	**Find** the best individual, q^*, of $P_{t'+1}$.
Step 11.	**If** (stopping criterion of DE is not met) **Then**
Step 12.	**Set** $t' \leftarrow t'+1$.
Step 13.	**Go To** Step 3.
Step 14.	**Else**
Step 15.	**Go To** Step 17.
Step 16.	**End If**
Step 17.	**Set** $(w, c_1, c_2) \leftarrow q^*$ and update S_t, V_t, with equations (15) and (16).
Step 18.	**End While**
Step 19.	**If** (stopping criterion of PSO is not met) **Then**
Step 20.	**Go To** Step 1.
Step 21.	**Else**
Step 22.	**Terminate** algorithm.
Step 23.	**End If**

Besides the increased efficiency of COMPSO, a very interesting property is observed in the results: the mean values assigned by the DE algorithm to the three PSO parameters are very close to the values recognized later in the stability analysis of Clerc and Kennedy as "optimal" for the constriction coefficient variant of PSO (Clerc & Kennedy, 2002). These values along with the corresponding test problems and dimensions are reported in Table 11. We refer the reader to the original paper by Parsopoulos and Vrahatis (2002a) for further details on the configuration of the algorithms.

In Table 11, we can see that the inertia weight assumed values close to the default one of the constriction coefficient PSO variant, i.e., $\chi = 0.729$. Also, the assigned values of c_1 and c_2 seem to be interrelated with their sum having an average equal to 3.3, while the corresponding sum for the original constriction coefficient PSO variant is almost 2.99 (the convergence regions of PSO were theoretically obtained for values higher than 2.916).

Moreover, the DE algorithm always assigns a larger value to c_1 than c_2. This tends to enforce the global perspective of COMPSO, i.e., to maintain the diversity of the swarm for a large number of iterations, and, consequently, avoid premature convergence to local minima. Parsopoulos and Vrahatis (2002a) also reported that the behavior of COMPSO was not significantly sensitive to different values of the parameters F and CR of DE.

COMPSO was shown to be able to provide a means for reliable control on the parameters of PSO. Of course, this comes at the cost of extra computational effort. However, it might be inevitable to pay this effort, since the determination of proper parameters in unexplored search spaces often requires a large number of preliminary experiments. Additionally, COMPSO controls parameters during optimization, changing them radically if needed, thereby preventing bad outcomes due to wrong parameter settings.

COMPSO has offered a good starting point for research in a field of general interest, namely the field of meta-strategies (also called *meta-heuristics*). DE can be replaced by any algorithm and numerous modifications can be made to the original COMPSO scheme. Thus, a lot of research is still needed, to reveal the potential of this apparently promising approach.

Table 11. Mean values of w, c_1, c_2, assigned by COMPSO to the inertia weight PSO variant through DE. All test problems are reported in Appendix A of the book at hand

Problem	Dim.	Mean w	Mean c_1	Mean c_2
TP_{UO-2}	2	0.7337	2.0421	1.1943
TP_{UO-3}	6	0.7027	2.0462	0.9881
TP_{UO-4}	2	0.7082	2.0171	1.3926
TP_{UO-4}	10	0.6415	2.1291	1.4534
TP_{UO-10}	2	0.7329	2.0197	1.3148
TP_{UO-11}	2	0.7153	2.1574	1.2706
TP_{UO-13}	2	0.7085	2.0378	1.2712
TP_{UO-15}	2	0.7023	2.0145	1.2098
TP_{UO-16}	3	0.7053	2.1062	1.1694
TP_{UO-17}	6	0.7179	1.9796	1.3812
TP_{UO-19}	3	0.7739	1.9722	1.3063

GUARANTEED CONVERGENCE PARTICLE SWARM OPTIMIZATION

The *guaranteed convergence PSO* (GCPSO) was introduced by Van den Bergh and Engelbrecht (2002), based on the finding that the inertia weight variant of PSO can be non-convergent even on local minimizers if velocities become very small (Van den Bergh, 2002). To address this problem, a modified update scheme was introduced for the overall best particle, based on the local search algorithm of Solis and Wets (1981).

More specifically, if the current and the best position of the *i*-th particle at iteration *t* coincide with the overall best position, i.e., $x_i(t) = p_i(t) = p_g(t)$, then the position update of x_i will depend only on the previous velocity term, $wv_i(t)$. Thus, if velocity is very close to zero, the particle will be almost immobilized. Since, in the long run, all particles are expected to approximate the global best position, it is possible that the aforementioned deficiency will hold for most, resulting in search stagnation (Van den Bergh & Engelbrecht, 2002).

The problem can be alleviated by introducing a different update scheme for the overall best particle. Let *g* be the index of the best particle in the inertia weight variant of PSO defined by equations (7) and (8) in Chapter Two. Then, according to GCPSO, the best particle is updated using the equations (Van den Bergh & Engelbrecht, 2002):

$$v_{gj}(t+1) = -x_{gj}(t) + p_{gj}(t) + wv_{gj}(t) + \rho(t)(1-2r), \tag{17}$$

$$x_{gj}(t+1) = x_{gj}(t) + v_{gj}(t+1), \tag{18}$$

$$j = 1, 2, \ldots, n,$$

where *n* is the dimension of the problem; *r* is a uniformly distributed random number in [0,1]; and $\rho(t)$ is a *scaling factor*. The rest of the particles are updated using the standard equations (7) and (8) in Chapter Two. Equations (17) and (18) can be combined by substitution, resulting in the equivalent scheme:

$$x_{gj}(t+1) = p_{gj}(t) + wv_{gj}(t) + \rho(t)(1-2r). \tag{19}$$

According to this update, the best particle will not stagnate, generating new candidate solutions randomly in an area that surrounds the overall best position, p_g (Van den Bergh & Engelbrecht, 2002).

In addition, GCPSO uses two counters to count the number of consecutive successes and failures in updating the best position, where *success* is defined as the improvement of the overall best function value, while *failure* is defined otherwise. Thus, if $M_{\text{SUC}}(t)$ denotes the number of consecutive successes and $M_{\text{FAIL}}(t)$ is the number of successive failures at iteration *t*, then the scaling factor $\rho(t)$ is updated as follows (Van den Bergh & Engelbrecht, 2002):

$$\rho(t+1) = \begin{cases} 2\rho(t), & \text{if } M_{\text{SUC}}(t) > s_{\text{SUC}}, \\ \dfrac{1}{2}\rho(t), & \text{if } M_{\text{FAIL}}(t) > s_{\text{FAIL}}, \\ \rho(t), & \text{otherwise}, \end{cases} \tag{20}$$

where s_{SUC} and s_{FAIL} are user-defined parameters. In addition to equation (20), the following rules apply on M_{SUC} and M_{FAIL} (Van den Bergh & Engelbrecht, 2002):

$$M_{SUC}(t+1) > M_{SUC}(t) \quad \Rightarrow \quad M_{FAIL}(t+1) = 0,$$
$$M_{FAIL}(t+1) > M_{FAIL}(t) \quad \Rightarrow \quad M_{SUC}(t+1) = 0,$$

such that only the number of consecutive successes and failures are preserved. The update of equation (20) resembles that of Solis and Wets (1981) for step update in random local search.

The adaptation of the scaling factor controls the sampling volume of random search, where consecutive successes increase volume to allow larger steps, while failures have the opposite effect, by reducing volume. The value, $\rho(0) = 1.0$, was experimentally shown to be a good initial value for the scaling factor. Regarding the rest of the parameters, the values $s_{FAIL} = 5$ and $s_{SUC} = 15$ are recommended by Van den Bergh and Engelbrecht (2002). This setting implies that GCPSO will penalize bad values of $\rho(t)$ faster than it awards its good values. Alternatively, dynamic control of these parameters was proposed as a mean to prevent the rapid oscillation of $\rho(t)$ (Van den Bergh & Engelbrecht, 2002). Nevertheless, a lower bound on the value of $\rho(t)$ may be beneficial for the algorithm, preventing it from taking almost zero values.

GCPSO has proved to converge on local minima (Van den Bergh, 2002), based on the convergence proof of Solis and Wets for the corresponding random local search algorithm (Solis & Wets, 1981). In parallel, Van den Bergh and Engelbrecht (2002) reported results from the application of GCPSO on 30-dimensional instances of test problems TP_{UO-1}, TP_{UO-3}, TP_{UO-6}, and TP_{UO-20}, described in Appendix A of the book. Table 12 reports the improvement percentages achieved by GCPSO against PSO on the aforementioned test problems, in terms of success in achieving the required accuracy over 50 independent experiments, as well as in the required number of function evaluations for the successful experiments, based on the results reported in (Van den Bergh & Engelbrecht, 2002). For the comparisons, the inertia weight PSO variant was used, although its parameter setting implied the constriction coefficient variant, i.e., $w = 0.72$, $c_1 = c_2 = 1.49$ (Van den Bergh & Engelbrecht, 2002).

In the unimodal test problems, TP_{UO-1} and TP_{UO-20}, GCPSO had superior performance than PSO, especially in the case of small swarm sizes, where the improvement in function evaluations was higher than 70%. Improvement was also observed in the number of successes, although only for the case of $N = 10$ particles. This is indicative of the efficiency of GCPSO in unimodal problems, which is the problem type that better suits it (since it is proven to be a local search algorithm).

In the rest (multimodal) problems, GCPSO had ambiguous performance. While in TP_{UO-6} there was a significant improvement in the number of successes, especially for larger swarms, in TP_{UO-3} inferior performance was observed. However, in both cases, GCPSO remained the fastest algorithm, requiring up to 30% less function evaluations in successful experiments.

In conclusion, GCPSO is a PSO variant perfectly suited for local optimization, accompanied by a convergence proof. This property renders it a suitable and efficient choice for hybrid PSO schemes that combine local and global optimization components, such as the memetic approaches described in the previous sections. The next section presents a different approach that promotes cooperation among particles to probe the search space.

Table 12. Improvement percentages achieved by GCPSO against PSO for two swarm sizes, N = 10, 20, in terms of success in achieving the required accuracy over 50 independent experiments. The required number of function evaluations in successful experiments is also reported. The numbers are based on the results provided in (Van den Bergh & Engelbrecht, 2002). Dash denotes that both algorithms were 100% successful, while negative values denote worse performance

Problem	N	GCPSO Improvement (%)	
		Success	Func. Eval.
TP_{UO-1}	10	4.16%	89.89%
	20	-	32.84%
TP_{UO-3}	10	-20.00%	22.53%
	20	-8.16%	30.38%
TP_{UO-6}	10	9.09%	24.44%
	20	24.32%	16.94%
TP_{UO-20}	10	31.57%	73.35%
	20	-	22.14%

COOPERATIVE PARTICLE SWARM OPTIMIZATION

Cooperation can be defined as the interaction of individuals for the exchange information that can enhance their search capabilities during the optimization procedure. This description is rather abstract and embraces different approaches. In some sense, even a simple evolutionary algorithm with a recombination operator can be considered as a cooperative algorithm with information being exchanged through recombination. Cobb (1992) considered this kind of cooperation for a GA-based scheme, while Potter and De Jong (2000) offer a comprehensive description of the main issues arising in cooperative evolutionary approaches.

Cooperation can be either *explicit*, through direct one-to-one communication among individuals, or *implicit*, through a shared memory for storing information. In the latter case, each individual can contribute information on the shared memory, placing it at the disposal of the rest (Clearwater *et al.*, 1992). The shared information can be either a complete or a partial solution of the problem. In the latter case, the shared information shall be combined with different parts of other individuals to form a complete candidate solution, i.e., the contribution of each individual pertains only to some of the coordinate directions of the problem.

Potter and De Jong (1994) proposed an approach based on GAs, where one population per coordinate direction of the solution vector was used to probe the search space. Thus, each population was performing a one-dimensional optimization, and cooperation was the communication channel among them that allowed the construction of complete solutions.

The same ideas were adopted by Van den Bergh and Engelbrecht (2004) for the development of *cooperative PSO* (CPSO) algorithms. More specifically, if n is the dimension of the problem, then, instead of using a single swarm, S, of N particles, CPSO employs n swarms, $S_1, S_2, ..., S_n$, of N_i, $i = 1, 2, ..., n$, particles each, and each swarm copes only with a single coordinate direction of the solution vector.

A fundamental question arises in such algorithms: how shall each particle be evaluated since a complete (n-dimensional) solution requires components encoded in other swarms? This question implies the following issues that need to be addressed:

1.   How shall particles be selected from each swarm to form complete (n-dimensional) candidate solutions?
2.   How shall a particle be penalized or awarded for its contribution in the quality of solutions?

The answers to these issues determine the special features of the algorithm, having significant impact on its performance. Therefore, they shall be carefully treated based on the available data and problem types. Van den Bergh and Engelbrecht (2004) proposed two different CPSO approaches, described in the following sections.

The CPSO-S$_K$ Algorithm

This CPSO approach uses the concept of the *context vector* to tackle the function evaluation problem. Since a direct evaluation of the one-dimensional particles of each swarm is impossible with an n-dimensional objective function, a context that provides the missing information, i.e., the remaining (n-1) coordinates, to each particle is used. This context is an n-dimensional vector where a particle contributes its coordinate value, while the rest (n-1) coordinates are fixed to the values of the best particles of the rest swarms. Van den Bergh and Engelbrecht (2004) denoted this approach as CPSO-S$_K$.

Putting it formally, let $x_i^{[k]}$ be the i-th (one-dimensional) particle in the k-th swarm, S_k, for $i = 1, 2,\ldots,$ N_i, and $k = 1, 2,\ldots, n$. Also, let $p_g^{[k]}$ be the (one-dimensional) overall best position of S_k. Then, the particle $x_i^{[k]}$ is evaluated with the n-dimensional objective function, $f(z)$, by constructing the n-dimensional context vector:

$$z_i^{[k]} = (p_g^{[1]}, p_g^{[2]},\ldots, p_g^{[k-1]}, x_i^{[k]}, p_g^{[k+1]},\ldots, p_g^{[n]})^{\mathrm{T}}, \tag{21}$$

and setting:

$$f(x_i^{[k]}) = f(z_i^{[k]}).$$

Then, each one-dimensional swarm is updated according to its PSO velocity and position update, which may not be identical for all swarms.

Van den Bergh and Engelbrecht (2004) used the inertia weight variant of PSO. The update of best positions of each swarm follows the standard PSO rules exactly, using the objective function value assigned to each particle with the aforementioned procedure. The objective function is re-evaluated each time one of the context vector components changes. The context vector that contains all the overall best positions of the swarms:

$$z^* = (p_g^{[1]}, p_g^{[2]},\ldots, p_g^{[n]})^{\mathrm{T}},$$

is, by definition, the overall best solution detected by CPSO-S$_k$ for the problem at hand.

CPSO-S$_K$ can be considered also with swarms of different dimensionality. Thus, instead of using n one-dimensional swarms, $S_1, S_2, \ldots, S_n$, one can use just $m < n$ swarms, such that:

$$\sum_{i=1}^{m} d_i = n, \text{ and } d_i \geq 1, \quad i = 1, 2, \ldots, m.$$

This modification is important especially in cases where correlations exist among the directions of the objective function. If correlated directions are assigned to different swarms, then their nonaligned changes will be detrimental for the algorithm (Van den Bergh & Engelbrecht, 2004). Unfortunately, these correlations are not always known to the user.

The performance of CPSO-S$_K$ depends on several factors related to the following issues:

1.   Selection of the underlying PSO variant per swarm
2.   Selection of the context vector building scheme

Regarding the first issue, the inertia weigh PSO variant was used in the approach of Van den Bergh and Engelbrecht (2004). Of course, any other PSO variant could be employed as well, and the convergence properties of the CPSO-S$_K$ algorithm would be then heavily dependent on the employed algorithm. In addition, different PSO variants can be used per swarm, combining their different properties. To the best of our knowledge, no such cooperative scheme has been proposed for CPSO up-to-date.

Regarding the second issue, the proposed context vector building scheme employed the best particle per swarm. Alternatively, a randomly selected best position could be used from each swarm, replacing equation (21) with:

$$z_i^{[k]} = (p_{r_1}^{[1]}, p_{r_2}^{[2]}, \ldots, p_{r_{k-1}}^{[k-1]}, x_i^{[k]}, p_{r_{k+1}}^{[k+1]}, \ldots, p_{r_n}^{[n]})^{\mathrm{T}}, \tag{22}$$

where $r_1, r_2, \ldots, r_n$, are randomly selected indices. Also, a combination of equations (21) and (22) could be used in a competitive manner by evaluating both and selecting the best one. This corresponds to a cooperation scheme that is more biased towards exhaustive search.

Preliminary experiments revealed performance deficiencies of CPSO-S$_K$, which appears to be prone to get stuck in local minima. Van den Bergh and Engelbrecht (2004) proposed a hybrid scheme that combines CPSO-S$_K$ with standard PSO to address these deficiencies. This scheme is described in the following section.

The CPSO-H$_K$ Algorithm

This hybrid scheme, denoted as CPSO-H$_K$, combines CPSO-S$_K$ with PSO (Van den Bergh & Engelbrecht, 2004). The main idea is the application of PSO as soon as the search of CPSO-S$_K$ stagnates. It would be ideal if one could identify when CPSO-S$_K$ stagnates to switch over the standard PSO. However, this is not always possible, dictating the necessity for a different scheme. To this end, the application of each algorithm alternately at each iteration was shown to constitute a promising approach. The search can be further enhanced by exchanging information between the active and the idle algorithm at the end of each iteration, in terms of the discovered solutions.

Such an information exchange scheme can be the replacement of some particles of the one algorithm with the overall best position of the other (Van den Bergh & Engelbrecht, 2004). More specifically, after the application of CPSO-S_K at iteration t, the produced context vector replaces a randomly selected particle of the standard PSO. Then, in the next iteration, $t+1$, the standard PSO is applied and updates its overall best position, whose direction components replace the best positions in randomly selected swarms of CPSO-S_K. Notice that the replacements shall not affect the overall best position detected by each algorithm, otherwise performance will most likely be reduced.

Moreover, Van den Bergh and Engelbrecht (2004) noticed that if information exchange takes place among all particles of CPSO-S_K and PSO, this can result in inferior performance of CPSO-H_K compared to more conservative schemes. Loss of diversity and deletion of good particles have been identified as the reasons behind this effect. Thus, using only part (about half) of the particles per algorithm to exchange information has been shown to be a more efficient scheme (Van den Bergh & Engelbrecht, 2004).

Van den Bergh and Engelbrecht (2004) applied CPSO-H_K and CPSO-S_K on the test problems TP_{UO-2}, TP_{UO-3}, TP_{UO-4}, TP_{UO-6}, and TP_{UO-20}, reported in Appendix A of the book at hand. Moreover, they compared these two approaches with the standard PSO, as well as with the cooperative co-evolutionary GA (CCGA) approach presented in (Potter & De Jong, 1994), providing a significant amount of experimental results. Both CPSO approaches exhibited competitive performance to the rest of the algorithms. Especially for small swarms (10 particles), CPSO approaches were shown to be very fast and obtained satisfactory results even for rotated versions of the test problems.

There seems to be a difference in performance between CPSO approaches with one-dimensional and higher-dimensional swarms when coordinate axes are rotated, with the latter exhibiting superior performance. On the other hand, in problems with equidistantly distributed local minima, one-dimensional swarms performed very well. For a detailed presentation and discussion of the results, the reader is referred to the original paper of Van den Bergh and Engelbrecht (2004).

Although several important issues need further investigation, CPSO approaches were shown to exhibit promising behavior, rendering them as an interesting alternative in cases where standard PSO fails to provide satisfactory results. In the next section, we describe a PSO variant that can find several solutions of the problem, concurrently, by using niches of particles.

NICHING PARTICLE SWARM OPTIMIZATION

Niching is a framework for developing evolutionary algorithms capable of locating several minimizers of the objective function (Horn, 1997). There are two types of niching: *parallel niching*, where several niches are recognized in a population and maintained simultaneously, and *sequential niching*, where niching is applied iteratively on the problem, while a procedure ensures that already detected solutions will not be detected again.

Niching PSO (NPSO) was proposed by Brits *et al.* (2002) as a PSO variant capable of locating several solutions of a problem, simultaneously. Hence, it can be categorized as a parallel niching approach. In the original paper, a number of subswarms following the GCPSO variant, described in a previous section, were considered (Brits *et al.*, 2002).

NPSO utilizes a swarm, *S*, of *N* particles. The swarm is initialized uniformly within the search space, $A \subset \mathbf{R}^n$. Initialization plays a crucial role, since uniform distribution of the particles is required at the beginning by the algorithm. For this purpose, Brits *et al.* (2002) used Faure sequences of random numbers in their implementation (Thiémard, 1998). Then, a number of iterations is performed using only the cognitive term in velocity update:

$$v_{ij}(t+1) = wv_{ij}(t) + c_1r_1(p_{ij}(t)-x_{ij}(t)). \tag{23}$$

This allows each particle to explore locally the search space without being influenced from the rest. The model of equation (23) is also called the *cognitive only model*.

Niches shall now be identified in the swarm. Brits *et al.* (2002) tracked the variance, σ_i, of the function value of each particle, x_i, for a number of iterations, t_σ. When σ_i falls under a threshold, δ, a new subswarm is created consisting of x_i and its closest neighbor. Claiming that it avoids the use of tunable parameters, Brits *et al.* (2002) proposed this approach as an alternative to the approach of Parsopoulos and Vrahatis (2001). However, in contrast to these claims, the proposed approach still uses a tunable parameter, δ, that requires user intervention.

Nevertheless, both approaches are useful, although in different cases. For example, in square-error objective functions (e.g., neural networks training, model identification problems etc.) where the global minimum is by definition equal to zero, the approach of Parsopoulos and Vrahatis (2001) is more suitable. On the other hand, if no information is known on the form of the objective function, the approach of Brits *et al.* (2002) can be very useful. Perhaps a combination of the two approaches may be superior from each one separately, a fact identified also by Brits *et al.* (2002).

In order to avoid problem dependencies, the parameter σ_i was normalized based on the upper and lower bounds of the particles. The closest neighbor of x_i is defined as follows:

$$x_{cl} = \arg\min_{j \neq i}\{\| x_i - x_j \|\}.$$

Let $S^{[j]}$ denote the j-th subswarm, and $x_i^{[j]}$ denote its i-th particle. Then, its radius is defined as follows:

$$R^{[j]} = \max_{i \neq g_j}\{\| x_{g_j}^{[j]} - x_i^{[j]} \|\},$$

where g_j is the index of the global best particle of the j-th subswarm. Notice that, the radius of a subswarm can become arbitrarily small, if its particles have all converged on the same solution.

If a particle x_i moves within the range of the j-th subswarm:

$$\| x_i - x_{g_j}^{[j]} \| \leq R^{[j]},$$

then it is absorbed by $S^{[j]}$. Moreover, two subswarms that intersect:

$$\| x_{g_{j_1}}^{[j_1]} - x_{g_{j_2}}^{[j_2]} \| < R^{[j_1]} + R^{[j_2]}, \tag{24}$$

can be merged, since they are expected to probe the same regions of the search space. In the special case where both subswarms have radius equal to zero, then a sufficiently small value, $\mu > 0$, can be used as a threshold in the right part of equation (24).

Each subswarm of the NPSO approach uses the GCPSO variant, which was proved to be locally convergent. In the approach of Brits *et al.* (2002), subswarms were let to consist of 2 particles only, while the variance of each particle was tracked for $t_\sigma = 3$ iterations. The pseudocode of NPSO is given in Table 13. Brits *et al.* (2002) applied the NPSO approach on four one-dimensional and one 2-dimensional test problems with very promising results. However, they identified a number of issues that need to be resolved, such as:

a. Sensitivity of the algorithm on the parameters σ_i and μ.
b. Correlations between the number of niches, the number of solutions, and swarm size.
c. Impact of using different algorithms than GCPSO in subswarms.

These issues, along with the need for further experimentation in high dimensional problems, opened the way for the development of more efficient NPSO approaches (Bird & Li, 2006).

The sequential niching approaches are also related to techniques presented in the next chapter; thus, we shall re-examine these issues later. The next section presents a PSO variant that attempts to minimize user intervention for parameter settings by using a parameter-free scheme

TRIBES

The *TRIBES* algorithm was proposed by Clerc (2006, Chapter 11) as a black-box PSO variant with capabilities of self-adaptation. Its foundation stone is the concept of *tribe*, which describes a special neighborhood connection scheme for exchanging information among particles, while the same structure and update rules with PSO are used.

Table 13. Pseudocode of the NPSO algorithm

Input:	Main swarm and GCPSO parameters.
Step 1.	**Initialize** the main swarm uniformly within the search space.
Step 2.	**Apply** one iteration of the model defined by equation (23).
Step 3.	**Evaluate** particles of the main swarm.
Step 4.	**Do** (number of subswarms)
Step 5.	**Apply** one iteration of GCPSO on the subswarm.
Step 6.	**Evaluate** the particles of the subswarm.
Step 7.	**Update** the radius of the subswarm.
Step 8.	**End Do**
Step 9.	**Merge** intersecting subswarms.
Step 10.	**Let** subswarms absorb particles from the main swarm, if they have entered their range.
Step 11.	**Check** condition for producing a new subswarm for each particle of the main swarm.
Step 12.	**If** (stopping criteria not met) **Then**
Step 13.	**Go To** Step 2.
Step 14.	**Else**
Step 15.	**Terminate** algorithm.
Step 16.	**End If**
Step 17.	**Report** solutions.

If a particle y shares its own memory, i.e., its best position, with another particle, x, then y is called an *informant* of x (Clerc, 2006), and we will denote this connection as:

$\inf(y, x) = 1$.

On the other hand, if y is not an informant of x, we will denote it as:

$\inf(y, x) = 0$.

In the PSO framework, it holds that $\inf(y, x) = \inf(x, y)$, i.e., if y is an informant of x, then x is also an vinformant of y. This compromise is adopted also in TRIBES. As we have already seen in neighborhood topologies of PSO, this information exchange channel is depicted with an arc between y and x. In most neighborhood topologies, only a number of particles are connected with a given particle x, i.e., the number of its informants is smaller than the swarm size, N. Now, we can give the following definition:

Definition 1. A *tribe* of a swarm, S, is a subset, $T \subseteq S$, of particles where any one of them is an informant for the rest in the same subset, i.e.:

$\inf(x_i, x_j) = 1$, for all i and j such that $x_i, x_j \in T$.

In a graph theoretical sense, a tribe is a symmetrical clique in the graph representing the neighborhood topology of the swarm (Clerc, 2006). Obviously, if $T_j, j = 1, 2, \ldots, J$, are tribes of a swarm, S, then it holds that:

$$S = \bigcup_{j=1}^{J} T_j,$$

where $J \leq N$.

Each tribe resembles a team that moves in the search space with complete cooperation among its members, while communicating with the rest of the tribes. This concept brings to mind the cooperative and niching techniques described in previous sections. Indeed, TRIBES can be considered as cooperative/niching approach in the sense that several teams cooperate to probe the search space simultaneously.

Communication among tribes is crucial for the operation of the algorithm. Clerc (2006) requires that each two particles, y and x, of the swarm must have a connecting path in the neighborhood graph, i.e., there shall exist a set of particles, $\{z_1, z_2, \ldots, z_C\}$, with $C < N$-1, such that:

$\inf(y, z_1) = \inf(z_1, z_2) = \cdots = \inf(z_C, x) = 1$.

Thus, each tribe must have at least one communication channel with another tribe. However, since the algorithm is self-adaptive, it can generate or delete some tribes; therefore, the connection channels are expected to change dynamically without losing the aforementioned interconnection property among the remaining tribes.

Regarding the particles of a tribe, quality measures are defined to assess their performance through time. If $x(t)$ is a particle at iteration t, and $p(t)$ is its best position, then:

$$x(t) \text{ is} \begin{cases} \text{neutral,} & \text{if } p(t) = p(t-1), \\ \text{good,} & \text{if } p(t) \text{ is better than } p(t-1), \\ \text{excellent,} & \text{if } p(t) \text{ is better than } p(t-1), \text{ and } p(t-1) \text{ is better than } p(t-2). \end{cases}$$

i.e., TRIBES considers rather a history of the best positions than only the last one, which is the case of standard PSO (Clerc, 2006). Within the context of a tribe, one can also define the *worst* and the *best* particle, as the ones with the worst and best function value, respectively.

Based on the quality of its particles, a tribe, T, of size, N_T, is characterized as *good* or *bad* as follows:

$$T \text{ is} \begin{cases} \text{good,} & \text{if } N_T^{[\text{good}]} > r, \\ \text{bad,} & \text{otherwise,} \end{cases}$$

where $N_T^{[\text{good}]} \leq N_T$ is the number of good particles of T, and $r \in \{0, 1, \ldots, N_T\}$ is a randomly selected integer. This probabilistic rule provides the space to build new tribes based on the following set of accompanying evolution rules (Clerc, 2006):

1. **Elimination of a particle:** A particle can be eliminated only from the best tribe, and it can be only its worst one. This constriction minimizes the risk of deleting a particle that carries crucial information regarding the actual solution of the problem. If a tribe consists of only one particle, then it will be deleted only if it has at least one informant with better performance. Deleting a particle will probably require a re-organization of the communication channels between particles and tribes. Thus, if the tribe T_j consists of more than one particles and its worst particle, $x_{\text{worst}}^{[T_j]}$, is connected to the k-th particle of another tribe, T_i, i.e.:

    $$\inf(x_{\text{worst}}^{[T_j]}, x_k^{[T_i]}) = 1,$$

    then, deleting the worst particle will result in the following connection change:

    $$\inf(x_{\text{best}}^{[T_j]}, x_k^{[T_i]}) = 1,$$

    i.e., the connections of the deleted particle are inherited by the best particle of its tribe. On the other hand, if T_j consists of one particle solely, connected with the k_i-th particle of tribe T_i as well as with the k_l-th particle of tribe T_l, i.e.:

    $$\inf(x^{[T_j]}, x_{k_i}^{[T_i]}) = 1 \text{ and } \inf(x^{[T_j]}, x_{k_l}^{[T_l]}) = 1,$$

    then, deleting this particle produces the following update:

    $$\inf(x_{k_i}^{[T_i]}, x_{k_l}^{[T_l]}) = 1,$$

    i.e., the tribes T_i and T_l are connected directly, while T_j vanishes. Notice that $x^{[T_j]}$ will be deleted only if at least one of its informants, $x_{k_i}^{[T_i]}$ or $x_{k_l}^{[T_l]}$, has better function value.

2. **Generation of new particles:** Each tribe that was characterized as "bad" generates two new particles; a *free* and a *confined* one. The free particle is generated randomly using a uniform distribution either within the whole search space or on one of its sides or vertices. On the other hand, the confined particle is generated randomly in a more restricted region. More specifically, let T_j be the tribe under consideration and g_j be the index of its best particle, with current position, x_{g_j}, and best position, p_{g_j}. Let also, z, be the best informant of x_{g_j}, and z_g be its best position. Then, the confined particle will be generated randomly and uniformly in a sphere of center z_g and radius $\| z_g\text{-}p_{g_j} \|$. Therefore, for each "bad" tribe, the free particle is initialized such that it has higher probability of discovering new promising regions, while the confined one is initialized in already identified promising regions. All particles generated from "bad" tribes form a new one that retains connections with its originating "bad" tribes.

The gain from the aforementioned evolution rules can be revealed only after a few iterations of the algorithm are performed. Therefore, it is not wise to perform adaptations very often. Clerc (2003) proposes the following rule: after each adaptation, the diameter of the relation graph among the particles, i.e., the number of arcs constituting the shortest path between any pair of particles of different tribes, is calculated. The maximum among the lengths of these shortest paths provides an estimation of the required number of iterations for transmitting information carried by any particle to any other particle. If this number is equal to L after an adaptation, the next adaptation shall take place after $L/2$ iterations.

In practice, the application of TRIBES requires a randomly generated initial particle, which constitutes the initial tribe. This particle is updated using the PSO rules (any PSO variant can be applied) and, if it is not improved in the first iteration, then a second particle is generated, producing a second tribe. If in the next iteration both tribes are bad, then two new particles are generated from each one. The four new particles constitute a new tribe, and the procedure continues in the same manner. Thus, as long as the algorithm does not exhibit an improvement, TRIBES will increase the number of particles, trying to equip the search mechanism with more search units. As soon as good positions start being detected, TRIBES will decrease the number of particles by deleting the worst among them, thereby retaining a trade-off between computational cost and efficient search (Clerc, 2003).

Clerc tested TRIBES under different PSO update schemes per particle, even per coordinate direction (Clerc, 2003, p. 145). He also proposed a generalized comparison scheme between particles, rendering the algorithm applicable even in spaces without a metric (Clerc, 2003, p. 146). Moreover, he reported results from the application of TRIBES on several benchmark problems and provided its source code freely on the web page: http://clerc.maurice.free.fr/pso/.

The popularity of TRIBES is still limited in literature, perhaps due to its more complex structure than PSO, while the reported results are still ambiguous, admitting significant criticism. Nevertheless, TRIBES is a very interesting and promising approach for future generation variants of PSO. The user needs to specify only the objective function, its search space, the desired accuracy, and a maximum number of function evaluations. Although a black-box algorithm that can adapt to any problem by responding to stimulations during the optimization procedure and taking all decisions without user intervention, may seem a very ambitious or misbegotten task, it still remains the holly grail of intelligent optimization and artificial intelligence. Thus, any effort towards this direction shall be embraced and improved to enrich our experience and provide hints that can lead to the design of new approaches closer to this goal. In our opinion, TRIBES constitutes a first step towards this direction.

QUANTUM PARTICLE SWARM OPTIMIZATION

Quantum PSO (QPSO) was introduced by Sun *et al.* (2004a). Although it is considered a PSO variant, its operation is placed on a rather different framework than standard PSO. Thus, if PSO follows a Newtonian approach for the movement of the particles, QPSO considers rather a quantum behavior of particles, based on laws of quantum mechanics.

In quantum mechanics, the time-dependent Schrödinger equation:

$$j\hbar \frac{\partial}{\partial t}\Psi(r,t) = \widehat{H}(r)\Psi(r,t),$$

(25)

has a dominant position, where:

$$\widehat{H}(r) = -\frac{\hbar^2}{2m}\nabla^2 + V(r),$$

(26)

is a time-dependent Hamiltonian operator; $\hbar$ is the Planck's constant; m is the mass of the particle; and $V(r)$ is a potential energy distribution function. The squared amplitude, $Q = |\Psi|^2$, of the wave function $\Psi(r,t)$ in equation (25) serves as a probability measure for the movement of the particle, under the normalization:

$$\iiint |\Psi(r,t)|^2 \, \mathrm{d}x\mathrm{d}y\mathrm{d}z = 1.0.$$

In QPSO, the swarm is considered as a quantum system where each particle has a quantum state based on the employed wave function, while it moves in a *Delta potential well* (DPW) towards a position p. Influenced by the analysis of Clerc and Kennedy (2002) presented in the previous chapter, Sun *et al.* (2004a) assumed that the position, p, of a particle, x_i, is defined as the weighted mean of its best position, p_i, and the overall best position of the swarm, p_g, as follows:

$$p_j = \frac{\varphi_1 p_{ij} + \varphi_2 p_{gj}}{\varphi_1 + \varphi_2}, \quad j = 1, 2, \ldots, n,$$

(27)

where $\varphi_1 = c_1 r_1$ and $\varphi_2 = c_2 r_2$, with c_1, c_2, being the cognitive and social parameter of PSO, respectively, and r_1, r_2, being uniformly distributed random numbers in $[0,1]$.

To illustrate the workings of QPSO, let us consider the simplest, one-dimensional case. Assuming that the center, p, of the potential is defined by equation (27), the DPW is defined as (Sun *et al.*, 2004a):

$$V(x) = -\gamma \, \delta(x-p) = -\gamma \, \delta(y),$$

(28)

where $y = p-x$. Through proper mathematical manipulations (Sun *et al.*, 2004a), we obtain the following wave function:

$$\Psi(y) = \frac{1}{\sqrt{L}}\exp\left(-\frac{|y|}{L}\right),$$

(29)

and, hence, a probability measure:

$$Q(y) = |\Psi(y)|^2 = \frac{1}{L}\exp\left(-2\frac{|y|}{L}\right),$$ (30)

where $L = \hbar^2/m\gamma$.

So far, we obtained a probability density of the particle positions. However, this is not adequate to serve as an algorithm, since the evaluation of a particle requires an exact position. Therefore, the position of the particle shall be gauged; this procedure is called *collapse* of the quantum state to the classical state.

Collapsing is possible through Monte Carlo simulation. More specifically, let s be a random number uniformly distributed in $(0, 1/L)$. Then, s can be written as:

$$s = \frac{1}{L}u,$$

where u is a random number uniformly distributed in $(0,1)$. Substituting $Q(y)$ in the left part of equation (30) with s, and solving for x, results in the following QPSO model (Sun *et al.*, 2004a):

$$x = p \pm \frac{L}{2}\ln\left(\frac{1}{u}\right).$$ (31)

Equation (31) provides two possible new positions of the particle, which are measurable with the objective function. Sun *et al.* (2004a) provided a convergence proof of this QPSO model on the position p. The parameter L is the only control parameter that appears in the update equations of QPSO. The lack of other user-defined parameters renders QPSO an interesting approach.

Alternatively to the DPW, different quantum field models can be used. Sun *et al.* (2004a) proposed different wave functions that result in different update schemes for the particles. Therefore, if S is a swarm of size N, x_i is the i-th particle, p_i is its best position, p_g is the overall best particle, and t is the iteration counter, we can describe the general form of QPSO as follows:

$$x_i(t) = p(t) + F(L, \pm u),$$ (32)

where,

$$p(t) = \frac{\varphi_1 p_i(t) + \varphi_2 p_g(t)}{\varphi_1 + \varphi_2},$$

where $i = 1, 2,..., N$; u is a uniformly distributed number in $(0,1)$; and F is a functional form obtained through the inversion of the probability density function, thereby depending on the employed quantum field model. Table 14 shows the pseudocode of QPSO, and Table 15 provides the three most popular potential well models (Mikki & Kishk, 2006; Sun *et al.*, 2004a; Sun *et al.*, 2004b).

Sun *et al.* (2004a) have recognized the sensitivity of QPSO on the parameter L (or equivalently on the parameter q in Table (15)). However, their experiments on widely used test problems revealed that QPSO can become an efficient approach under proper fine-tuning, as most of the PSO variants. The different philosophy than the rest PSO variants, as well as its susceptibility to improvements (Coelho, 2008; Xi *et al.*, 2008) and its interesting applications (Coelho & Mariani, 2008; Liu *et al.*, 2009) rendered QPSO a worth-noting approach.

CHAPTER SYNOPSIS

We presented some of the most established and recently proposed variants of PSO with dominant position in PSO literature. The presentation detail was kept at a low level to make the underlying ideas attainable by readers with different scientific backgrounds. Pseudocode accompanied the algorithms, providing a sketch for their implementation in any programming language.

Our aim was the exposition of the main ideas and features that constitute the core of research in the development of modern PSO variants, without focusing on specific problem types or instances. The presented variants served as prototypes for accomplishing our goal. However, they shall not be considered as the unique PSO variants for solving optimization problems efficiently. In fact, over-specialized PSO approaches may effectively exploit the inherent properties of a specific problem type, producing even better results than the exposed methods. Such approaches were out of the scope of this chapter, although the interested reader can easily access the relative literature, starting from sources reported herein.

Table 14. Pseudocode of the QPSO algorithm

Input:	Swarm S, swarm size N.
Step 1.	**Initialize** swarm and best positions.
Step 2.	**Find** the index g of the best particle.
Step 3.	**While** (stopping condition not met)
Step 4.	**Do** ($i = 1 \ldots N$)
Step 5.	**Compute** position $p(t)$ using $p_i(t)$ and $\mathrm{p_g}(t)$.
Step 6.	**Draw** a random number $u \sim U(0,1)$.
Step 7.	**Set** $L = L(q, u, |x_i(t)-p(t)|)$.
Step 8.	**Draw** a random number $R \sim U(0,1)$.
Step 9.	**If** ($R > 0.5$) **Then**
Step 10.	$x_i(t+1) = x_i(t) + F(L, u)$
Step 11.	**Else**
Step 12.	$x_i(t+1) = x_i(t) + F(L, -u)$
Step 13.	**End If**
Step 14.	**End Do**
Step 15.	**End While**
Step 16.	**Report** solutions

Table 15. The QPSO update equations for different potential energy models

Potential well	QPSO update equation
Delta potential well	$x_i(t+1) = p(t) \pm \dfrac{\ln(1/u)}{2q\ln\sqrt{2}}\left\|x_i(t) - p(t)\right\|$
Harmonic oscillator	$x_i(t+1) = p(t) \pm \dfrac{\sqrt{\ln(1/u)}}{0.47694q}\left\|x_i(t) - p(t)\right\|$
Square well	$x_i(t+1) = p(t) + \dfrac{0.6574}{\xi q}\cos^{-1}(\pm\sqrt{u})\left\|x_i(t) - p(t)\right\|$

REFERENCES

Angeline, P. J. (1995). Adaptive and self–adaptive evolutionary computations. In M. Palaniswami, Y. Attikiouzel, R. Marks, D. Fogel & T. Fukuda (Eds.), *Computational Intelligence: A Dynamic Systems Perspective* (pp. 152-163). Washington, DC: IEEE Press.

Belew, R. K. (1990). Evolution, learning and culture: computational metaphors for adaptive algorithms. *Complex Systems*, *4*, 11–49.

Belew, R. K., McInerny, J., & Schraudolph, N. N. (1991). Evolving networks: using the genetic algorithm with connectionist learning. In C. Langton, C. Taylor, J. Farmer & S. Rasmussen (Eds.), *Artificial Life II* (pp. 511-547). New York: Addison-Wesley.

Bird, S., & Li, X. (2006). Adaptively choosing niching parameters in a PSO. In *Proceedings of the 2006 Genetic and Evolutionary Computation Conference (GECCO'06), Seattle (WA), USA* (pp. 3-9).

Brits, R., Engelbrecht, A. P., & Van den Bergh, F. (2002). A niching particle swarm optimizer. In *Proceedings of the 4th Asia-Pacific Conference on Simulated Evolution and Learning (SEAL 2002), Singapore* (pp. 692-696).

Burke, E. K., Gustafson, S., & Kendall, G. (2004). Diversity in genetic programming: an analysis of measures and correlation with fitness. *IEEE Transactions on Evolutionary Computation*, *8*(1), 1098–1107. doi:10.1109/TEVC.2003.819263

Clearwater, S. H., Hogg, T., & Huberman, B. A. (1992). Cooperative problem solving. In *Computation: The micro and macro view* (pp. 33-70). Singapore: World Scientific.

Clerc, M. (2006). *Particle swarm optimization*. London: ISTE Ltd.

Clerc, M., & Kennedy, J. (2002). The particle swarm - explosion, stability, and convergence in a multidimensional complex space. *IEEE Transactions on Evolutionary Computation*, *6*(1), 58–73. doi:10.1109/4235.985692

Cobb, H. G. (1992). Is the genetic algorithm a cooperative learner? In *Foundations of genetic algorithms 2* (pp. 277-296). San Mateo, CA: Kaufmann.

Coelho, L. S. (2008). A quantum particle swarm optimizer with chaotic mutation operator. *Chaos, Solitons, and Fractals, 37*(5), 1409–1418. doi:10.1016/j.chaos.2006.10.028

Coelho, L. S., & Mariani, V. C. (2008). Particle swarm approach based on quantum mechanics and harmonic oscillator potential well for economic load dispatch with valve-point effects. *Energy Conversion and Management, 49*(11), 751–759.

Cui, X., Li, M., & Fang, T. (2001). Study of population diversity of multiobjective evolutionary algorithm based on immune and entropy principles. In *Proceedings of the 2001 IEEE Congress on Evolutionary Computation (CEC'01), Seoul, Korea* (pp. 1316–1321).

Dawkins, R. (1976). *The selfish gene*. New York: Oxford University Press.

Goldberg, D. E. (1989). *Genetic algorithms in search, optimization and machine learning*. Reading, MA: Addison Wesley.

Hart, W. E. (1994). *Adaptive global optimization with local search*. PhD thesis, University of California, San Diego, USA.

Hinton, G. E., & Nowlan, S. J. (1987). How learning can guide evolution. *Complex Systems, 1*, 495–502.

Hooke, R., & Jeeves, T. A. (1961). Direct search solution of numerical and statistical problems. *Journal of the ACM, 8*, 212–229. doi:10.1145/321062.321069

Hoos, H. H., & Stützle, T. (2004). *Stochastic local search: foundations and applications*. San Mateo, CA: Kaufmann.

Horn, J. (1997). *The nature of niching: genetic algorithms and the evolution of optimal, cooperative populations*. PhD thesis, University of Illinois, Illinois Genetic Algorithm Lab, Urbana, USA.

Keesing, R., & Stork, D. G. (1990). Evolution and learning in neural networks: the number and distribution of learning trials affect the rate of evolution. In R. Lippmann, J. Moody, & D. Touretzky (Eds.), *Proceedings of the 1990 Conference on Advances in Neural Information Processing Systems (NIPS 3), Denver, Colorado* (pp. 804–810).

Krasnogor, N. (2002). *Studies on the theory and design space of memetic algorithms*. PhD thesis, University of the West of England, Bristol, UK.

Land, M. W. S. (1998). *Evolutionary algorithms with local search for combinatorial optimization*. PhD thesis, University of California, San Diego, USA.

Liu, L., Sun, J., Zhang, D., Du, G., Chen, J., & Xu, W. (2009). Culture conditions optimization of hyaluronic acid production by Streptococcus zooepidemicus based on radial basis function neural networks and quantum-behaved particle swarm optimization algorithm. *Enzyme and Microbial Technology, 44*(1), 24–32. doi:10.1016/j.enzmictec.2008.09.015

Matyas, J. (1965). Random optimization. *Automatization and Remote Control, 26,* 244–251.

Merz, P. (1998). *Memetic algorithms for combinatorial optimization (fitness landscapes and effective search strategies).* PhD thesis, Department of Electrical Engineering and Computer Science, University of Siegen, Germany.

Mikki, S. M., & Kishk, A. A. (2006). Quantum particle swarm optimization for electromagnetics. *IEEE Transactions on Antennas and Propagation, 54*(10), 2764–2775. doi:10.1109/TAP.2006.882165

Moscato, P. (1989). *On evolution, search, optimization, genetic algorithms and martial arts. Towards memetic algorithms* (Tech. Rep. C3P Report 826). Caltech Concurrent Computation Program, California, USA.

Muhlenbein, M., Gorges-Schleiter, M., & Kramer, O. (1988). Evolution algorithms in combinatorial optimization. *Parallel Computing, 7,* 65–85. doi:10.1016/0167-8191(88)90098-1

Parsopoulos, K. E., Tasoulis, D. K., & Vrahatis, M. N. (2004). Multiobjective optimization using parallel vector evaluated particle swarm optimization, In *Proceedings of the 2004 IASTED International Conference on Artificial Intelligence and Applications (AIA 2004), Innsbruck, Austria* (Vol. 2, pp. 823-828).

Parsopoulos, K. E., & Vrahatis, M. N. (2001). Modification of the particle swarm optimizer for locating all the global minima. In V. Kurkova, N.C. Steele, R. Neruda, & M. Karny (Eds.), *Artificial Neural Networks and Genetic Algorithms* (pp. 324-327). Wien: Springer.

Parsopoulos, K. E., & Vrahatis, M. N. (2002a). Recent approaches to global optimization problems through particle swarm optimization. *Natural Computing, 1*(2-3), 235–306. doi:10.1023/A:1016568309421

Parsopoulos, K. E., & Vrahatis, M. N. (2002b). Particle swarm optimization method in multiobjective problems. In *Proceedings of the 2002 ACM Symposium on Applied Computing (SAC 2002), Madrid, Spain* (pp. 603-607).

Parsopoulos, K. E., & Vrahatis, M. N. (2004). UPSO: a unified particle swarm optimization scheme. In T. Simos & G. Maroulis (Eds.), *Lecture Series on Computer and Computational Sciences* (Vol. 1, pp. 868-873). Zeist, The Netherlands: VSP International Science Publishers.

Parsopoulos, K. E., & Vrahatis, M. N. (2007). Parameter selection and adaptation in unified particle swarm optimization. *Mathematical and Computer Modelling, 46*(1-2), 198–213. doi:10.1016/j.mcm.2006.12.019

Petalas, Y. G., Parsopoulos, K. E., Papageorgiou, E. I., Groumpos, P. P., & Vrahatis, M. N. (2007a). Enhanced learning in fuzzy simulation models using memetic particle swarm optimization. In *Proceedings of the 2007 IEEE Swarm Intelligence Symposium (SIS'07), Honolulu (HI), USA* (pp. 16-22).

Petalas, Y. G., Parsopoulos, K. E., & Vrahatis, M. N. (2007b). Memetic particle swarm optimization . *Annals of Operations Research, 156*(1), 99–127. doi:10.1007/s10479-007-0224-y

Petalas, Y. G., Parsopoulos, K. E., & Vrahatis, M. N. (2007c). Entropy-based memetic particle swarm optimization for computing periodic orbits of nonlinear mappings. In *Proceedings of the 2007 IEEE Congress on Evolutionary Computation (CEC'07), Singapore* (pp. 2040-2047).

Potter, M. A., & De Jong, K. A. (1994). A cooperative coevolutionary approach to function optimization. In *The Third Parallel Problem Solving From Nature* (pp. 249-257). Berlin: Springer.

Potter, M. A., & De Jong, K. A. (2000). Cooperative coevolution: an architecture for evolving coadapted subcomponents. *Evolutionary Computation, 8*(1), 1–29. doi:10.1162/106365600568086

Rao, S. S. (1992). *Optimization: theory and applications.* New York: Wiley Eastern.

Rosca, J. P. (1995). Entropy–driven adaptive representation. In *Proceedings of the Workshop on Genetic Programming: From Theory to Real–World Applications, Tahoe City (CA), USA* (pp. 23–32).

Shannon, C. E. (1964). *The mathematical theory of communication.* Champaign: University of Illinois Press.

Solis, F., & Wets, R. (1981). Minimization by random search techniques. *Mathematics of Operations Research, 6,* 19–30. doi:10.1287/moor.6.1.19

Storn, R., & Price, K. (1997). Differential evolution - a simple and efficient heuristic for global optimization over continuous spaces. *Journal of Global Optimization, 11,* 341–359. doi:10.1023/A:1008202821328

Sun, J., Feng, B., & Xu, W. (2004a). Particle swarm optimization with particles having quantum behavior. In *Proceedings of the 2004 IEEE Congress on Evolutionary Computation (CEC'04), Portland (OR), USA* (pp. 325-331).

Sun, J., Xu, W., & Feng, B. (2004b). A global search strategy for quantum-behaved particle swarm optimization. In *Proceedings of the 2004 IEEE Conference on Cybernetics and Intelligent Systems, Singapore* (pp. 111-116).

Tang, J., Lim, M. H., & Ong, Y. S. (2006). Adaptation for parallel memetic algorithm based on population entropy. In *Proceedings of the 2006 Genetic and Evolutionary Computation Conference (GECCO'06), Seattle (WA), USA* (pp. 575–582).

Thiémard, E. (1998). *Economic generation of low-discrepancy sequences with a b-ary gray code.* (Tech. Rep. EPFL-DMA-ROSO, RO981201). Department of Mathematics, Ecole Polytechnique Fédérale de Lausanne, Lausanne, Switzerland.

Trelea, I. C. (2003). The particle swarm optimization algorithm: convergence analysis and parameter selection. *Information Processing Letters, 85*(6), 317–325. doi:10.1016/S0020-0190(02)00447-7

Van den Bergh, F. (2002). *An analysis of particle swarm optimizers.* PhD thesis, Department of Computer Science, University of Pretoria, South Africa.

Van den Bergh, F., & Engelbrecht, A. P. (2002). A new locally convergent particle swarm optimizer. In *Proceedings of the 2002 IEEE International Conference on Systems, Man and Cybernetics (SMC'02), Hammamet, Tunisia* (Vol. 3, pp. 96-101).

Van den Bergh, F., & Engelbrecht, A. P. (2004). A cooperative approach to particle swarm optimization. *IEEE Transactions on Evolutionary Computation, 8*(3), 225–239. doi:10.1109/TEVC.2004.826069

Xi, M., Sun, J., & Xu, W. (2008). An improved quantum-behaved particle swarm optimization algorithm with weighted mean best position. *Applied Mathematics and Computation, 205*(2), 751–759. doi:10.1016/j.amc.2008.05.135

Chapter 5
Performance–Enhancing Techniques

This chapter presents techniques that have proved to be very useful in enhancing the performance of PSO in various optimization problem types. They consist of transformations of either the objective function or the problem variables, enabling PSO to alleviate local minimizers, detect multiple minimizers, handle constraints, and solve integer optimization problems. The chapter begins with a short discussion on the filled functions approach, and then presents the stretching technique as an alternative for alleviating local minimizers. Next, we present the deflection and repulsion techniques, as a means for detecting multiple global minimizers with PSO, followed by a penalty function approach for constraint handling. The chapter closes with the description of two rounding schemes that enable the continuous, real-valued PSO to solve integer programming problems. All techniques are thoroughly described and graphically illustrated whenever possible.

INTRODUCTION

The alleviation of local minimizers and the detection of several (global or local) ones has been a topic of ongoing research for many years in global optimization (Torn & Žilinskas, 1989). For this purpose, various techniques have been developed and incorporated in the context of global optimization algorithms.

The simplest approach to tackle the aforementioned problem is the *multistart* technique, where the algorithm is restarted from different initial conditions every time it has converged to a minimizer. However, this technique does not include a mechanism for preventing the algorithm from converging anew to the same or a worst minimizer in subsequent restarts. This deficiency limits the applicability of multistart, unless it is combined with different techniques that guide the algorithm towards more promising regions of the search space than those detected (Torn & Žilinskas, 1989).

DOI: 10.4018/978-1-61520-666-7.ch005

Copyright © 2010, IGI Global. Copying or distributing in print or electronic forms without written permission of IGI Global is prohibited.

Alternatively, one can modify the objective function after the detection of a minimizer, in such a way that its new form excludes already detected minimizers. In this context, Goldstein and Price (1971) proposed an efficient algorithm for the minimization of algebraic functions, which requires high-order derivatives of the involved polynomials. Later, they generalized their technique to non-polynomial problems, using a transformation that makes use of the Hessian of the objective function. However, the required derivatives and Hessian are not always available nor can they always be estimated accurately, thereby restricting the applicability of this method to problems with nice mathematical properties. Shusterman (1979) proposed a similar approach but the produced objective function becomes very flat after the detection of a small number of minimizers, setting obstacles on most optimization algorithms.

Vilkov *et al.* (1975) proposed the popular *tunneling* method for one-dimensional functions. This method can direct the search point away from a detected minimizer, towards the region of attraction of a better one. The algorithm was generalized in multi-dimensional problems by Montalvo (1979) and Gómez and Levy (1982). However, the constructed hypersurface becomes increasingly flat with the number of detected minimizers, imposing the same difficulties as the previous methods on the algorithms.

Filled functions constitute another popular approach, developed by Ge (1987, 1990). They consist of a transformation of the objective function after the detection of a minimizer. More specifically, let x^* be a detected (local) minimizer of the objective function, $f(x)$. Then, $f(x)$ is transformed as follows:

$$T(x;r,p) = f(x) + \frac{a}{r + f(x)} \exp\left(-\frac{\|x - x^*\|^2}{p^2}\right),$$ (1)

where a, r, and p, are user-defined parameters that control the magnitude of the transformation. The term:

$$\frac{a}{r + f(x)},$$

inverts the objective function, transforming the local minimizer, x^*, to a local maximizer. The exponential term:

$$\exp\left(-\frac{\|x - x^*\|^2}{p^2}\right),$$

imposes an exponentially increasing penalty to each point x, based on its distance from x^*. The parameter a determines the magnitude of the produced local maximum at x^*, with larger values producing higher local maxima. The parameter r is used to avoid division by zero in equation (1), and it shall be selected taking into consideration both the magnitude of the objective function as well as the value of a. Finally, p is perhaps the most crucial parameter, since it determines the scope of the transformation. Small values of p result in more local effects than (even slightly) higher values.

The parameters a, r, and p, have a crucial impact on the shape of the filled function. An erroneous choice can be detrimental for the optimization procedure, while even small differences can produce significantly different landscapes. We illustrate this parameter sensitivity with the example of Fig. 1, where the filled function of equation (1) (illustrated with a dashed line) is applied on the one-dimensional

Figure 1. The filled function (dashed line) of equation (1) applied on the one-dimensional instance of TP$_{UO-3}$ (Rastrigin function - solid line) defined in Appendix A, for the local minimizer x* = -1, a = 50, r = 0.1, and p = 0.01, 0.1, and 1.0

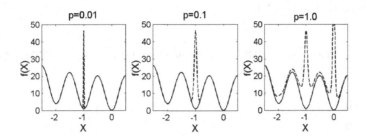

instance of TP$_{UO-3}$ (Rastrigin function), defined in Appendix A of the book at hand, and illustrated with a solid line. The transformation is applied for the local minimizer $x^* = -1$, fixed parameters, $a = 50$, $r = 0.1$, and three different values of p, namely 0.01, 0.1, and 1.0.

As we can see, the value $p = 0.01$ produces a very local effect around the detected minimizer, leaving the rest of the function unaffected. Increasing the value of p to 0.1, produces wider alteration around x^*, although the effect remains local. However, a further increase to 1.0 has a tremendous effect on the function, especially for the global minimizer that lies at zero, on the right of x^*. The effect becomes even stronger if we further vary the rest of the parameters. Nevertheless, it is expected that an appropriate set of parameters shall be problem-dependent. Indeed, Parsopoulos and Vrahatis (2004, p. 13) provided graphical evidence for a polynomial function, where the setting, $a = 200$ (and $a = 400$), $r = 0$, and $p = 1$, was promising, while the same setting is rather baneful for the function illustrated in Fig. 1. Finally, one can notice that the filled function introduces new local minima around the detected minimizer; an effect also known as *mexican hat*.

The aforementioned deficiencies dictate the cautious use of filled functions, suggesting the necessity for a remarkable effort to discover an appropriate configuration for the problem at hand. More recent methods, based on ideas similar to the concept of filled functions, have been proposed and used successfully with PSO for detecting several global minimizers and alleviating local ones. These techniques can also be considered as sequential niching schemes, described in the previous chapter, and they are exposed in the following sections.

THE STRETCHING TECHNIQUE FOR ALLEVIATING LOCAL MINIMIZERS

Stretching was introduced to PSO as a means for alleviating local minima, in Parsopoulos *et al.* (2001). It works similarly to filled functions, i.e., after the detection of a local minimizer a transformation is applied on the objective function such that the higher local minima vanish, while the lower minima (including the global one) remain unaffected. Stretching has a wide effect on the whole range of the objective function and is properly suited for alleviating local minimizers, especially in highly multimodal problems.

More specifically, if $f(x)$ is the objective function and x^* is a detected local minimizer, stretching is defined as the following two-stage transformation:

$$G(x) = f(x) + \gamma_1 \| x - x^* \| [1 + sign(f(x) - f(x^*))], \tag{2}$$

$$H(x) = G(x) + \gamma_2 \frac{1 + sign(f(x) - f(x^*))}{\tanh(\mu(G(x) - G(x^*)))}, \tag{3}$$

where γ_1, γ_2, and μ, are user-defined parameters, while $sign(z)$ is the three-valued sign function, defined as:

$$sign(z) = \begin{cases} +1, & \text{if } z > 0, \\ 0, & \text{if } z = 0, \\ -1, & \text{if } z < 0. \end{cases}$$

Thus, after the detection of x^*, all new candidate solutions will be evaluated using equations (2) and (3), instead of the original objective function $f(x)$. If a new local minimizer with lower function value than $f(x^*)$ is detected, then it replaces x^* in equations (2) and (3). Thus, stretching is applied on a sequence of local minimizers with monotonically decreasing function values, which, under proper assumptions on the number of local minima, can lead the algorithm to the global one.

The first transformation, $G(x)$, stretches the function upwards, covering higher local minima, while the second one, $H(x)$, transforms the detected minimizer to a maximizer. The whole procedure leaves all minimizers with lower function values unaffected. This is illustrated in Fig. 2 for the same objective function with Fig. 1, i.e., the one-dimensional instance of TP_{UO-3} (solid line), and for the local minimizer $x^* = -1$, with parameters $\gamma_1 = \gamma_2 = 20$, and $\mu = 0.1$. The first transformation, $G(x)$, is illustrated with a dotted line, while the final form, $H(x)$, of the stretched objective function is illustrated with a dashed line. We can clearly see the valley shaped on the right of the stretched minimizer, around the position of the global one at zero, as well as the maximum (central peak) introduced at the (previously local minimizer) $x^* = -1$. It is intuitively evident that an algorithm applied on the transformed landscape will have a higher probability of convergence on the global minimizer.

We can state the following significant remarks regarding the application of stretching:

Remark 1. *Stretching can also be applied on non-minimizers.* Indeed, it can be applied to any point except the global minimizer, in order to simplify the landscape by reducing the number of local minima.

Remark 2. *Stretching shall not be applied on the global minimizer.* Since it eliminates all minimizers with the same or higher values, it will destroy all remaining global and local minima. Thus, stretching is not appropriate for detecting several global minimizers of a function.

Remark 3. *Stretching introduces new local minima.* We can clearly distinguish the two new local minima introduced in both sides of the stretched local minimizer, $x^* = -1$, in Fig. 2. This is the Mexican hat effect that was also verified for the filled functions in Fig. 1.

Remark 4. *Stretching is not suitable for detecting all local minimizers.* Its application on a local minimizer will destroy all higher local minimizers.

Remark 5. *Stretching is sensitive to the values of its parameters.* Similarly to the filled functions, stretching requires a proper parameter set to work efficiently. Although the negative effects of a bad parameter set are not as extensive as in filled functions, it can produce transformations of

Figure 2. The stretching transformation applied on the one-dimensional instance of TP$_{UO-3}$ (Rastrigin function - solid line) defined in Appendix A, for the local minimizer x* = -1, and parameters, $\gamma_1 = \gamma_2 = 20$, $\mu = 0.1$

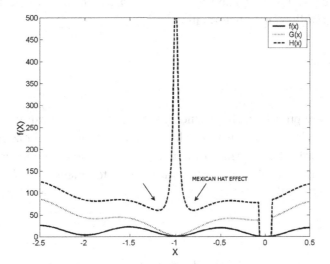

inferior quality. Parsopoulos and Vrahatis (2004) have illustrated this sensitivity for different parameter settings.

Summarizing, stretching can be a very useful technique for alleviating local minima, combined easily with any evolutionary algorithm. It has been used with PSO on several problems with very promising results (Parsopoulos *et al.*, 2001; Parsopoulos & Vrahatis, 2002a, 2004). However, its use requires some precautions such as a nice parameter setting, which can be determined through a preprocessing phase prior to the application of PSO. In parallel, a mechanism for distinguishing between local and global minimizers can be useful to avoid destructing all global minimizers of the objective function. The latter issue can be easily addressed in functions with known global minima, e.g., error functions where the global minimum is known to be equal to zero or bounded below.

Stretching can be applied anew to an arbitrary number of points, always using the original objective function, $f(x)$, in equations (2) and (3) for each new minimizer. An improved variant that claims to address the Mexican hat effect has been reported in Wang and Zhang (2007). Alternatively, the undesirable effect can also be tackled by using a *repulsion* technique that keeps the algorithm away from the newly introduced local minima. This technique is discussed in a later section, after the presentation of deflection, an intimately related technique for detecting several minimizers.

THE DEFLECTION TECHNIQUE FOR DETECTING SEVERAL MINIMIZERS

Deflection was introduced by Magoulas *et al.* (1997) as a technique for detecting several (local or global) minimizers of the error function in neural network training problems. Later, it was combined with PSO for solving widely used optimization problems (Parsopoulos & Vrahatis, 2004). Deflection works similarly to the previously presented approaches, i.e., it transforms the objective function so that already detected minimizers are converted to maximizers. However, it has only one additional constraint; it can be applied only on non-negative functions.

Let $f(x) > 0$ be the objective function, and x_i^*, $i = 1, 2,\ldots, m$, be the already detected minimizers. Then, deflection consists of the following transformation:

$$D(x) = f(x) \times \prod_{i=1}^{m} T_i(x; x_i^*, \lambda_i)^{-1}, \tag{4}$$

where T_i, $i = 1, 2,\ldots, m$, are proper functions, and λ_i, $i = 1, 2,\ldots, m$, are positive relaxation parameters. The functions T_i shall be selected such that $D(x)$ has exactly the same minimizers with $f(x)$ except x_i^*, $i = 1, 2,\ldots, m$. In other words, any sequence of points converging to a minimizer x_i^* shall not produce a minimum at $D(x_i^*)$. A category of functions that satisfy the aforementioned *deflection property* is the following (Magoulas *et al.*, 1997):

$$T_i(x; x_i^*, \lambda_i) = \tanh(\lambda_i \|x - x_i^*\|). \tag{5}$$

The effect of the deflection procedure is intended to be local around the position of the detected minimizer, such that the rest remain unaffected.

In contrast to stretching, deflection uses all detected minimizers to transform the objective function. Thus, if a new minimizer is detected, then it rather adds a new term in the product of equation (4) instead of replacing the previous minimizer as in equations (2) and (3). Thus, deflection is especially suited for detecting a multitude of (local or global) minimizers of the objective function. Similarly to stretching, deflection can be applied on any point of the search space regardless of its qualification as a minimizer, although without altering its surrounding minimizers. Therefore, in contrast to stretching, it can also be used on global minimizers.

The requirement of a positive objective function lies in the fact that, if $f(x^*) < 0$ for a given minimizer, x^*, then the transformation of equation (4) will retain all minimizers (including x^*), shifting their values towards $-\infty$. On the other hand, if $f(x^*) = 0$ then also $D(x^*) = 0$. Thus, a global minimum equal to zero cancels the deflection transformation, raising an issue that can be addressed through the use of a non-positive function. In this case, deflection is applied on the shifted objective function:

$$f'(x) = f(x) + c, \tag{6}$$

where $c > 0$ is a suitable positive constant such that, $f'(x) > 0$, for all x. The shifted objective function $f'(x)$ has exactly the same minimizers with $f(x)$, although all its minima are shifted upwards by c.

The workings of deflection on the same problem with Fig. 2, are illustrated in Fig. 3. More specifically, the one-dimensional instance of TP_{UO-3} (Rastrigin function), defined in Appendix A and illustrated with a solid line, is used, and deflection (dashed line) is applied on its local minimizer, $x_1^* = -1$, as well as on the global minimizer, $x_g^* = 0$, with parameters $\lambda_1 = \lambda_g = 1$. The parameter, $c = 1$, was used in equation (6) to shift the objective function and enable the deflection of the global minimizer $x_g^* = 0$. Otherwise, the deflection would have an effect only on the local minimizer, which has an original function value equal to $f_1^* = 1$.

Equation (4) implies that the deflected value of any given point x depends on three factors: the magnitude of its value, $f(x)$; its distance from the deflected minimizers, $\|x-x_i\|$; as well as on the corresponding parameters, λ_i. Regarding the first factor, it is clear that $D(x)$ increases proportionally to $f(x)$ for fixed values of the functions $T_i(x; x_i^*, \lambda_i)$. For this reason, the peaks produced in the regions of deflected minimizers with higher values are expected to also be higher than those of minimizers with smaller values. This effect is also observed in Fig. 3, where the region of the global minimizer positioned at zero assumes lower values than that of the local minimizer at -1. Regarding the second factor, points in adequate distance from the deflected minimizers will assume values close to their original ones, while points that lie close to a deflected minimizer will be more affected by the transformation, assuming higher values. Finally, keeping x and x_i fixed in equation (5) while increasing λ_i, results in monotonically increasing values of $T_i(x; x_i^*, \lambda_i)$ that approximate 1. Thus, the deflected value of x will be close to its original one. On the other hand, decreasing λ_i towards zero, produces values of $T_i(x; x_i^*, \lambda_i)$ close to zero. In this case, the value of x will become very large due to the transposition of the functions T_i in equation (4).

Figure 4 illustrates the effect of different values of λ_i on deflection. The same objective function with Fig. 3 is used, while deflection is applied on the local minimizers, $x_1^* = -2$, $x_2^* = -1$, as well as on the global one, $x_g^* = 0$, with parameter values $\lambda_i = 0.8$ (dash-dotted line), $\lambda_i = 1.0$ (dashed line), and $\lambda_i = 2.0$ (dotted line). Apparently, increasing values of λ_i have a milder effect on the objective function (a shifted function as in Fig. 3 was assumed also in Fig. 4).

The Mexican hat effect, which accompanies filled functions and stretching, is present also in deflection, as we observe in Figs. 3 and 4. Local minimizers introduced by this effect can still trap an algorithm, although they have remarkably higher values than the originally deflected minimizers. To alleviate this drawback, Parsopoulos and Vrahatis (2004) proposed the repulsion technique described in the next section.

Figure 3. The deflection transformation applied on the one-dimensional instance of TP$_{UO-3}$ (Rastrigin function - solid line) defined in Appendix A, on the local minimizer $x_1^* = -1$, as well as on the global minimizer $x_g^* = 0$, with parameters $\lambda_l = \lambda_g = 1$

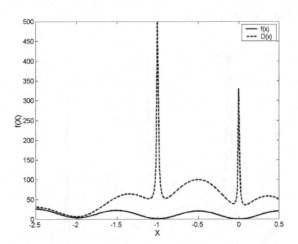

THE REPULSION TECHNIQUE

The *repulsion* technique was introduced by Parsopoulos and Vrahatis (2004) for preventing the attraction of PSO on the local minimizers artificially introduced by stretching and deflection. The underlying idea is intuitively appealing and straightforwardly realizable: after the detection of a minimizer and the application of a (stretching or deflection) transformation on it, a *repulsion area* is defined around it and any particle that falls into this, is repelled away.

To put it formally, let $X^* = \{x_j^*; j = 1, 2,..., m\}$ be the set of already detected minimizers and $S = \{x_1, x_2,..., x_N\}$ be the swarm at a given iteration. After the position update of particles using any PSO variant, the distance between each particle and detected minimizer, which is defined as (Parsopoulos & Vrahatis, 2004):

$$d_{ij} = d(x_i, x_j^*) = \|x_i - x_j^*\|, \quad i = 1, 2,..., N, \quad j = 1, 2,..., m, \tag{7}$$

is computed to check if it lies in the repulsion area of the minimizer. In general, this area is defined for each minimizer as a spherical region, centered at the minimizer, with radius r_{ij}, $i = 1, 2,..., N, j = 1, 2,..., m$. Hence, if $d_{ij} \leq r_{ij}$ then x_i shall be repelled away from x_j^*. This is achieved with a correction in the particle position, as follows (Parsopoulos & Vrahatis, 2004):

$$x_i = x_i + \rho_{ij} z_{ij}, \tag{8}$$

where ρ_{ij} is a fixed parameter determining the *repulsion strength*, and z_{ij} is the unitary vector with direction from x_j^* towards x_i, defined as:

Figure 4. The deflection transformation applied on the one-dimensional instance of TP$_{UO-3}$ (Rastrigin function - solid line) defined in Appendix A, on the local minimizers $x_1^* = -2$, $x_2^* = -1$, as well as on the global minimizer $x_g^* = 0$, for different parameter values, $\lambda_i = 0.8$ (dash-dotted line), $\lambda_i = 1.0$ (dashed line), and $\lambda_i = 2.0$ (dotted line), for all i

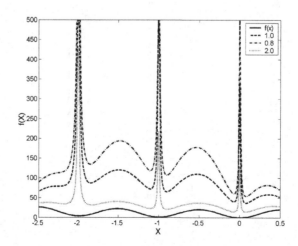

$$z_{ij} = \frac{(x_i - x_j^*)}{d_{ij}}.$$

(9)

For simplicity, r_{ij} and ρ_{ij} are selected to be equal for all particles and detected minimizers, although, if needed, the user can choose different settings per case. In unexplored problems, small values of r_{ij} are preferred to avoid the inclusion of an undiscovered minimizer in the repulsion area of a detected one. On the other hand, ρ_{ij} shall be adequately large to ensure the proper repulsion away from the detected minimizer. The repulsion procedure is described in pseudocode in Table 1.

Parsopoulos and Vrahatis (2004) illustrated the workings of repulsion in conjunction with the deflection technique on test problem TP_{UO-21}, defined in Appendix A of the book at hand, which has the shape of an egg holder, as depicted in the left part of Fig. 5. Their experiment aimed at the detection of all 12 global minimizers within the range $[-5,5]^2$. As they report, the plethora of equally attractive regions of the search space results in a rambling movement of the particles, which is illustrated in the right part of Fig. 5. Moreover, there is no guarantee that after an (even large) number of restarts, the algorithm will be able to detect all global minimizers. This problem was addressed by applying deflection with repulsion after the detection of each global minimizer.

The experiments used the constriction coefficient variant, PSO[co], as well as the inertia weight variant, PSO[in]. The corresponding parameter setting is reported in Table 2, while the results are reported in Table 3. Both variants were able to detect all 12 minimizers, with PSO[co] being more efficient than PSO[in]. The values of the parameters r_{ij} and ρ_{ij} are crucial, since they may result in the inclusion of neighboring minimizers in the repulsion area of a detected one. For example, this could happen if $r_{ij} = 3$ was used in the example of Fig. 5. Hence, in the application of repulsion, an estimation of the relative distance among the desirable minimizers can be more than useful. Otherwise, a preprocessing phase of preliminary experiments using different parameter settings can offer the necessary information.

Parsopoulos and Vrahatis (2004) also reported a plethora of experimental results on the problem of detecting periodic orbits of nonlinear mappings, where all techniques described so far find a rich field of application. We postpone the discussion of these results until a later chapter devoted to the applications

Table 1. Pseudocode of the repulsion procedure

Input:	Set of detected minimizers, X*; swarm, S; parameters r_{ij} and ρ_{ij} for all i and j.
Step 1.	**Do** $(i = 1\ldots N)$
Step 2.	**If** $(X^* \neq \varnothing)$ **Then**
Step 3.	**Do** $(j = 1\ldots m)$
Step 4.	**Compute** d_{ij} using equation (7).
Step 5.	**If** $(d_{ij} \leq r_{ij})$ **Then**
Step 6.	**Compute** z_{ij} using equation (9).
Step 7.	**Update** particle position x_i using equation (8).
Step 8.	**End If**
Step 9.	**End Do**
Step 10.	**End If**
Step 11.	**End Do**

of PSO in dynamical systems. The next section discusses the penalty function technique for tackling constrained optimization problems with PSO.

THE PENALTY FUNCTION TECHNIQUE FOR CONSTRAINED OPTIMIZATION PROBLEMS

In Chapter 1, we defined the constrained optimization problem as follows:

$$\min_{x \in A} f(x), \quad \text{subject to} \quad C_i(x) \leq 0, \quad i = 1, 2, ..., k, \tag{10}$$

where k is the number of constraints. The form of constraints in relation (10) is not restrictive, since different forms can be represented equivalently as follows:

$$C_i(x) \geq 0 \Leftrightarrow -C_i(x) \leq 0,$$

Figure 5. (Left) plot of test problem TP_{UO-21} (egg holder), defined in Appendix A of the book at hand. In the range $[-5,5]^2$, it has 12 global minimizers at the points $(k_1\pi/2, k_2\pi)^T$, for $k_1 = \pm1, \pm2$, and $k_2 = 0, \pm1$. (Right) contour plot of TP_{UO-21} and the rambling movement of a particle for 30 iterations. Darker contour lines denote the regions of the global minimizers

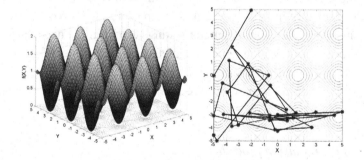

Table 2. Parameters of the constriction coefficient and inertia weight PSO variants for detecting all global minimizers of test problem TP_{UO-21}, illustrated in Fig. 5, using both deflection and repulsion

Parameter	Value
Swarm size	20
Accuracy	10^{-4}
c_1, c_2	2.05
w	$1 \rightarrow 0.1$ (linearly decreasing)
χ	0.729
v_{max}	5
r_{ij}	0.5
ρ_{ij}	0.8

Table 3. Results of the constriction coefficient and inertia weight PSO variants for detecting all global minimizers of test problem TP_{UO-21}, using the parameter setting of Table 2

PSO variant	Number of detected minimizers	Mean total number of function evaluations	Mean number of function evaluations per minimizer
PSO[co]	12	6300	524
PSO[in]	12	50660	4222

$C_i(x) = 0 \Leftrightarrow C_i(x) \leq 0$ and $-C_i(x) \leq 0$.

One of the most popular approaches for tackling constrained problems is the use of penalty functions. In general, a *penalty function* is defined as:

$$f_p(x) = f(x) + P(x), \quad x \in A \subset \mathbf{R}^n,$$

where $f(x)$ is the original objective function and $P(x)$ is a penalty term. Obviously, $P(x)$ shall be selected such that:

$$P(x) = \begin{cases} 0, & \text{if } x \text{ is a feasible point,} \\ a > 0, & \text{otherwise,} \end{cases}$$

in order to penalize only infeasible solutions. Also, $P(x)$ can be either fixed to a prescribed value for all infeasible solutions or proportional to the number of violated constraints and degree of violation.

Recently, the following penalty function was used with evolutionary algorithms and PSO, exhibiting very promising results (Parsopoulos & Vrahatis, 2002b; Yang *et al.*, 1997):

$$f_p(x) = f(x) + h(t)H(x), \quad x \in A \subset \mathbf{R}^n, \tag{11}$$

where $f(x)$ is the original objective function; $h(t)$ is a penalty value, dynamically changing with the iteration number, t; and $H(x)$ is a penalty factor defined as follows:

$$H(x) = \sum_{i=1}^{k} \theta(q_i(x)) q_i(x)^{\gamma(q_i(x))}, \tag{12}$$

where $q_i(x) = \max\{0, C_i(x)\}$, $i = 1, 2, \ldots, k$; $\theta(q_i(x))$ is a multi-stage assignment function; and $\gamma(q_i(x))$ is the power of the penalty function (Homaifar *et al.*, 1994).

The aforementioned penalty function takes all constraints into consideration, based on their corresponding degree of violation, while the user can manipulate each one independently, based on its level of importance. Also, the penalty value, $h(t)$, adds an additional degree of freedom, allowing the dynamic change of penalty magnitude during the optimization procedure.

An alternative penalty function is defined in Coello Coello (1999), as follows:

$$f_p(x) = f(x) + H(x), \tag{13}$$

with,

$$H(x) = w_1 H_{NVC}(x) + w_2 H_{SVC}(x), \tag{14}$$

where $H_{NVC}(x)$ is the number of violated constraints, and $H_{SVC}(x)$ is the sum of violated constraints, defined as follows:

$$H_{SVC}(x) = \sum_{i=1}^{k} \max\{0, C_i(x)\}.$$

The weights w_1 and w_2 permit the user to determine the importance of each constraint based on the problem at hand.

Parsopoulos and Vrahatis (2002b) investigated the performance of PSO equipped with the penalty function of equations (11) and (12) on several widely used constrained optimization test problems. They also investigated unified PSO (UPSO) on constrained engineering design problems, using the penalty function of equations (13) and (14) (Parsopoulos & Vrahatis, 2005). We postpone the discussion of their results until a later chapter that deals with applications of PSO on this type of problems. The next section closes this chapter with a discussion of rounding schemes that allow continuous-based PSO approaches to work efficiently on integer optimization problems.

ROUNDING TECHNIQUES FOR INTEGER OPTIMIZATION

PSO was originally introduced for continuous optimization problems. Both its philosophy and operation imply the existence of continuous variables. This property raises a rational question regarding its efficiency on integer subspaces of the n-dimensional Euclidean space.

A common approach for applying continuous optimization methods on integer problems is the transformation of the integer problem to the corresponding continuous problem. Then, the continuous problem is solved and integer solutions are obtained by truncating the detected continuous ones. Such approaches have been used with traditional branch and bound optimization algorithms, which transform the integer problem to a corresponding continuous one, and apply quadratic programming techniques to solve it (Lawler & Wood, 1966; Manquinho *et al.*, 1997).

Laskari *et al.* (2002) considered the aforementioned approach in the context of PSO for solving integer programming problems. More specifically, they applied the standard PSO update rules, while each particle component was rounded to the nearest integer value. Hence PSO retained its dynamics, while the produced solutions were integer.

To put it more formally, let $x_i = (x_{i1}, x_{i2}, \ldots, x_{in})^T$ be the i-th particle of the swarm, with its components $x_{ij} \in \mathbf{R}, j = 1, 2, \ldots, n$. Then, its rounded counterpart, $z_i = (z_{i1}, z_{i2}, \ldots, z_{in})^T$, defined as:

$$z_{ij} = \lfloor x_{ij} + 0.5 \rfloor, \quad j = 1, 2, \ldots, n, \tag{15}$$

replaces x_{ij} in the swarm at each iteration. Therefore, the swarm always consists of integer particles, although their velocities can be real-valued.

Obviously, following this approach may result in multiple appearances of the same integer particle in the swarm, since an integer vector corresponds to an infinite number of different real-valued vectors rounded with equation (15). The original real-valued trajectory of a single particle (solid line) and its corresponding integer trajectory (dashed-dotted line) are denoted in Fig. 6 for the 2-dimensional instance of test problem TP_{UO-1} defined in Appendix A of the book at hand, for 10 iterations.

Intuitively, the repeated rounding at each iteration may result in search stagnation, especially in cases where particles cluster close to each other. To tackle this potential deficiency, Laskari *et al.* (2002) also investigated a different scheme with gradual truncation of the particles, in order to retain the dynamics of PSO, while gradually pushing the particles towards integer values. According to this scheme, particles are truncated to their first k_1 decimal digits in the first T iterations, to $k_2 < k_1$ decimal digits in the next T iterations, and so on, until k_i reaches zero. Typically, one can consider $k_{i+1} = k_i - 2$, i.e., two decimal digits are eliminated from the particle components after every T iterations of the algorithm. A more aggressive scheme would be $k_{i+1} = k_i/2$, where the transformation of the particles to integers is rapid. The aforementioned schemes require that the initial number of decimal digits is even, while different schemes with dynamically changing values of k_i can be also considered.

Laskari *et al.* (2002) applied the aforementioned approaches in the context of PSO on several integer programming test problems reported in Appendix A. We postpone the discussion of their results until a later chapter of the book, which considers the application of PSO on such problems.

CHAPTER SYNOPSIS

This chapter was devoted to the presentation of techniques that enhance PSO performance, rendering it efficient in demanding applications. All the presented techniques are based on transformations of the objective function and/or problem variables. Additional parameters are introduced in most cases, which

Figure 6. Original (real-valued) trajectory of a single particle (solid line) and its corresponding integer trajectory (dashed-dotted line) for 10 iterations, produced with the rounding scheme of equation (15), on the contour plot of the 2-dimensional instance of test problem TP_{UO-1} defined in Appendix A of the book at hand

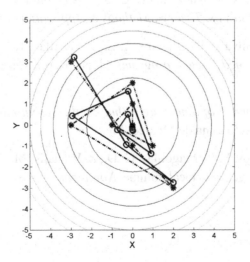

require the interference of the user. A preprocessing phase of preliminary experiments can be very useful for determining the proper values of these parameters. In some cases, wrong values can be detrimental for the algorithm; thus, the user should pay special attention when applying these techniques to avoid undesirable effects.

REFERENCES

Coello, C. A. C. (1999). Self-adaptive penalties for GA-based optimization. In *Proceedings of the 1999 IEEE Congress on Evolutionary Computation (CEC'99), Washington (DC), USA* (pp. 573-580).

Ge, R. P. (1987). The theory of the filled function method for finding a global minimizer of a nonlinearly constrained minimization problem. *Journal of Computational Mathematics, 5,* 1–9.

Ge, R. P. (1990). A filled function method for finding a global minimizer of a function of several variables. *Mathematical Programming, 46,* 191–204. doi:10.1007/BF01585737

Goldstein, A. A., & Price, J. F. (1971). On descent from local minima. *Mathematics of Computation, 25*(3), 569–574. doi:10.2307/2005219

Gómez, S., & Levy, A. V. (1982). The tunneling method for solving the constrained global optimization problem with several nonconnected feasible regions. In *Lecture Notes in Mathematics, Vol. 909* (pp. 34-47). Berlin: Springer.

Homaifar, A., Lai, A. H.-Y., & Qi, X. (1994). Constrained optimization via genetic algorithms. *Simulation, 2*(4), 242–254. doi:10.1177/003754979406200405

Laskari, E. C., Parsopoulos, K. E., & Vrahatis, M. N. (2002). Particle swarm optimization for integer programming. *In Proceedings of the 2002 IEEE Congress on Evolutionary Computation (CEC'02), Honolulu (HI), USA* (pp. 1582-1587).

Lawler, E. L., & Wood, D. W. (1966). Branch and bound methods: a survey. *Operations Research, 14,* 699–719. doi:10.1287/opre.14.4.699

Magoulas, G. D., Vrahatis, M. N., & Androulakis, G. S. (1997). On the alleviation of local minima in backpropagation. *Nonlinear Analysis, Theory . Methods & Applications, 30*(7), 4545–4550.

Manquinho, V. M., Marques Silva, J. P., Oliveira, A. L., & Sakallah, K. A. (1997). *Branch and bound algorithms for highly constrained integer programs* (Tech. Rep.). Cadence European Laboratories, Portugal.

Montalvo, A. (1979). *Development of a new algorithm for the global minimization of functions.* PhD thesis, Universidad National Autónoma de Mexico, Mexico.

Parsopoulos, K. E., Plagianakos, V. P., Magoulas, G. D., & Vrahatis, M. N. (2001). Objective function "stretching" to alleviate convergence to local minima. *Nonlinear Analysis, Theory . Methods & Applications, 47*(5), 3419–3424.

Parsopoulos, K. E., & Vrahatis, M. N. (2002a). Recent approaches to global optimization problems through particle swarm optimization. *Natural Computing, 1*(2-3), 235–306. doi:10.1023/A:1016568309421

Parsopoulos, K. E., & Vrahatis, M. N. (2002b). Particle swarm optimization method for constrained optimization problems. In P. Sincak, J. Vascak, V. Kvasnicka, & J. Pospichal (Eds.), *Intelligent Technologies-Theory and Applications: New Trends in Intelligent Technologies (Frontiers in Artificial Intelligence and Applications series, Vol. 76)* (pp. 214-220). IOS Press.

Parsopoulos, K. E., & Vrahatis, M. N. (2004). On the computation of all global minimizers through particle swarm optimization. *IEEE Transactions on Evolutionary Computation, 8*(3), 211–224. doi:10.1109/TEVC.2004.826076

Parsopoulos, K. E., & Vrahatis, M. N. (2005). Unified particle swarm optimization for solving constrained engineering optimization problems. In [LNCS]. *Lecture Notes in Computer Science, 3612*, 582–591.

Shusterman, L. B. (1979). The method of successive elimination in search for the global optimum of multiextremal algebraic functions. *Radioelektronika, 79*(6), 58–63.

Torn, A., & Žilinskas, A. (1989). *Global optimization*. Heidelberg, Germany: Springer.

Vilkov, A. V., Zhidkov, N. P., & Schedrin, B. M. (1975). A method of search for the global minimum of a function of one variable. *Journal of Computational Mathematics and Mathematical Physiscs, 75*, 1040–1043.

Wang, Y.-J., & Zhang, J.-S. (2007). A new constructing auxiliary function method for global optimization. *Mathematical and Computer Modelling, 47*(11-12), 1396–1410. doi:10.1016/j.mcm.2007.08.007

Yang, J.-M., Chen, Y.-P., Horng, J.-T., & Kao, C.-Y. (1997). Applying family competition to evolution strategies for constrained optimization. [Berlin: Springer.]. *Lecture Notes in Computer Science, 1213*, 201–211. doi:10.1007/BFb0014812

Section 2
Applications of Particle Swarm Optimization

Chapter 6
Applications in Machine Learning

This chapter presents the fundamental concepts regarding the application of PSO on machine learning problems. The main objective in such problems is the training of computational models for performing classification and simulation tasks. It is not our intention to provide a literature review of the numerous relative applications. Instead, we aim at providing guidelines for the application and adaptation of PSO on this problem type. To achieve this, we focus on two representative cases, namely the training of artificial neural networks, and learning in fuzzy cognitive maps. In each case, the problem is first defined in a general framework, and then an illustrative example is provided to familiarize readers with the main procedures and possible obstacles that may arise during the optimization process.

INTRODUCTION

Machine learning is the field of artificial intelligence that deals with algorithms that render computational models capable of learning and adapting to their environment. From an abstract viewpoint, machine learning is the procedure of extracting information in the form of patterns or rules from data. This purpose requires the use of computational methods. Human interaction can also be beneficial within a collaborative framework to the algorithms. However, the elimination of this necessity still remains the main challenge in the development of intelligent systems.

There are different types of machine learning procedures, based on the desired outcome as well as on the degree of human intervention:

a. **Supervised learning:** The algorithm builds a mapping between a set of presented input data and a set of desired output. This is possible by altering the parameters of the computational model so that the produced error between input and output is minimized.

DOI: 10.4018/978-1-61520-666-7.ch006

Copyright © 2010, IGI Global. Copying or distributing in print or electronic forms without written permission of IGI Global is prohibited.

b. **Unsupervised learning:** The algorithm tunes the computational model to regularities of the available data without a task-oriented measure of quality. This is possible through competition among the modules of the computational model.

c. **Semi-supervised learning:** This is a combination of the two previous approaches that employs both learning with explicit input-output examples, as well as non-labeled examples.

d. **Reinforcement learning:** The algorithm learns an input-output mapping by continuously interacting with the environment, which in turn admits an impact from every taken step, providing feedback to the model.

Several additional learning subtypes, which are outside the scope of the book at hand, can be distinguished.

Supervised learning constitutes a very prosperous application field for evolutionary algorithms and PSO, due to the existence of explicit performance measures. These measures usually come in the form of objective functions in the parameters of the computational model. Thus, training procedures aim at the detection of parameter values that minimize the model's error in learning a set of presented examples.

In this context, PSO has been applied for training artificial neural networks and fuzzy cognitive maps. The rest of this chapter is dedicated to an overview of these applications.

TRAINING ARTIFICIAL NEURAL NETWORKS WITH PSO

In the following sections, we briefly present the problem of neural network *training* for the most common case of feedforward neural networks. We also present an illustrative example for the logical XOR classification task. Further applications are also reported.

The Multi-Layer Perceptron Model

Artificial neural networks (NNs) are computational models based on the operation of biological neural networks, which constitute the information processing mechanism of the human brain. Their structure is based on the concept of the *artificial neuron*, which resembles biological neurons, as their main processing unit. The artificial neuron constitutes a nonlinear mapping between a set of input and a set of output data. Thus, if the input data are represented as a vector, $Q = (q_1, q_2, ..., q_m)^{\mathrm{T}}$, the artificial neuron implements a function:

$$y = F\left(b_i + \sum_{i=1}^{m} w_i q_i \right),$$

where F is the *transfer function*; w_i, $i = 1, 2, ..., m$, are the *weights*; and b_i is a *bias*. In order to retain a compact notation, we will henceforth represent the bias, b_i, as a weight, w_0, with an auxiliary constant input, $q_0 = 1$.

The training of the neuron to learn an input-output pair, $\{Q, y\}$, is the procedure of detecting proper weights so that the output y is obtained if Q is presented to the neuron. Obviously, this procedure can be modeled as an error minimization problem:

$$\min_{w_i} \left(y - F\left(\sum_{i=0}^{m} w_i q_i \right) \right)^2.$$

If more than one input vectors, $Q_1, Q_2, \ldots, Q_K$, are to be learned by the neuron, then the objective function is augmented with a separate square-error term for each input vector:

$$\min_{w_i} \sum_{k=1}^{K} \left(y_k - F\left(\sum_{i=0}^{m} w_i q_{ki} \right) \right)^2,$$

where q_{ki} is the i-th component of the k-th input vector, $Q_k = (q_{k1}, q_{k2}, \ldots, q_{km})^{\mathrm{T}}$ (recall that w_0 is the bias of the neuron, with $q_{k0} = 1$ for all k). This kind of training procedure is also known as *batch training*, as all input vectors are presented to the neuron prior to any change of its weights.

Obviously, the dimension of the minimization problem is equal to the number, m, of the weights and biases, which is in direct correlation with the dimension of the input vector. The transfer function, $F(x)$, is selected based on the desired properties that shall be attributed to the model. Although linear transfer functions can be used, nonlinear functions equip the model with enhanced classification capabilities. Thus, nonlinear transfer functions are most commonly used, with *sigmoid* functions defined as:

$$F(x, \lambda) = \frac{1}{1 + \exp(-\lambda x)},$$

being the most common choice. In addition, alternative transfer functions have been proposed in the literature (Magoulas & Vrahatis, 2006).

The gathering of artificial neurons, interconnected and structured in layers, produces various NN models. The most popular one is the *multi-layer perceptron*. The first layer in such a structure is called the *input layer*. Its neurons are commissioned simply to forward the input vector to the neurons of the next layer through their weighted interconnections. Thus, they do not perform any computation and, hence, do not posses biases or transfer functions. The output neurons constitute the *output layer*, while all intermediate layers are called *hidden layers*, and their neurons admit biases and transfer functions. Since each layer forwards its output strictly to the neurons of the next one, these NN models are also called *feedforward NNs*. Figure 1 illustrates the structure of such a NN architecture.

Let L be the number of layers, including the input and output layer, and let $N_1, N_2, \ldots, N_L$, be the number of neurons per layer. Then, the number of weights between the l-th and the $(l+1)$-th layer is equal to $N_l(1+N_{l+1})$, including biases. Thus, the total number of weights and biases of the whole NN is equal to:

$$n = N_L - N_1 + \sum_{l=1}^{L-1} N_l(1 + N_{l+1}). \tag{1}$$

Let $\{Q_k, y_k\}$, $k = 1, 2, \ldots, K$, be the training patterns, i.e., input-output pair examples, with $Q_k = (q_{k1}, q_{k2}, \ldots, q_{km})^{\mathrm{T}}$ and $y_k = (y_{k1}, y_{k2}, \ldots, y_{kD})^{\mathrm{T}}$. Obviously, it must hold that $N_1 = m$ and $N_L = D$. Also, let $w_{ij}^{[l]}$ denote the weight of the interconnection between the i-th neuron of the l-th layer with the j-th neuron of the $(l+1)$-th layer, with $l = 1, 2, \ldots, L$, $i = 0, 1, 2, \ldots, N_l$, and $j = 1, 2, \ldots, N_{l+1}$ (the value $i = 0$ stands for the biases of the neurons in the l-th layer for $l = 2, \ldots, L$). Then, the training of the network is equivalent to the following minimization problem (Magoulas & Vrahatis, 2006):

Figure 1. A multi-layer perceptron model

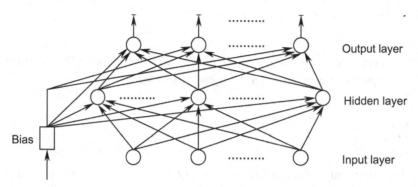

$$\min_{w_{ij}^{[l]}} \sum_{k=1}^{K} \sum_{d=1}^{D} \left(Y_{kd} - y_{kd} \right)^2, \tag{2}$$

where Y_{kd} is the actual output of the *d*-th output neuron when the input vector Q_k is presented to the network, while y_{kd} is the corresponding desired output. This error function is not the only possible choice for the objective function. A variety of distance functions are available in the literature, such as the *Minkowsky, Mahalanobis, Camberra, Chebychev, quadratic, correlation, Kendall's rank corellation* and *chi-square* distance metrics; the *context-similarity measure*; the *contrast model*; *hyperrectangle* distance functions and others (Magoulas & Vrahatis, 2006; Wilson & Martinez, 1997).

The minimization problem defined in equation (2) can be straightforwardly solved using PSO by encoding the network weights as particle components. Thus, each particle constitutes an individual weight configuration, evaluated in the context of equation (2). The underlying optimization problem is high-dimensional and can be highly nonlinear under sigmoid transfer functions. All PSO concepts and properties still hold in the case of NN training, since the formulation of the problem as a problem of global optimization adheres to the general optimization framework that governs all PSO developments discussed in the first part of the book at hand.

Learning the eXclusive OR: A Simple Example

Let us now illustrate the application of PSO in NN training with a simple example, namely the eXclusive OR (XOR) problem. The main goal is to train a NN to learn the logical XOR operation, which is defined with the truth table of Table 1. A properly trained NN shall admit a two dimensional input vector from the set {(0,0), (0,1), (1,0), (1,1)} and produce the correct output reported in Table 1. A NN with a single hidden layer of two neurons and a logistic transfer function are used to achieve this (Haykin, 1999, p. 175).

Setting the problem in the framework of the previous section, there will be three layers, i.e., $L = 3$. The first (input) layer consists of $N_1 = 2$ neurons, since the input vectors are 2-dimensional. The hidden layer consists of $N_2 = 2$ neurons, and the third (output) layer consists of $N_3 = 1$ neuron, since the output can be either "0" or "1". There is a total number of 6 weights in the interconnections among all neurons, plus 3 biases of the neurons in the hidden and output layer. Hence, there is a total of 9 parameters that need to be specified, i.e., the corresponding optimization problem is 9-dimensional, as it is also derived

Table 1. Truth table of the logical XOR operation

A	B	A XOR B
0	0	0
0	1	1
1	0	1
1	1	0

from equation (1). The corresponding objective function for the specific problem can be also given in a closed-form expression, as reported in Parsopoulos and Vrahatis (2002).

PSO can now be applied using a swarm of 9-dimensional particles. Each particle corresponds to a different weight configuration of the NN, and it is evaluated based on equation (2). However, the problem is highly nonlinear with a plethora of local minima as well as flat regions. This can impose difficulties on the detection of the global minimizer to any optimization algorithm. In this case, the stretching technique of Chapter Five can be used.

Parsopoulos and Vrahatis (2002) applied the inertia weight PSO variant on this problem, using a swarm size equal to 80, while the inertia weight was decreased from 1.0 towards 0.4; $c_1 = c_2 = 0.5$; and the weights and biases were initialized randomly and uniformly in the range $[-1,1]^9$. As the global minimum of the objective function is by definition equal to zero, one can identify whether the algorithm has detected the global minimum or not. Parsopoulos and Vrahatis (2002) performed 100 experiments and, if PSO failed to converge to a global minimizer, then stretching was applied and the algorithm continued its operation on the transformed function. The obtained results revealed that stretching was able to increase efficiency of PSO from 77% to 100%, in terms of the number of successful experiments, at the cost of an increased mean number of iterations as reported in Table 2.

Alternatively to stretching, the deflection procedure, which has a more local effect on the objective function, can be used, as presented in Chapter Five. In general, our experience shows that these two techniques can be valuable tools for tackling machine learning problems with PSO.

FURTHER APPLICATIONS

In the XOR problem, as well as in the general framework presented so far in this chapter, human intervention was needed to determine crucial properties of the NN, such as the number of hidden layers and neurons. These requirements can be reduced by encoding the corresponding parameters, $L, N_2, \ldots, N_{L-1}$, as components of the particles of PSO. The number of neurons in the input and output layers, N_1 and N_L, respectively, are excluded since they depend solely on the dimension of the input and output vectors. Thus, the number of hidden layers, neurons, weights, and biases, can all be considered as parameters to be determined by PSO. In this case, special care shall be taken since the produced particles will not necessarily be of the same size.

For example, in the XOR problem, a network with 1 hidden layer and 3 hidden neurons would be modeled with a 15-dimensional particle, instead of the 9-dimensional case of 2 hidden neurons presented in the previous section. This particle would consist of the number of layers, $L = 3$, the number of hidden neurons in the hidden layer, $N_2 = 3$, and 13 weights and biases. Similarly, a network with 2 hidden layers

Table 2. Results for the XOR problem with and without stretching

Statistic	Standard PSO	Stretched cases only	PSO with stretching
Success	77%	23%	100%
Mean	1459.7	29328.6	7869.6
St.D.	1143.1	15504.2	13905.4

and 2 hidden neurons in each, would result in an 18-dimensional particle containing, L, N_2, N_3, and 15 weights and biases. Clearly, if the swarm consists of particles with different dimensions, the standard PSO update equations must be properly modified. This is possible by truncating or augmenting each particle; such approaches are reported in (Binos, 2002; Zhang *et al.*, 2000).

Moreover, instead of feedforward NN, different network types have also been also tackled with different PSO variants. For instance, Ismail and Engelbrecht (2000) used PSO to train product unit NNs, while Van den Bergh and Engelbrecht (2000) applied cooperative PSO for feedforward NNs with promising results. Al-kazemi and Mohan (2002) proposed a multi-phase PSO approach to train feedforward NNs using many swarms that combine different search criteria and hill-climbing. A hybrid approach that combines GAs with PSO for designing recurrent NNs to solve a dynamic plant control problem is proposed by Juang (2004). In this case, PSO and the genetic algorithm produce an equal portion of a population of recurrent NN designs at each iteration, while PSO additionally refines the elite individuals of the population. Finally, very interesting applications for analyzing the levels of pollution in central Hong Kong and the stage prediction of the Shing Mun river are reported in Lu *et al.* (2002) and Chau (2006), respectively.

FUZZY COGNITIVE MAPS LEARNING WITH PSO

Learning in fuzzy cognitive maps constitutes an active research field with many significant applications in industry, engineering, and bioinformatics. In the following sections, we present the latest developments on the application of PSO on this machine learning task.

Fuzzy Cognitive Maps

Fuzzy cognitive maps (FCMs) are simulation models that combine concepts from artificial neural networks and fuzzy logic. The neuro-fuzzy representation equips FCMs with an inherent ability for abstraction in knowledge representation and adaptation, rendering them a very useful tool for modeling and studying complex systems. To date, FCMs have been used in a plethora of applications in diverse scientific fields, including social and organizational systems (Craiger *et al.*, 1996; Taber, 1991, 1994), circuit design (Styblinski & Meyer, 1988), industrial process control (Stylios *et al.*, 1999), supervisory control systems (Groumpos & Stylios, 2000; Stylios & Groumpos, 1998; Stylios *et al.*, 1999), and bioinformatics (Georgopoulos *et al.*, 2003; Parsopoulos *et al.*, 2004).

FCMs were originally introduced by Kosko (1986) as directional graphs with feedback. Their representation is similar to that of causal concept maps, consisting of nodes that represent key concepts

Figure 2. A simple FCM with 5 concepts (nodes) and 7 weights (arcs)

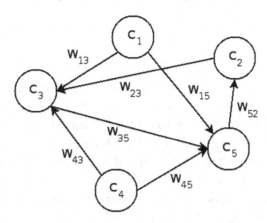

of the simulated system, and links among them that represent their causal relationships. The degree of causality between two concepts is represented with a numerical weight on their interconnecting link. A simple FCM with 5 concepts and 7 weights is depicted in Fig. 2.

To put it formally, let M be the number of concepts (nodes), C_i, $i = 1, 2,..., M$, of the FCM. Each concept assumes a numerical value, $A_i \in [0,1]$, $i = 1, 2,..., M$, that quantifies C_i or its effect. An edge with direction from C_i to another concept, C_j, denotes a causality relationship between them. The link is weighted with a numerical value, $w_{ij} \in [-1,1]$. Positive weights denote positive causality; hence, an increase in the value, A_i, of the concept C_i triggers an increase in the value, A_j, of C_j. On the other hand, negative causality is expressed with negative weights, and an increase in A_i results in decrease of A_j and vice versa. The weights of an FCM can be represented with a matrix:

$$W = \begin{pmatrix} w_{11} & w_{12} & \cdots & w_{1M} \\ w_{21} & w_{22} & \cdots & w_{2M} \\ \vdots & \vdots & \ddots & \vdots \\ w_{M1} & w_{M2} & \cdots & w_{MM} \end{pmatrix},$$

where the i-th row includes the causality relationships between C_i and the rest of the concepts. Zero weights, $w_{ij} = 0$, are used to represent the absence of causality (and therefore interconnection) between C_i and C_j.

The initial setting of an FCM, including both its design and initial values of weights and concepts, is determined by a group of experts with in-depth knowledge of the considered problem (Stylios *et al.*, 1999; Stylios & Groumpos, 2000). In order to avoid the error-prone procedure of assigning directly numerical values on concepts and weights, linguistic modifiers are used and then converted to fuzzy functions. Thus, numerical values are finally assigned through a fuzzification-defuzzification procedure. If needed, the experts can impose strict bounds on these values to retain their physical meaning.

After the initial configuration, the FCM behaves similarly to a discrete dynamical system. Keeping the weights fixed, the concept values are let to converge to a stable state by applying the following iterative update rule (Kosko, 1997; Stylios & Groumpos, 2004):

$$A_i(t+1) = F\left(A_i(t) + \sum_{\substack{k=1 \\ k \neq i}}^{M} w_{ki} A_k(t)\right), \tag{3}$$

where t stands for the iteration counter; $A_i(t+1)$ is the value of concept C_i at iteration $t+1$; $A_k(t)$ is the value of C_k at iteration t; and w_{ki} is the weight of the link from C_k to C_i. The function F is usually a sigmoid, similarly to the case of feedforward neural networks described in the previous sections.

After its convergence, which usually requires a small number of iterations of the equation (3), the FCM shall be capable of simulating the underlying system accurately, and desirable values shall be assumed by the concepts. Unfortunately, this is not always possible. Wide opinion variations among the experts are translated into weight values that are incapable of leading the system to desirable states. In such cases, a *learning algorithm* is needed to modify the weights further within their bounds, so that desirable steady states can be achieved.

There are just a few established learning algorithms for FCMs and they can be classified in two major categories. The first one consists of algorithms based on rules for unsupervised training of artificial neural networks (Kosko, 1997). The second category consists of evolutionary algorithms (Khan *et al.*, 2004; Koulouriotis *et al.*, 2001) and PSO-based approaches (Papageorgiou *et al.*, 2004, 2005; Parsopoulos *et al.*, 2003, 2004; Petalas *et al.*, 2007, 2009). The latter approaches are described in the following section.

A Learning Approach Based on PSO

The learning procedure is similar to some extent with that of neural network training. Let M be the number of concepts, C_i, and $A_i \in [0,1]$, $i = 1, 2,..., M$, be their values. Let also, $M^* \leq M$, be the number of concepts whose output is of interest, i.e., their values are crucial for the operation of the simulated system. These concepts are called *output concepts*, and we denote them as $C_{o1}, C_{o2},..., C_{oM^*}$. Their values, $A_{o1}, A_{o2},..., A_{oM^*}$, are monitored and used as performance criterion for the learning procedure. Hence, the user is interested in detecting a weight matrix, $W = [w_{ij}]$, $i, j = 1, 2,..., M$, so that the converged FCM attains a desirable steady state, while its weights retain their physical meaning.

A desirable steady state for a given weight matrix, W, shall provide output concept values that lie within prespecified bounds:

$$A_{oi}^{[\min]} \leq A_{oi} \leq A_{oi}^{[\max]}, \quad i = 1, 2,..., M^*,$$

considered to be crucial for the proper operation of the simulated system. A proper objective function that guarantees this property shall be defined for the learning procedure. Papageorgiou *et al.* (2005) proposed the following objective function:

$$f(W) = \sum_{i=1}^{M^*} \left[\left|A_{oi}^{[\min]} - A_{oi}\right| \times H\left(A_{oi}^{[\min]} - A_{oi}\right)\right] + \sum_{i=1}^{M^*} \left[\left|A_{oi} - A_{oi}^{[\max]}\right| \times H\left(A_{oi} - A_{oi}^{[\max]}\right)\right], \tag{4}$$

where H is the Heaviside function:

$$H\left(A_{oi}^{[min]} - A_{oi}\right) = \begin{cases} 0, & \text{if } A_{oi}^{[min]} - A_{oi} < 0, \\ 1, & \text{otherwise,} \end{cases} \quad \text{and} \quad H\left(A_{oi} - A_{oi}^{[max]}\right) = \begin{cases} 0, & \text{if } A_{oi} - A_{oi}^{[max]} < 0, \\ 1, & \text{otherwise,} \end{cases}$$

and A_{oi}, $i = 1, 2, \ldots, M^*$, are the steady state values of output concepts obtained by applying the iterative rule of equation (3), using the weight matrix W.

The objective function of equation (4) is actually a penalty function. Its global minimizers are weight matrices that produce output concept values within the prespecified bounds. Any other weight matrix is penalized by an amount proportional to the degree of bound violation. In contrast to gradient-based approaches, the non-differentiability of $f(W)$ does not constitute an obstacle to the application of PSO. Possible further requirements implied by the problem at hand can be easily incorporated in equation (4) to achieve a desirable weight matrix.

PSO is applied straightforwardly using the objective function of equation (4). Each particle of the swarm is a weight matrix, encoded as a vector by considering its rows in turn:

$$W = [\underbrace{w_{12}, \ldots, w_{1M}}_{\text{row 1}}, \underbrace{w_{21}, \ldots, w_{2M}}_{\text{row 2}}, \ldots, \underbrace{w_{M1}, \ldots, w_{M,M-1}}_{\text{row } M}]^{\text{T}}.$$

The elements, w_{11}, $w_{22}, \ldots$, w_{MM}, of the main diagonal of W are omitted in the vectorial representation, as, by definition, FCMs have no self-feedback in their nodes, and therefore the corresponding weights are all equal to zero, $w_{ii} = 0$, $i = 1, 2, \ldots, M$. The dimension of particles for an FCM with M concepts will be at maximum equal to $M \times (M-1)$, although in many applications it is significantly smaller due to sparsity of the weight matrices.

Each particle (weight matrix W) is evaluated with $f(W)$, by using the weight matrix W in the FCM and letting it converge to a steady state to obtain the required concept values involved in equation (4). Subsequently, the PSO update equations are used to produce new weight matrices (particles). Initialization of each weight is performed randomly and uniformly either in [-1,0], if it is specified as negative by the experts, or in [0,1], otherwise. Obviously, there is no restriction regarding the employed PSO variant, while the objective function can be modified accordingly to fit the framework of different problems. A flowchart of the proposed PSO-based learning procedure is depicted in Fig. 3. In the next section, an illustrative example of the learning procedure is provided for an industrial process control problem.

Industrial Process Control: An Illustrative Example

This problem was addressed by Papageorgiou *et al.* (2005), and it is ideal to illustrate the PSO-based learning procedure. The problem, previously considered by Stylios and Groumpos (1998), consists of the simulation of a simple process control problem from industry. The system, which is illustrated in Fig. 4, consists of a tank and three valves, denoted as V1, V2, and V3, which control the amount of liquid in the tank. Valves V1 and V2 pour two liquid chemicals, whose chemical reaction produces a new liquid, into the tank. A sensor (gauger) is sunk into the tank and gauges the specific gravity, G, of the produced liquid. If G attains a value within a desirable range, $[G_{min}, G_{max}]$, then V3 opens and empties the tank. There is also a security limit on the height, T, of the liquid in the tank, which shall not exceed lower and upper bounds, T_{min} and T_{max}, respectively. Therefore, the main goal of this simple control process is the preservation of G and T within the desirable limits:

Figure 3. Flowchart of the PSO-based learning procedure

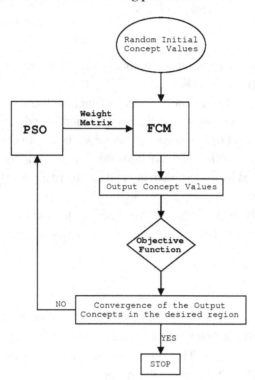

Figure 4. The industrial process control system

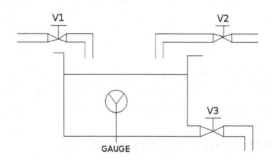

$$G_{min} \leq G \leq G_{max}, \quad T_{min} \leq T \leq T_{max}.$$

A group of experts designed an FCM that simulates the system, following the procedure described in the previous sections. First, they decided on the number of concepts and their interactions, which are described as "negative", "positive", and "no influence", based on the causality relationships among them. The corresponding FCM is depicted in Fig. 5 and consists of the following five concepts:

1. **Concept C_1:** Height of liquid in tank. It depends on the state of valves V1, V2, and V3.
2. **Concept C_2:** State of valve V1 (open, closed, or partially open).

Figure 5. The FCM that simulates the system of Fig. 4

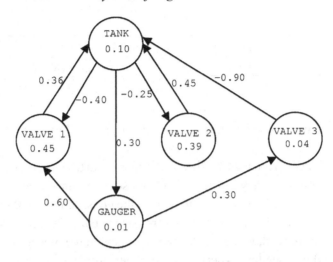

3. **Concept C_3:** State of valve V2 (open, closed, or partially open).
4. **Concept C_4:** State of valve V3 (open, closed, or partially open).
5. **Concept C_5:** Specific gravity of the produced liquid.

The concepts are connected with eight links. A consensus among experts was attained regarding the directions of the links. Also, the initial weights were determined by assigning linguistic variables, such as "weak", "strong" etc., along with the corresponding fuzzy sets, as defined in Cox (1999). The linguistic variables are combined to a single linguistic weight using the widely used SUM technique (Lin & Lee, 1996). This weight is transformed to a numerical value using the *center of area* defuzzification method (Kosko, 1992; Lin & Lee, 1996).

All experts agreed on the same range for the weights w_{21}, w_{31} and w_{41}, while most of them agreed for w_{12} and w_{13} (Papageorgiou *et al.*, 2005). However, no agreement was attained for w_{15}, w_{52}, and w_{54}, where their opinions varied significantly. The ranges as implied by the fuzzy regions of the weights are reported in Table 3.

PSO was applied on the corresponding 8-dimensional optimization problem for the detection of a weight setting that retains the output concepts, C_1 and C_5, within the following bounds assigned by the experts:

$$0.68 \leq A_1 \leq 0.70, \quad 0.78 \leq A_5 \leq 0.85. \tag{5}$$

Thus, the objective function of equation (4) becomes:

$$f(W) = |0.68 - A_1| H(0.68 - A_1) + |A_1 - 0.70| H(A_1 - 0.70) +$$
$$|0.78 - A_5| H(0.78 - A_5) + |A_5 - 0.85| H(A_5 - 0.85).$$

Table 3. Ranges of the weights as implied by their fuzzy regions for the industrial process control problem of Fig. 4

$-0.5 \leq w_{12} \leq -0.3,$	$0.2 \leq w_{15} \leq 0.4,$	$0.4 \leq w_{31} \leq 0.5,$	$0.5 \leq w_{52} \leq 0.7,$
$-0.4 \leq w_{13} \leq -0.2,$	$0.3 \leq w_{21} \leq 0.4,$	$-1.0 \leq w_{41} \leq -0.8,$	$0.2 \leq w_{54} \leq 0.4,$

Additionally, each weight was constrained in the range [-1,0] or [0,1] to avoid physically meaningless weights.

Papageorgiou *et al.* (2005) considered two different scenarios. In the first, they admitted all constraints posed by the experts. In the second, they considered only those constraints where a unanimous agreement among the experts was achieved. In both cases they used the constriction coefficient PSO variant with ring neighborhood topology of radius $r = 3$, and a swarm size equal to 20. Also, the default parameters, $\chi = 0.729$, $c_1 = c_2 = 2.05$, were used, while the desired accuracy on the objective function value was set to 10^{-8}. For each case, 100 independent experiments were performed and their results were statistically analyzed. The most important observations and conclusions are reported in the following sections.

Scenario One: All Constraints are Retained

In the first set of experiments, all constraints defined by the experts in Table 3 were used. However, as reported in Papageorgiou *et al.* (2005), no solution that fulfills the relations (5) was found in 100 experiments. This is a strong indication that the provided weight bounds are not proper and, consequently, they cannot lead the FCM to a desirable steady state. The best of the obtained weight matrices was the following:

$$W^* = \begin{pmatrix} 0.00 & -0.35 & -0.20 & 0.00 & 0.40 \\ 0.40 & 0.00 & 0.00 & 0.00 & 0.00 \\ 0.50 & 0.00 & 0.00 & 0.00 & 0.00 \\ -0.80 & 0.00 & 0.00 & 0.00 & 0.00 \\ 0.00 & 0.75 & 0.00 & 0.20 & 0.00 \end{pmatrix},$$

which leads the FCM to the following steady state:

$$A_1 = 0.6723, \quad A_2 = 0.7417, \quad A_3 = 0.6188, \quad A_4 = 0.6997, \quad A_5 = 0.7311.$$

This steady state clearly violates both constraints in relation (5).

Assuming that the constraints for the weights w_{15}, w_{52}, and w_{54}, for which the opinions of experts varied significantly, were the reason for the disability of the algorithm to achieve a solution, Papageorgiou *et al.* (2005) attempted to solve the problem by omitting them. Thus, they experimented by omitting the constraints, initially one by one, and subsequently in pairs. Despite their effort, again, no solution was detected.

Finally, they omitted all three constraints, allowing the three weights to assume values in the whole range [0,1] (they all have positive signs). In this case, proper weight matrices were obtained, although in substantially different ranges than those determined by the experts (Papageorgiou *et al.*, 2005). Thus, the algorithm was able to correct the inconsistencies of the experts in a very efficient manner, requiring at most 620 function evaluations to find a solution with the desirable accuracy. Of course, due to the interactions among weights and concepts, there is a multitude of different optimal matrices. Indicatively, we report one of the obtained optimal matrices:

$$W^* = \begin{pmatrix} 0.00 & -0.45 & -0.20 & 0.00 & 0.84 \\ 0.40 & 0.00 & 0.00 & 0.00 & 0.00 \\ 0.50 & 0.00 & 0.00 & 0.00 & 0.00 \\ -0.80 & 0.00 & 0.00 & 0.00 & 0.00 \\ 0.00 & 0.99 & 0.00 & 0.10 & 0.00 \end{pmatrix},$$

which leads the FCM to the desirable steady state:

$$A_1 = 0.6805, \quad A_2 = 0.7798, \quad A_3 = 0.6176, \quad A_4 = 0.6816, \quad A_5 = 0.7967.$$

Scenario Two: Only Unanimously Agreed Constraints are Retained

In the second scenario, Papageorgiou *et al.* (2005) retained only the unanimously agreed constraints, i.e., those of the weights w_{21}, w_{31}, and w_{41}. The rest of the weights were let to move unrestricted in [-1,0] or [0,1], depending on their sign.

In this case, it was observed that the three constrained weights assumed values in remarkably narrower ranges than those proposed by the experts. This implies that PSO can be also used for the further refinement of the assigned bounds. Numerous experiments were performed with the three weights being fixed or moving in their corresponding ranges (Papageorgiou *et al.*, 2005). Some of the obtained optimal weight matrices were the following:

$$W^* = \begin{pmatrix} 0.00 & -0.44 & -0.10 & 0.00 & 1.00 \\ 0.40 & 0.00 & 0.00 & 0.00 & 0.00 \\ 0.50 & 0.00 & 0.00 & 0.00 & 0.00 \\ -0.81 & 0.00 & 0.00 & 0.00 & 0.00 \\ 0.00 & 1.00 & 0.00 & 0.13 & 0.00 \end{pmatrix},$$

which leads the FCM to the steady state:

$$A_1 = 0.6805, \quad A_2 = 0.7872, \quad A_3 = 0.6390, \quad A_4 = 0.6898, \quad A_5 = 0.8172;$$

$$
W^* = \begin{pmatrix}
0.00 & -0.27 & -0.20 & 0.00 & 1.00 \\
0.40 & 0.00 & 0.00 & 0.00 & 0.00 \\
0.50 & 0.00 & 0.00 & 0.00 & 0.00 \\
-0.81 & 0.00 & 0.00 & 0.00 & 0.00 \\
0.00 & 1.00 & 0.00 & 0.10 & 0.00
\end{pmatrix},
$$

which leads the FCM to the steady state:

$$A_1 = 0.6816, \quad A_2 = 0.8090, \quad A_3 = 0.6174, \quad A_4 = 0.6822, \quad A_5 = 0.8174;$$

and

$$
W^* = \begin{pmatrix}
0.00 & -0.23 & -0.13 & 0.00 & 0.86 \\
0.40 & 0.00 & 0.00 & 0.00 & 0.00 \\
0.50 & 0.00 & 0.00 & 0.00 & 0.00 \\
-0.81 & 0.00 & 0.00 & 0.00 & 0.00 \\
0.00 & 0.92 & 0.00 & 0.12 & 0.00
\end{pmatrix},
$$

which leads the FCM to the steady state:

$$A_1 = 0.6817, \quad A_2 = 0.7985, \quad A_3 = 0.6323, \quad A_4 = 0.6860, \quad A_5 = 0.8007.$$

Although different, all the obtained weight matrices lead the FCM to desirably steady states. The weights w_{13}, w_{15}, w_{52}, and w_{54}, converged to regions significantly different than those suggested by the experts, while w_{21}, w_{31}, and w_{41}, remained almost fixed at values close to the initial ones suggested by the experts. Also, the weight w_{12} deviates slightly from its initial region.

These interesting observations constitute a small fraction of the information derived from the application of the PSO-based learning procedure on the industrial process control problem. However it suffices to reveal the potential of the PSO-based learning algorithm to significantly increase our intuition on the problem dynamics, and help towards the further refinement of expert knowledge. Moreover, it can provide robust solutions in cases where the experts disagree or doubt on their decisions. The next section briefly reports further developments and applications of PSO-based learning procedures.

Further Developments on PSO-Based Learning Algorithms

The PSO-based learning procedure was introduced in Parsopoulos *et al.* (2003) and further evaluated in Papageorgiou *et al.* (2005) for the industrial process control problem presented in the previous section. Its efficiency and apparent practical value prompted its further use in different applications, such as modeling of radiation therapy systems (Parsopoulos *et al.*, 2004), and more complex industrial process control problems (Papageorgiou *et al.*, 2004).

Concurrently, different PSO variants were assessed on the FCM learning problem. One of the most efficient approaches was recently proposed by Petalas *et al.* (2007, 2009). This approach utilizes the memetic PSO (MPSO) algorithm presented in Chapter Four. More specifically, MPSO approaches, equipped with either the Hook and Jeeves or the Solis and Wets local search, were investigated on interesting FCM learning problems, including a complex industrial process control problem, the simulation of a radiation therapy system, a heat exchanger problem, and the simulation of an ecological industrial park problem (Petalas *et al.*, 2009). Moreover, MPSO was compared favorably against the DE and GA algorithm on these problems. The reader is referred to the original paper for the complete presentation of these applications. Besides the form of the equation (4), different objective functions can be used, depending on the problem at hand. In the following, we briefly describe the radiation therapy problem, where such an objective function is used.

Radiotherapy is a popular means of cancer treatment. It is a complex process that involves a large number of treatment variables. The main objective of radiotherapy is the delivery of the highest possible amount of radiation to the tumor, while minimizing the exposure of healthy tissue and critical organs to the radiation. Hence, treatment planning and doctor-computer interaction is required prior to the actual final treatment (Parsopoulos *et al.*, 2004; Petalas *et al.*, 2009).

The radiation therapy process is modeled by a supervisor-FCM, which is constructed by experts and consists of the following six concepts:

1. **Concept C_1:** Tumor localization.
2. **Concept C_2:** Dose prescribed for the treatment planning.
3. **Concept C_3:** Machine-related factors.
4. **Concept C_4:** Human-related factors.
5. **Concept C_5:** Patient positioning and immobilization.
6. **Concept C_6:** Final dose received by the targeted tumor.

The corresponding FCM is depicted in Fig. 6.

The main objective in this problem is the maximization of the final dose received by the tumor, which is described by C_6, as well as the maximization of the dose, C_2, prescribed by the treatment planning as defined by the AAPM and ICRP protocols for the determination of acceptable dose per organ and part of the human body (Khan, 1994; Wells & Niederer, 1998; Willoughby *et al.*, 1996). Thus, the objective function is modeled so that the values of the two aforementioned concepts are maximized, instead of restricting them in bounds as in equation (4), and it is defined as:

$$f(W) = -A_2 - A_6,$$

where the negative sign is used to transform the maximization of the positive values A_2 and A_6 to an equivalent minimization problem.

This objective function was easily addressed with PSO as reported in Parsopoulos *et al.* (2004). In their experiments, both the inertia weight and the constriction coefficient PSO variant were investigated and compared against two DE variants. The standard parameter setting, $\chi = 0.729$, $c_1 = c_2 = 2.05$, and a decreasing inertia weight from 1.2 to 0.1 were employed, while the desirable bounds for the output concepts were determined by the experts as follows:

$$0.80 \leq A_2 \leq 0.95, \, 0.90 \leq A_6 \leq 0.95.$$

Figure 6. The supervisor-FCM for the radiotherapy problem

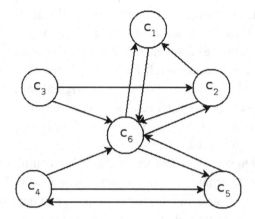

All algorithms detected the same solution matrix (Parsopoulos *et al.*, 2004):

$$W^* = \begin{pmatrix} 0.0 & 0.0 & 0.0 & 0.0 & 0.0 & 0.5 \\ 0.4 & 0.0 & 0.0 & 0.0 & 0.0 & 0.7 \\ 0.0 & -0.2 & 0.0 & 0.0 & 0.0 & -0.1 \\ 0.0 & 0.0 & 0.0 & 0.0 & -0.2 & -0.2 \\ 0.0 & 0.0 & 0.0 & -0.6 & 0.0 & 0.8 \\ 0.4 & 0.9 & 0.0 & 0.0 & 0.9 & 0.0 \end{pmatrix},$$

which leads the FCM to the steady state:

$$A_1 = 0.819643, \quad A_2 = 0.819398, \quad A_3 = 0.659046, \quad A_4 = 0.501709, \quad A_5 = 0.824788, \quad A_6 = 0.916315.$$

An interesting point is the high positive influence of concept C_6 (final dose) on concept C_2 (dose prescribed from the treatment planning), as well as on concept C_5 (patient positioning). This means that if we succeed in delivering the maximum dose to the target volume, then the initial calculated dose from treatment planning is the desired; the same happens with patient positioning (Parsopoulos *et al.*, 2004).

The convenient manipulation and easy adaptation of the PSO-based learning scheme to different problems renders it a valuable tool in machine learning tasks. Using these developments as a starting point, new simulation models were also introduced. Such a model is the *interval cognitive map* introduced by Petalas *et al.* (2005), which is based on modeling the concepts and weights with interval numbers. The presentation of the interval arithmetic methodology for learning in interval FCM models is outside the scope of the book at hand, and thus omitted.

CHAPTER SYNOPSIS

Fundamental concepts of applying PSO on machine learning problems were presented and analyzed. The problems were restricted to the representative cases of training feedforward artificial neural networks and learning in fuzzy cognitive maps. For each case, an illustrative example was used to introduce the reader to the workings of the corresponding PSO-based approach. Thus, the training of a feedforward neural network with one hidden layer for learning the XOR logical operation, as well as the simulation of an industrial process control problem with a fuzzy cognitive map, were analyzed. Moreover, elements for further acquisition were provided in the reported literature, along with brief references to relative applications.

REFERENCES

Al-kazemi, B., & Mohan, C. K. (2002). Training feedforward neural networks using multi-phase particle swarm optimization. In L. Wang, J.C. Rajapakse, K. Fukushima, S.-Y. Lee & X. Yao (Eds.), *Proceedings of the 9th International Conference on Neural Information Processing (ICONIP'02), Singapore* (pp. 2615-2619).

Binos, T. (2002). *Evolving neural network architecture and weights using an evolutionary algorithm.* MSc thesis, RMIT University, Melbourne, Australia.

Chau, K. W. (2006). Particle swarm optimization training algorithm for ANNs in stage prediction of Shing Mun river. *Journal of Hydrology (Amsterdam)*, *329*, 363–367. doi:10.1016/j.jhydrol.2006.02.025

Cox, E. (1999). *The fuzzy systems handbook.* Cambridge, MA: Academic Press.

Craiger, J. P., Goodman, D. F., Weiss, R. J., & Butler, A. (1996). Modeling organizational behavior with fuzzy cognitive maps. *International Journal of Computation Intelligent Organization*, *1*, 120–123.

Georgopoulos, V., Malandraki, G., & Stylios, C. (2003). A fuzzy cognitive map approach to differential diagnosis of specific language impairment. *Journal of Artificial Intelligence in Medicine*, *29*(3), 261–278. doi:10.1016/S0933-3657(02)00076-3

Groumpos, P. P., & Stylios, C. D. (2000). Modelling supervisory control systems using fuzzy cognitive maps. *Chaos, Solitons, and Fractals*, *11*, 329–336. doi:10.1016/S0960-0779(98)00303-8

Haykin, S. (1999). *Neural networks: a comprehensive foundation.* Upper Saddle River, NJ: Prentice Hall.

Ismail, A., & Engelbrecht, A. P. (2000). Global optimization algorithms for training product unit neural networks. In *Proceedings of the 2000 IEEE International Joint Conference on Neural Networks (IJCNN'00), Como, Italy* (pp. 132-137).

Juang, C.-F. (2004). A hybrid of genetic algorithm and particle swarm optimization for recurrent network design. *IEEE Transactions on Systems, Man, and Cybernetics. Part B, Cybernetics*, *34*(2), 997–1006. doi:10.1109/TSMCB.2003.818557

Khan, F. (1994). *The physics of radiation therapy.* Baltimore: Williams & Wilkins.

Khan, M. S., Khor, S., & Chong, A. (2004). Fuzzy cognitive maps with genetic algorithm for goal-oriented decision support. *International Journal of Uncertainty . Fuzziness and Knowledge-Based Systems*, *12*, 31–42. doi:10.1142/S0218488504003028

Kosko, B. (1986). Fuzzy cognitive maps. *International Journal of Man-Machine Studies, 24*, 65–75. doi:10.1016/S0020-7373(86)80040-2

Kosko, B. (1992). *Neural networks and fuzzy systems*. Upper Saddle River, NJ: Prentice Hall.

Kosko, B. (1997). *Fuzzy engineering*. Upper Saddle River, NJ: Prentice Hall.

Koulouriotis, D. E., Diakoulakis, I. E., & Emiris, D. M. (2001). Learning fuzzy cognitive maps using evolution strategies: a novel schema for modeling and simulating high-level behavior. In *Proceedings of the 2001 IEEE Congress on Evolutionary Computation (CEC'01), Seoul, Korea* (pp. 364-371).

Lin, C. T., & Lee, C. S. (1996). *Neural fuzzy systems: a neuro-fuzzy synergism to intelligent systems*. Upper Saddle River, NJ: Prentice Hall.

Lu, W. Z., Fan, H. Y., Leung, A. Y. T., & Wong, J. C. K. (2002). Analysis of pollutant levels in central Hong Kong applying neural network method with particle swarm optimization. *Environmental Monitoring and Assessment, 79*, 217–230. doi:10.1023/A:1020274409612

Magoulas, G. D., & Vrahatis, M. N. (2006). Adaptive algorithms for neural network supervised learning: a deterministic optimization approach. *International Journal of Bifurcation and Chaos in Applied Sciences and Engineering, 16*(7), 1929–1950. doi:10.1142/S0218127406015805

Papageorgiou, E. I., Parsopoulos, K. E., Groumpos, P. P., & Vrahatis, M. N. (2004). Fuzzy cognitive maps learning through swarm intelligence. In *Lecture Notes in Artificial Intelligence (LNAI), Vol. 3070* (pp. 344-349). Berlin: Springer.

Papageorgiou, E. I., Parsopoulos, K. E., Stylios, C. D., Groumpos, P. P., & Vrahatis, M. N. (2005). Fuzzy cognitive maps learning using particle swarm optimization. *Journal of Intelligent Information Systems, 25*(1), 95–121. doi:10.1007/s10844-005-0864-9

Parsopoulos, K. E., Papageorgiou, E. I., Groumpos, P. P., & Vrahatis, M. N. (2003). A first study of fuzzy cognitive maps learning using particle swarm optimization. In *Proceedings of the 2003 IEEE Congress on Evolutionary Computation (CEC'03), Canberra, Australia* (pp. 1440-1447).

Parsopoulos, K. E., Papageorgiou, E. I., Groumpos, P. P., & Vrahatis, M. N. (2004). Evolutionary computation techniques for optimizing fuzzy cognitive maps in radiation therapy systems. In [LNCS]. *Lecture Notes in Computer Science, 3102*, 402–413.

Parsopoulos, K. E., & Vrahatis, M. N. (2002). Recent approaches to global optimization problems through particle swarm optimization. *Natural Computing, 1*(2-3), 235–306. doi:10.1023/A:1016568309421

Petalas, Y. G., Papageorgiou, E. I., Parsopoulos, K. E., Groumpos, P. P., & Vrahatis, M. N. (2005). Interval cognitive maps. In *Proceedings of the International Conference of Numerical Analysis and Applied Mathematics (ICNAAM 2005), Rhodes, Greece* (pp. 882-885).

Petalas, Y. G., Parsopoulos, K. E., Papageorgiou, E. I., Groumpos, P. P., & Vrahatis, M. N. (2007). Enhanced learning in fuzzy simulation models using memetic particle swarm optimization. In *Proceedings of the 2007 IEEE Swarm Intelligence Symposium (SIS'07)* (pp. 16-22).: Washington, D.C.: IEEE.

Petalas, Y. G., Parsopoulos, K. E., & Vrahatis, M. N. (2009). Improving fuzzy cognitive maps learning through memetic particle swarm optimization. *Soft Computing, 13*(1), 77–94. doi:10.1007/s00500-008-0311-2

Styblinski, M. A., & Meyer, B. D. (1988). Fuzzy cognitive maps, signal flow graphs, and qualitative circuit analysis. In *Proceedings of 2ⁿᵈ IEEE International Conference on Neural Networks, San Diego (CA), USA* (pp. 549-556).

Stylios, C. D., Georgopoulos, V., & Groumpos, P. P. (1999). Fuzzy cognitive map approach to process control systems. *Journal of Advanced Computational Intelligence and Intelligent Informatics, 3*(5), 409–417.

Stylios, C. D., & Groumpos, P. P. (1998). The challenge of modelling supervisory systems using fuzzy cognitive maps. *Journal of Intelligent Manufacturing, 9*, 339–345. doi:10.1023/A:1008978809938

Stylios, C. D., & Groumpos, P. P. (2000). Fuzzy cognitive maps in modeling supervisory control systems. *Journal of Intelligent & Fuzzy Systems: Applications in Engineering and Technology, 8*(2), 83–98.

Stylios, C. D., & Groumpos, P. P. (2004). Modeling complex systems using fuzzy cognitive maps. *IEEE Transactions on Systems . Man and Cybernetics: Part A, 34*(1), 159–165.

Taber, R. (1991). Knowledge processing with fuzzy cognitive maps. *Expert Systems with Applications, 2*, 83–87. doi:10.1016/0957-4174(91)90136-3

Taber, R. (1994). Fuzzy cognitive maps model social systems. *AI Expert, 9*, 8–23.

Van den Bergh, F., & Engelbrecht, A. P. (2000). Cooperative learning in neural networks using particle swarm optimizers. *South African Computer Journal, 26*, 84–90.

Wells, D., & Niederer, J. (1998). A medical expert system approach using artificial neural networks for standardized treatment planning. *International Journal of Radiation Oncology, Biology, Physics, 41*(1), 173–182. doi:10.1016/S0360-3016(98)00035-2

Willoughby, T., Starkschall, G., Janjan, N., & Rosen, I. (1996). Evaluation and scoring of radiotherapy treatment plans using an artificial neural network. *International Journal of Radiation Oncology, Biology, Physics, 34*(4), 923–930. doi:10.1016/0360-3016(95)02120-5

Wilson, D. R., & Martinez, T. R. (1997). Improved heterogeneous distance functions. *Journal of Artificial Intelligence Research, 6*, 1–34.

Zhang, C., Shao, H., & Li, Y. (2000). Particle swarm optimisation for evolving artificial neural network. In *Proceedings of the 2000 IEEE International Conference on Systems, Man, Cybernetics, Nashville (TN), USA* (pp. 2487-2490).

Chapter 7
Applications in Dynamical Systems

This chapter is devoted to the application of PSO in dynamical systems. The core subject of the chapter is the problem of detecting periodic orbits of nonlinear mappings. This problem is very interesting and significant, as the study of periodic orbits can reveal several crucial properties of a dynamical system. Traditional root-finding algorithms, such as the Newton-family methods, are widely applied on such problems. However, obstacles arise as soon as non-differentiable or discontinuous mappings come under investigation. In such cases, PSO has been shown to be a very useful and efficient alternative. The chapter aims at presenting fundamental ideas and specific application issues. We thoroughly discuss the transformation of the original problem to a corresponding global optimization task. The application of the deflection technique, presented in Chapter Five, for computing several periodic orbits is analyzed and the algorithm is illustrated on well known benchmark problems. Finally, we present and discuss a very significant application, i.e., the detection of periodic orbits in 3-dimensional galactic potentials.

INTRODUCTION

Nonlinear mappings are widely used to model conservative or dissipative dynamical systems (Birkhoff, 1917; Bountis & Helleman, 1981; Greene, 1979; Polymilis *et al.*, 1997, 2000, 2003; Skokos, 2001a; Skokos *et al.*, 1997; Vrahatis, 1995; Vrahatis & Bountis, 1994; Vrahatis *et al.*, 1993, 1996, 1997; Verhulst, 1990). Points that are invariant under the mapping possess a central role in its analysis. Such points are called *fixed points* or *periodic orbits* of the mapping (Verhulst, 1990). To put it formally, let $\Phi(x)$ be a nonlinear mapping, defined as:

$$\Phi(x) = (\Phi_1(x), \Phi_2(x),\ldots, \Phi_n(x))^{\mathrm{T}}: \mathbf{R}^n \to \mathbf{R}^n. \tag{1}$$

DOI: 10.4018/978-1-61520-666-7.ch007

Copyright © 2010, IGI Global. Copying or distributing in print or electronic forms without written permission of IGI Global is prohibited.

Then, an *n*-dimensional point:

$$x = (x_1, x_2, \ldots, x_n)^{\mathrm{T}} \in \mathbf{R}^n,$$

is called a *fixed point* of $\Phi(x)$, if it holds that:

$$\Phi(x) = x,$$

i.e., the image of *x* through the mapping $\Phi(x)$ remains equal to *x*. In addition, *x* is called a *fixed point of order p* or *periodic orbit of period p* of $\Phi(x)$, if it holds that:

$$\Phi^p(x) \equiv \underbrace{\Phi(\Phi(\cdots \Phi(x) \cdots))}_{p \text{ times}} = x, \tag{2}$$

i.e., the image of *x* after *p* subsequent applications of the mapping $\Phi(x)$ is equal to *x*.

The numerical detection of periodic orbits is a very challenging task, as analytical expressions are rarely available and only for polynomial mappings of low degree and low periods (Helleman, 1977; Hénon, 1969). Classical efficient methods constitute a typical tool for tackling the problem numerically. However, these algorithms often fail; a failure attributed to the nonexistence (or poor behavior) of partial derivatives in the neighborhood of the fixed points or to their sensitivity to large values assumed by the mapping in the neighborhood of saddle-hyperbolic periodic orbits, which are unstable in the linear approximation.

These pathological cases can be remedied by using different methods, such as topological degree-based methods, named as *characteristic bisection methods* (or *sign methods*) (Vrahatis, 1995; Vrahatis & Bountis, 1994; Vrahatis *et al.*, 1993, 1996, 1997). These methods require only the accurate knowledge of the sign of the components of the considered function. Furthermore, evolutionary computation and swarm intelligence offer an abundance of alternative tools. Parsopoulos and Vrahatis (2003, 2004) offered the first studies of PSO on the detection of periodic orbits of nonlinear mappings. In parallel, they combined PSO with the deflection technique, described in Chapter Five, introducing a scheme capable of detecting multiple periodic orbits of the same period. Their results on widely used benchmark problems justified the efficiency of this approach, rendering it a valuable tool.

An important application of this methodology was published by Skokos *et al.* (2005), where the aforementioned PSO approach was applied for the detection of periodic orbits in 3-dimensional galactic potentials. The obtained results were very promising, enriching the available knowledge on the studied models by detecting new and unknown periodic orbits. The methodology was further improved with the introduction of an entropy-based memetic PSO approach by Petalas *et al.* (2007). All these advances are described in detail in the following sections.

DETECTION OF PERIODIC ORBITS OF NONLINEAR MAPPINGS USING PSO

The application of PSO for the detection of periodic orbits of nonlinear mappings requires the development of a proper optimization framework. This framework is needed to transform the nonlinear system-solving problem of equation (2) to a proper minimization task, suitable for the application of

PSO. This can be accomplished by proper mathematical manipulations of equation (2) as described in the following paragraphs.

Let $\Phi(x)$ be an n-dimensional nonlinear mapping defined as in equation (1) and $O_n = (0, 0,\ldots, 0)^T$ be the origin of the n-dimensional Euclidean space, $\mathbf{R}^n$. Also, let p be an integer denoting the desired period. Then, by definition, we can derive the following:

$$\Phi^p(x) = x \Rightarrow \Phi^p(x) - x = O_n \Rightarrow \begin{pmatrix} \Phi_1^p(x) \\ \Phi_2^p(x) \\ \vdots \\ \Phi_n^p(x) \end{pmatrix} - \begin{pmatrix} x_1 \\ x_2 \\ \vdots \\ x_n \end{pmatrix} = \begin{pmatrix} 0 \\ 0 \\ \vdots \\ 0 \end{pmatrix} \Rightarrow \begin{cases} \Phi_1^p(x) - x_1 = 0, \\ \Phi_2^p(x) - x_2 = 0, \\ \vdots \\ \Phi_n^p(x) - x_n = 0. \end{cases} \tag{3}$$

All solutions of this nonlinear system are also periodic orbits of period p of $\Phi(x)$. We can now define the following objective function:

$$f(x) = \sum_{i=1}^{n} (\Phi_i^p(x) - x_i)^2. \tag{4}$$

This function is non-negative and its global minimizers are also solutions of the nonlinear system defined by equation (3), i.e., they are periodic orbits of period p of $\Phi(x)$. Therefore, the original problem is transformed to the equivalent global optimization task of minimizing the objective function $f(x)$, which measures the distance of a point, x, from its image, $\Phi^p(x)$, using the ℓ_2-norm (square-error). The selection of this norm was based mainly on its wide applicability with gradient-based methods. Of course, different and probably nondifferentiable norms, such as ℓ_1 and ℓ_∞, can be alternatively used with PSO. Moreover, if x^* is a periodic orbit of period p, then the rest $(p-1)$ periodic orbits of the same period and type can be obtained with $(p-1)$ subsequent iterations of the mapping on x^*, while its stability can be determined through established techniques (Howard & Mackay, 1987; Skokos, 2001b).

The properties of the objective function defined in equation (4) are heavily dependent on the form of the mapping. Continuous and differentiable mappings produce objective functions that can be easily minimized using gradient-based methods. However, in many applications, discontinuities are introduced in the mappings, resulting in objective functions that lack nice mathematical properties. Moreover, even in continuous and differentiable cases, the shape of the objective function may include very narrow valleys, and flat or very steep regions, resulting in a declining efficiency of the traditional optimization algorithms. In such cases, where gradient-based approaches have controversial performance, stochastic optimization algorithms such as PSO can be very useful.

Let us illustrate the transformation of the original problem to a corresponding optimization one with a few examples of widely used nonlinear mappings. For this purpose, we will use the 2-dimensional Hénon, the standard, and the Gingerbreadman mappings, defined as TP_{NM-1}, TP_{NM-2}, and TP_{NM-3}, respectively, in Appendix A of the book at hand. TP_{NM-1} is defined as:

$$\Phi(x) = \begin{pmatrix} \cos a & -\sin a \\ \sin a & \cos a \end{pmatrix} \begin{pmatrix} x_1 \\ x_2 - x_1^2 \end{pmatrix} \Leftrightarrow \begin{cases} \Phi_1(x) = x_1 \cos a - (x_2 - x_1^2)\sin a, \\ \Phi_2(x) = x_1 \sin a - (x_2 - x_1^2)\cos a, \end{cases}$$

Figure 1. Phase plot (left) of TP$_{NM-1}$ (2-dimensional Hénon mapping) for cosa = 0.24 and contour plot (right) of the corresponding objective function, as defined in equation (4), for period p = 5. Darker contour lines denote lower function values

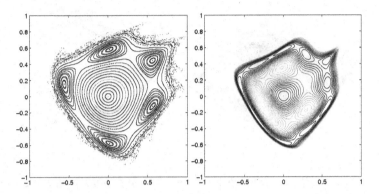

Figure 2. Phase plot (left) of TP$_{NM-1}$ (2-dimensional Hénon mapping) for cosa = 0.8 and contour plot (right) of the corresponding objective function, as defined in equation (4), for period p = 1. Darker contour lines denote lower function values

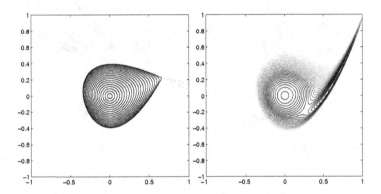

where $a \in [0,\pi]$ is the rotation angle. Following the definition of equation (4), the corresponding objective function for period $p = 1$ is defined as:

$$f(x) = \sum_{i=1}^{2} (\Phi_i^1(x) - x_i)^2 = \left((x_1 \cos a - (x_2 - x_1^2)\sin a) - x_1 \right)^2 + \left((x_1 \sin a - (x_2 - x_1^2)\cos a) - x_2 \right)^2.$$

Figure 1 depicts the phase plot (left) of TP$_{NM-1}$ for cos$a$ = 0.24 and the contour plot (right) of the corresponding objective function defined in equation (4), for period $p = 5$, with darker lines denoting lower function values. The same is depicted in Fig. 2 for cos$a$ = 0.8 and period $p = 1$. Figure 3 depicts both a 3-dimensional plot (left) and the contour plot (right) of the corresponding objective function of TP$_{NM-2}$ for period $p = 5$. Finally, Fig. 4 depicts the phase plot (left) of TP$_{NM-3}$ and the corresponding contour plot (right) for period $p = 5$.

Clearly, in all cases, the produced landscape depends on the mapping's nonlinearity. For example, TP$_{NM-2}$ (Fig. 3) produces a rougher landscape than TP$_{NM-1}$ (Figs. 1 and 2); this is due to discontinuities

Figure 3. The 3-dimensional plot (left) of the objective function produced for TP$_{NM-2}$ (Standard map) and its contour plot (right) for period p = 5. Darker contour lines denote lower function values

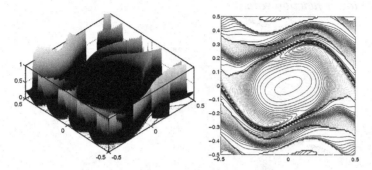

Figure 4. Phase plot (left) of TP$_{NM-3}$ (Gingerbreadman map) and contour plot (right) of the corresponding objective function, as defined in equation (4), for period p = 5. Darker contour lines denote lower function values

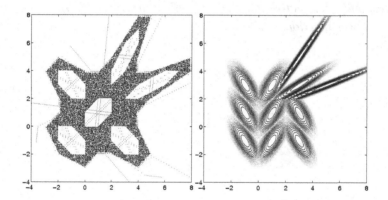

introduced by its modular form. Obviously, in this landscape a gradient-based algorithm would have questionable performance. Even in the case of TP$_{NM-1}$ for cos$a$ = 0.8 and period p = 1, a gradient-based approach is expected to easily attain the periodic orbit that lies at the origin. However, it will face serious problems with the second periodic orbit that lies in the very narrow valley on the right side of the contour plot, which, in Fig. 2, is shown as an upwards tail. Moreover, in the case of TP$_{NM-3}$ (Fig. 4), initial conditions will be a crucial factor for the performance of traditional algorithms due to the high number of global minimizers.

Obviously, periodic orbits of period p are also orbits of any other period p' = kp, for any positive integer k. Hence, a periodic orbit of period p = 1 will also have any other period p > 1. This property raises a problem of distinguishing periodic orbits of period p = 1 from orbits of higher periods. Another problem is the detection of periodic orbits of same period but different stability types, since subsequent applications of the algorithm cannot guarantee that already detected orbits will not be obtained repeatedly.

These problems can be addressed by using the deflection technique presented in Chapter Five. More specifically, as soon as a periodic orbit x^* is detected by PSO, it undergoes a test of having period p = 1. The test consists of a simple application of the mapping on x^*. If the result is equal to x^*, then the detected solution has period $p = 1$ and deflection is applied on it. After deflection, the algorithm is re-applied on the deflected objective function, which prohibits convergence to x^*. Of course, the precautions analyzed in Chapter Five shall accompany the application of deflection to avoid undesirable effects on neighboring minimizers due to wrong parameter setting.

Parsopoulos and Vrahatis (2003) applied the PSO algorithm with deflection on the nonlinear mappings TP_{NM-1}-TP_{NM-8}, defined in Appendix A of the book at hand. Their approach is described by the pseudocode of Table 1. In their experiments, the constriction coefficient PSO variant was used due to its fast convergence properties. The default PSO parameter values, $\chi = 0.729$, $c_1 = c_2 = 2.05$, were used, while the desirable accuracy was set to 10^{-10}. The requirement for high accuracy is quite common in such problems, since even the slightest deviation from an unstable periodic orbit can lead to chaotic behavior. In all experiments, a swarm size equal to 150 was used (Parsopoulos & Vrahatis, 2003).

The analysis of deflection in Chapter Five mentioned that functions with global minimum equal to zero shall be lifted prior to the application of the deflection transformation. For this purpose, Parsopoulos and Vrahatis (2003) used a bias (lift) equal to $b = 1$. Thus, the objective function for each mapping in their experiments was defined as:

$$f(x) = 1 + \sum_{i=1}^{n} (\Phi_i^p(x) - x_i)^2,$$

and its global minimum was equal to $f^* = 1$.

Several of the detected periodic orbits reported in Parsopoulos and Vrahatis (2003), along with the required number of iterations, are reported in Table 2. Deflection, as described above, was used for the

Table 1. Pseudocode of the PSO algorithm with deflection, applied for the detection of periodic orbits of nonlinear mappings

Input:	$\Phi(x)$ (mapping), p (desirable period), K (maximum number of deflections).
Step 1.	Set $SF \leftarrow$ "FALSE" and $k \leftarrow 0$.
Step 2.	**While** ($SF =$ "FALSE")
Step 3.	**Apply** PSO on the produced optimization problem.
Step 4.	**If** (PSO detected a solution x_1) **Then**
Step 5.	**Compute** all points, $x_2, \ldots, x_p$, of the same type and period by iterating (p-1) times the mapping $\Phi(x)$ on x_1.
Step 6.	**If** ($k < K$) **Then**
Step 7.	**Apply** deflection on $x_1, \ldots, x_p$, and set $k \leftarrow k+1$.
Step 8.	**Else**
Step 9.	Set $SF \leftarrow$ "TRUE".
Step 10.	**End If**
Step 11.	**Else**
Step 12.	**Write** "No Further Solution Detected".
Step 13.	Set $SF \leftarrow$ "TRUE".
Step 14.	**End If**
Step 15.	**End While**
Step 16.	**Report** all detected solutions (if any).

detection of different periodic orbits of the same period. The algorithm was able to detect periodic orbits of different periods after a small number of iterations in all cases. The computational burden appeared to be intimately related to the form of the mapping rather than the desirable period p. For example, in the Hénon map (TP_{NM-1}), there are small deviations between the mean number of iterations required for periodic orbits of period $p = 1$ and $p = 149$, as seen in Table 2. On the other hand, for the standard map (TP_{NM-2}), orbits of period $p = 5$ required an increased computational burden by a factor larger than 2, compared to orbits of period $p = 1$. Finally, an expected increase in the number of required iterations is observed when the dimensionality of the mapping increases, as observed in Table 2 for the 3-dimensional mappings TP_{NM-5} and TP_{NM-6}.

The apparent efficiency of the algorithm has increased interest in similar approaches. Petalas *et al.* (2007) proposed a similar scheme that employs the entropy-based memetic PSO (E-MPSO) approach, presented in Chapter Four of the book at hand. Their experiments focused on the test problems TP_{NM-1}, TP_{NM-2}, TP_{NM-7}, and TP_{NM-8}, for accuracy equal to 10^{-16}. Table 3 reports the achieved improvement of this approach against the standard PSO, in terms of the required function evaluations, for different periods and swarm sizes. For the standard PSO, only successful experiments were taken into consideration in the reported results (Petalas *et al.*, 2007).

In most cases, E-MPSO achieved a significant improvement both in the number of function evaluations and the number of successful experiments (Petalas *et al.*, 2007). Only in a few cases, it required a slightly higher number of function evaluations; however, in these cases, E-MPSO was completely successful in contrast to the standard PSO, which failed to converge in some experiments. The reader is referred to the original paper for a detailed presentation of these results (Petalas *et al.*, 2007).

The presented approaches place PSO and its variants at a conspicuous place among alternatives for detecting periodic orbits of nonlinear mappings, especially in cases that appear to be pathological for

Table 2. Selected periodic orbits reported in (Parsopoulos & Vrahatis, 2003)

Mapping	Period	Periodic Orbits			Iterations
TP_{NM-1}	1		$(0.6666755407,$	$0.2222243088)^T$	56
$(\cos a = 0.8)$	1		$(0.0000147632,$	$0.0000051785)^T$	46
	1		$(0.0000064371,$	$0.0000043425)^T$	55
$(\cos a = 0.24)$	5		$(0.5696231776,$	$0.1622612843)^T$	54
	5		$(0.5672255008,$	$-0.1223401431)^T$	47
	43		$(0.2576802556,$	$0.0196850254)^T$	61
	97		$(0.2310634711,$	$0.3622185202)^T$	59
	149		$(0.2232720401,$	$0.2588270953)^T$	68
TP_{NM-2}	1		$(-0.5000026466,$	$0.0000120344)^T$	59
	5		$(-0.2924003042,$	$0.1541390546)^T$	140
TP_{NM-3}	1		$(1.0000108347,$	$1.0000061677)^T$	73
	6		$(3.0000036048,$	$0.9999987183)^T$	82
TP_{NM-4}	1		$(0.7263613548,$	$0.0000047179)^T$	69
	2		$(0.8756170073,$	$-0.0000020728)^T$	61
TP_{NM-5}	1	$(0.0000000203,$	$-0.0000001703,$	$-0.0000026306)^T$	135
TP_{NM-6}	1	$(-0.0132365558,$	$-0.0165516360,$	$0.0297928545)^T$	115

Table 3. Improvement gained by using the E-MPSO approach instead of standard PSO, in terms of the number of required function evaluations. Negative values indicate worse performance

Mapping	Period	Swarm Size	Improvement
TP_{NM-1}	5	20	16.2%
		3	8.9%
		4	3.7%
TP_{NM-1}	17	100	36.3%
		20	22.9%
		30	31.9%
TP_{NM-2}	5	20	44.2%
		3	24.7%
		4	8.7%
TP_{NM-2}	17	100	56.3%
		20	19.6%
		30	20.3%
TP_{NM-7}	5	40	-1.1%
		6	4.8%
		8	1.9%
TP_{NM-7}	11	100	39.9%
		20	-1.6%
		30	1.6%
TP_{NM-8}	3	100	47.2%
		20	19.3%
		30	15.0%
TP_{NM-8}	7	100	29.6%
		20	17.4%
		30	20.9%

gradient-based approaches. This claim is also supported by the application of the PSO-based methodology on the 3-dimensional galactic potential problem presented in the next section.

DETECTION OF PERIODIC ORBITS IN 3-DIMENSIONAL GALACTIC POTENTIALS USING PSO

The study of periodic orbits of barred potentials can provide useful information regarding the structure of galactic bars (Athanassoula, 1984; Contopoulos, 1980; Contopoulos & Grosbøl, 1989; Patsis, 2005; Pfenniger, 1996; Sellwood & Wilkinson, 1993). In 2-dimensional models, the galactic bar is supported by regular orbits trapped around the "x1" periodic orbits, which are elongated along the major axis of the bar (Contopoulos & Grosbøl, 1989). Based on the property that these orbits do not extend beyond the corotation resonance, Contopoulos (1980) predicted that bars should end at, or before corotation.

On the other hand, in 3-dimensional models, the planar "x1" family has in general large unstable parts, rendering its orbits insufficient for building the bar. However, other families of periodic orbits that bifurcate from "x1" have large stable parts and they can support the bar. These families of orbits constitute the "x1-tree" (Skokos *et al.*, 2002a, 2002b), which is associated with morphological features observed in real galaxies (Patsis *et al.*, 2002, 2003a, 2003b).

Basic families of periodic orbits that belong to "x1-tree" can be usually detected without heavy computational burden. However, this does not hold as we approach corotation, where tracing periodic orbits is affected by high-order resonances. This region hosts a plethora of unstable periodic orbits that lie close to each other. Thus, the phase space is characterized by chaotic behavior and the detection of small islands of stability in this region is a demanding task. Once a periodic orbit is detected, its whole family can be obtained through changes to the system parameter. In this way, the morphological evolution and stability transitions of the family can be monitored, inferring their importance for the system.

Skokos *et al.* (2005) introduced a PSO-based algorithmic scheme for detecting initial conditions of periodic orbits in Hamiltonian systems with given parameter sets. Based on the methodology described in the previous section, the problem is transformed into a minimization task where the global minimizers of a particular objective function, which is defined on the Poincaré surface section of the system, correspond to the periodic orbits of the Hamiltonian. Then, PSO is applied, in combination with deflection, to locate periodic orbits and follow them as the system parameter changes. This approach was applied successfully on a 3-dimensional galactic potential of a Ferrers bar.

In the following paragraphs, the galactic potential is described and the corresponding objective function is derived to illustrate the application of the PSO-based scheme on this interesting problem. The algorithm's performance and its contribution to detection of new periodic orbits are also reported and discussed.

The Galactic Potential

Skokos *et al.* (2005) used the 3-dimensional barred galaxy model described in Skokos *et al.*, (2002a). This model consists of a Miyamoto disc, a Plummer bulge, and a Ferrers bar. The potential of the Miyamoto disc is defined as (Miyamoto & Nagai, 1975):

$$V_D = -\frac{GM_D}{\sqrt{x^2 + y^2 + \left(A + \sqrt{B^2 + z^2}\right)^2}}, \tag{5}$$

where M_D is the total mass of the disc; A and B are scale lengths, such that the ration B/A provides a measure of the model flatness; and G is the gravitational constant. The bulge is a Plummer sphere, hence its potential is given by:

$$V_S = -\frac{GM_S}{\sqrt{x^2 + y^2 + z^2 + \varepsilon_S^2}}, \tag{6}$$

where ε_S^2 is the bulge scale length, and M_S is its total mass. The last component of the model is a triaxial Ferrers bar with density defined as:

$$\rho(m) = \begin{cases} \dfrac{105 M_B}{32\pi abc}(1 - m^2)^2, & for \ \ m \le 1, \\ 0, & for \ \ m > 1, \end{cases} \tag{7}$$

where,

$$m^2 = \frac{y^2}{a^2} + \frac{x^2}{b^2} + \frac{z^2}{c^2}, \qquad a > b > c \tag{8}$$

and a, b, c, are the principal semi-axes, while M_B denotes the mass of the bar component. The corresponding potential, V_B, along with the forces are given in Pfenniger (1984) in a closed form, which is proper for numerical manipulations.

The parameter values used by Skokos *et al.* (2005) were, $A = 3$, $B = 1$, $a = 6$, $b = 1.5$, and $c = 0.6$. The three masses should satisfy the relation:

$$G \times (M_D + M_S + M_B) = 1.$$

Under this restriction, proper values were used such that, $GM_D = 0.82$, $GM_S = 0.08$, $GM_B = 0.10$, and $\varepsilon_S = 0.4$. Length unit of 1 kpc; time unit of 1 Myr; and mass unit 2×10^{11} M$_\odot$, were also assumed (Skokos *et al.*, 2005). The bar rotates with a pattern speed, $\Omega_b = 0.054$, around the z-axis, which corresponds to 54 km s^{-1} kpc^{-1}, and places corotation at 6.13 kpc. The governing Hamiltonian for the motion of a test particle can be given in the form:

$$H = \frac{1}{2}\left(p_x^2 + p_y^2 + p_z^2\right) + V_D + V_S + V_B - \Omega_b(xp_y - yp_x), \tag{9}$$

with,

$$p_x = \dot{x} - \Omega_b y, \qquad p_y = \dot{y} - \Omega_b x, \qquad p_z = \dot{z},$$

being the canonical momenta. The numerical value of H is reported as the *energy*, E_j, of the system. The next section describes the construction of the corresponding objective function for the equivalent optimization problem.

Construction of the Objective Function

The *Poincaré surface of section* (PSS) is a well-known method for studying the stability of periodic orbits in the 6-dimensional phase space of the Hamiltonian system defined in equation (9) (Lieberman & Lichtenberg, 1992, p. 17). Instead of following the temporal evolution of an orbit, Skokos *et al.* (2005) confined their study to an appropriate subspace, namely $(x, z, \dot{x}, \dot{z})$, which is defined by the conditions $y = 0$ and $\dot{y} > 0$. The major axis of the bar lies along the y-axis. Hence, for a given energy value, E_j, an orbit with initial conditions, $X_0 = (x_0, z_0, \dot{x}_0, \dot{z}_0)^T$, on the PSS is fully defined by assuming $y = 0$, while $\dot{y}$ can be obtained as the positive solution of equation (9). Thus, only the four initial conditions of an orbit on the PSS are required to identify it.

The temporal evolution of the orbit is derived by solving the Hamiltonian equations of motion. The next intersection of the periodic orbit with the PSS defined by $y = 0$ and $\dot{y} > 0$, is denoted as (Skokos *et al.*, 2005):

$$\Phi(X_0) = \left(\Phi_x(X_0), \Phi_z(X_0), \Phi_{\dot{x}}(X_0), \Phi_{\dot{z}}(X_0) \right)^{\mathrm{T}} : \mathbf{R}^4 \rightarrow \mathbf{R}^4, \tag{10}$$

which is obviously a point belonging to the 4-dimensional PSS of the system. Following closely the mathematical derivations of equations (3) and (4) in the first section of the present chapter, the initial condition, X, of a periodic orbit of period p of the system satisfies the following system of equations:

$$\begin{pmatrix} \Phi_x^p(X) \\ \Phi_z^p(X) \\ \Phi_{\dot{x}}^p(X) \\ \Phi_{\dot{z}}^p(X) \end{pmatrix} - \begin{pmatrix} x \\ z \\ \dot{x} \\ \dot{z} \end{pmatrix} = \begin{pmatrix} 0 \\ 0 \\ 0 \\ 0 \end{pmatrix}. \tag{11}$$

Thus, the problem of detecting the initial conditions of a periodic orbit of period p is equivalent to solving the system defined by equation (11), which in turn is equivalent to the global minimization of the following objective function:

$$f(X) = \left(\Phi_x^p(X) - x \right)^2 + \left(\Phi_z^p(X) - z \right)^2 + \left(\Phi_{\dot{x}}^p(X) - \dot{x} \right)^2 + \left(\Phi_{\dot{z}}^p(X) - \dot{z} \right)^2. \tag{12}$$

This objective function can be solved with the methodology described in the first section of the present chapter. Skokos *et al.* (2005) employed this methodology, combining PSO with deflection in the same manner as Parsopoulos and Vrahatis did for nonlinear mappings (Parsopoulos & Vrahatis, 2003, 2004). More details on their experimental setting, as well as several of the detected periodic orbits, are reported in the next section.

Experimental Setting and Results

Skokos *et al.* (2005) reported a multitude of periodic orbits of the Hamiltonian of equation (9). Besides the known orbits of period $p = 1$, several 2-dimensional and 3-dimensional orbits were detected. The orbits were also followed under energy changes, revealing their stability properties and locating their bifurcation. Periodic orbits of period $p = 1$ were mostly considered at the region between the radial 4:1 and corotation resonances. For this purpose, a subspace, S, of the PSS was used, where the values of the initial conditions lie within suitable intervals. Also, instead of considering all four variables, x, z, $\dot{x}$, $\dot{z}$, which results in a 4-dimensional search space, Skokos *et al.* (2005) followed a different approach.

More specifically, they first located the planar (2-dimensional) orbits of the system on the equatorial plane of the 4-dimensional space, in two phases. In the first phase, the 2-dimensional orbits that start perpendicular to the $y = 0$ axis were located. These orbits have initial conditions in the form of $(x, 0, 0, 0)$, thereby resulting in an one-dimensional search. In this phase, a constriction coefficient PSO with a small swarm size, $N = 5$, and ring topology with radius $r = 1$ was used. The detected periodic orbits of the first phase were deflected to avoid their detection anew in the next phase.

The second phase considered 2-dimensional periodic orbits not perpendicular to the x-axis, with initial conditions of the form $(x, 0, \dot{x}, 0)$. Thus, the search space became 2-dimensional in this set of experiments; this change was accompanied by an increased swarm size and neighborhood radius, $N = 10$ and $r = 2$, respectively. Since periodic orbits with $\dot{x} = 0$ were deflected in the first phase, the condition $\dot{x} \neq 0$ held for all new orbits.

After the detection of the 2-dimensional periodic orbits in the previous two phases, pure 3-dimensional orbits were considered. This search was also conducted in multiple phases. At start, orbits with initial conditions of the common forms, $(x, z, 0, 0)$ and $(x, 0, 0, \dot{z})$, were considered. Since the search space remained 2-dimensional in these experiments, the same swarm size and neighborhood radius with the second phase were used. Having detected the physically most important orbits, the 4-dimensional case with initial conditions of the form $(x, z, \dot{x}, \dot{z})$ was considered. In this case, the swarm size was increased to $N = 20$ and the neighborhood's radius was set to $r = 3$.

The higher swarm sizes that accompanied high-dimensional search spaces obviously aimed at enhancing the search (exploration) capability of PSO. However, larger swarms under the ring topology usually require longer convergence times. This effect was extenuated by increasing also neighborhood radius to promote the faster circulation of information among particles. Also, the detection of periodic orbits with gradually increased dimensionality, in combination with the consecutive deflections, worked beneficially for the algorithm, since a lower dimension of the search space corresponded to easier computational tasks.

The orbits detected by Skokos *et al.* (2005) are presented in the *characteristic diagram* (Contopoulos, 2002, Section 2.4.3) depicted in Fig. 5. This diagram illustrates the x-coordinate of the initial conditions of periodic orbits as a function of the energy, E_j. The depicted points in Fig. 5 define completely initial conditions of the corresponding periodic orbits, since only planar orbits that start perpendicular to the x-axis (i.e., $z = \dot{z} = 0$ and $\dot{x} = 0$) with $y = 0$ and $\dot{y} > 0$, are depicted in the characteristic diagram. Also, black curves represent stable periodic orbits, while gray curves represent unstable periodic orbits.

Figure 5. (a) Characteristic diagram of the families "x1", "f", "s", "e", and "te". The dash-dotted curve is the section of zero velocity surface with the (E_j,x) plane. (b) Magnification of the rectangular region in the upper right part of diagram (a). Black curves represent stable periodic orbits while gray curves represent unstable periodic orbits

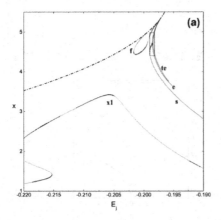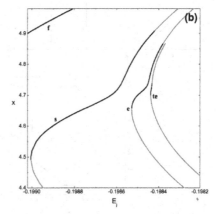

Besides the depicted orbits of two known families, namely the "x1" family (of period 1) (Contopoulos & Papayannopoulos, 1980), and the 2-dimensional "f" and "s" families (Patsis *et al.*, 2003b), two new families of periodic orbits were discovered by the PSO approach of Skokos *et al.* (2005). Their characteristic curves lie very close to each other as well as to the characteristic curve of family "s", and they are denoted as "e" and "te" in the right part of Fig. 5. The close distance of the different families constitutes a big obstacle in their detection with traditional gradient-based algorithms, requiring initial conditions very close to them. This justifies the usefulness of the PSO-based method in such problems. The new "e" and "te" families detected by PSO exist for energies higher than a minimum energy threshold, similarly to the "f" and "s" families. As they are close to the minimum energy thresholds, the morphology of the two families is influenced by 8:1 and 10:1 resonances.

The "e" family exists for $E_j > -0.19853$, and its characteristic curve has two branches with the stable orbits occupying its upper branch, as depicted in Fig. 5 (right). Figure 6 depicts orbits of the "e" family for increasing values of the energy. Orbits (a), (b), and (c), belong to the upper branch, while (d) and (e) belong to the lower branch of the characteristic curve. Also, a new 2-dimensional family, named "er1", is born by bifurcation from the "e" family for $E_j \approx -0.19847$, while a 3-dimensional family, named as "ev1", is born for $E_j \approx -0.19849$ (Skokos *et al.*, 2005). A detailed presentation of these families is out of the scope of this book. The interested reader can refer to the original paper of Skokos *et al.*, (2005) for a complete presentation of all detected periodic orbits.

The "te" family exists for $E_j > -0.198438$, and its evolution exhibits a similar behavior with the "e" family. The upper branch of its characteristic contains stable orbits, such as the ones depicted in Fig. 7 (orbits (a) and (b)). On the other hand, all orbits at the lower branch are unstable. Also, two families,

Figure 6. Periodic orbits of the 2-dimensional "e" family obtained using the PSO method. Figures (a)-(c) illustrate the evolution of the orbital morphology along the upper branch as energy increases, while (d) and (e) illustrate the evolution along the lower branch of the characteristic curve depicted in Fig. 5 (left). Orbits (a) and (b) are stable, while (c), (d), and (e), are simple unstable

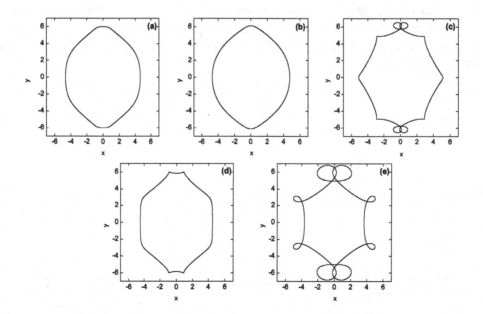

a 2-dimensional (named "ter1") and a 3-dimensional (named "tev1"), are born by bifurcation from the "te" family for $E_j \approx -0.198434$, and $E_j \approx -0.198435$, respectively. A detailed presentation of these orbits is provided in Skokos *et al.* (2005).

The presented results justify the usefulness of the PSO-based methodology with deflection. Its advantages against traditional gradient-based algorithms spring from its ability to locate periodic orbits in regions of the search space that contain a multitude of orbits close to each other, even in cases where many of them are unstable. Additionally, deflection provides a means of avoiding the detection of the same periodic orbit repeatedly in subsequent applications of the algorithm. This renders the PSO-based methodology a valuable tool that complements existing approaches for detecting periodic orbits.

CHAPTER SYNOPSIS

We presented the application of PSO on the problem of detecting periodic orbits of nonlinear mappings. The original problem is transformed to a global optimization problem, where global minimizers are periodic orbits of the mapping. The optimization problem is solved with PSO in combination with deflection. This offers an advantage against traditional gradient-based approaches, since PSO neither requires continuity or differentiability of the mapping nor depends crucially on the initial conditions. Moreover, the incorporation of the deflection technique renders it able to detect different orbits in subsequent runs. The algorithm was illustrated on benchmark problems, as well as on an interesting application in galactic potentials. In the latter, PSO was able to detect new families of periodic orbits, justifying its merit, and occupying a prominent position among the available approaches for addressing such problems.

Figure 7. Periodic orbits of the 2-dimensional "te" family obtained using the PSO method. Figures (a)-(c) illustrate the evolution of the orbital morphology along the upper branch as energy increases, while (d) and (e) illustrate the evolution along the lower branch of the characteristic curve depicted in Fig. 5 (left). Orbits (a) and (b) are stable; (c) and (e) are double unstable; and (d) is simple unstable

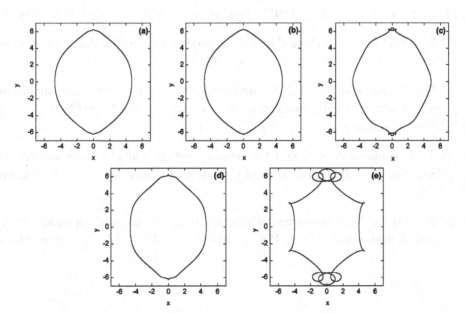

REFERENCES

Athanassoula, E. (1984). The spiral structure of galaxies. *Physics Reports, 114*(5-6), 319–403. doi:10.1016/0370-1573(84)90156-X

Birkhoff, G. (1917). Dynamical systems with two degrees of freedom. *Transactions of the American Mathematical Society, 18*, 199–300. doi:10.2307/1988861

Bountis, T., & Helleman, R. (1981). On the stability of periodic orbits of two-dimensional mappings. *Journal of Mathematical Physics, 22*(9), 1867–1877. doi:10.1063/1.525159

Contopoulos, G. (1980). How far do bars extend. *Astronomy & Astrophysics, 81*(1-2), 198–209.

Contopoulos, G. (2002). *Order and chaos in dynamical astronomy*. Berlin: Springer.

Contopoulos, G., & Grosbøl, P. (1989). Orbits in barred galaxies. *The Astronomy and Astrophysics Review, 1*, 261–289. doi:10.1007/BF00873080

Contopoulos, G., & Papayannopoulos, T. (1980). Orbits in weak and strong bars. *Astronomy & Astrophysics, 92*(1-2), 33–46.

Greene, J. (1979). A method for determining a stochastic transition. *Journal of Mathematical Physics, 20*, 1183–1201. doi:10.1063/1.524170

Helleman, R. (1977). On the iterative solution of a stochastic mapping. In U. Landman (ed.) *Statistical Mechanics and Statistical Methods in Theory and Applications* (pp. 343-370). New York: Plenum.

Hénon, M. (1969). Numerical study of quadratic area-preserving mappings. *Quarterly of Applied Mathematics, 27*, 291–312.

Howard, J., & Mackay, R. (1987). Linear stability of symplectic maps. *Journal of Mathematical Physics, 28*(5), 1036–1051. doi:10.1063/1.527544

Lieberman, M. A., & Lichtenberg, A. J. (1992). *Regular and chaotic dynamics*. New York: Springer.

Miyamoto, M., & Nagai, R. (1975). Three-dimensional models for the distribution of mass in galaxies. *Astronomical Society of Japan . Publications, 27*(4), 533–543.

Parsopoulos, K. E., & Vrahatis, M. N. (2003). Computing periodic orbits of nondifferentiable/ discontinuous mappings through particle swarm optimization. In *Proceedings of the 2003 IEEE Swarm Intelligence Symposium (SIS 2003), Indianapolis (IN), USA* (pp. 34-41).

Parsopoulos, K. E., & Vrahatis, M. N. (2004). On the computation of all global minimizers through particle swarm optimization. *IEEE Transactions on Evolutionary Computation, 8*(3), 211–224. doi:10.1109/TEVC.2004.826076

Patsis, P. A. (2005). On the relation between orbital structure and observed bar morphology. *Monthly Notices of the Royal Astronomical Society, 358*(2), 305–315. doi:10.1111/j.1365-2966.2005.08768.x

Patsis, P. A., Skokos, Ch., & Athanassoula, E. (2002). Orbital dynamics of three-dimensional bars-III. Boxy/peanut edge-on profiles. *Monthly Notices of the Royal Astronomical Society, 337*(2), 578–596. doi:10.1046/j.1365-8711.2002.05943.x

Patsis, P. A., Skokos, Ch., & Athanassoula, E. (2003a). Orbital dynamics of three-dimensional bars-IV. Boxy isophotes in face-on views. *Monthly Notices of the Royal Astronomical Society, 342*(1), 69–78. doi:10.1046/j.1365-8711.2003.06511.x

Patsis, P. A., Skokos, Ch., & Athanassoula, E. (2003b). On the 3D dynamics and morphology of inner rings. *Monthly Notices of the Royal Astronomical Society, 346*(4), 1031–1040. doi:10.1111/j.1365-2966.2003.07168.x

Petalas, Y. G., Parsopoulos, K. E., & Vrahatis, M. N. (2007). Entropy-based memetic particle swarm optimization for computing periodic orbits of nonlinear mappings. In *Proceedings of the 2007 IEEE Congress on Evolutionary Computation (CEC'07), Singapore* (pp. 2040-2047).

Pfenniger, D. (1984). The 3D dynamics of barred galaxies. *Astronomy & Astrophysics, 134*(2), 373–386.

Pfenniger, D. (1996). Stellar dynamics and the 3D structure of bars. In R. Buta, D.A. Crocker & B.G. Elmegreen (Eds.), *Barred galaxies (Astronomical Society of the Pacific Conference Series, Volume 91)*, (p. 273).

Polymilis, C., Servizi, G., & Skokos, C. (1997). A quantitative bifurcation analysis of Hénon-like 2D maps. *Celestial Mechanics and Dynamical Astronomy, 66*(4), 365–385. doi:10.1007/BF00049378

Polymilis, C., Servizi, G., Skokos, C., Turchetti, G., & Vrahatis, M. N. (2003). Topological degree theory and local analysis of area preserving maps. *Chaos (Woodbury, N.Y.), 13*(1), 94–104. doi:10.1063/1.1539011

Polymilis, C., Skokos, C., Kollias, G., Servizi, G., & Turchetti, G. (2000). Bifurcations of beam-beam like maps. *Journal of Physics. A. Mathematical Nuclear and General, 33*(5), 1055–1064. doi:10.1088/0305-4470/33/5/316

Sellwood, J. A., & Wilkinson, A. (1993). Dynamics of barred galaxies. *Reports on Progress in Physics, 56*(2), 173–256. doi:10.1088/0034-4885/56/2/001

Skokos, C. (2001a). Alignment indices: a new, simple method for determining the ordered or chaotic nature of orbits. *Journal of Physics. A. Mathematical Nuclear and General, 34*(47), 10029–10043. doi:10.1088/0305-4470/34/47/309

Skokos, C. (2001b). On the stability of periodic orbits of high dimensional autonomous Hamiltonian systems. *Physica D. Nonlinear Phenomena, 159*(3-4), 155–179. doi:10.1016/S0167-2789(01)00347-5

Skokos, C., Contopoulos, G., & Polymilis, C. (1997). Structures in the phase space of a four dimensional symplectic map. *Celestial Mechanics and Dynamical Astronomy, 65*(3), 223–251. doi:10.1007/BF00053508

Skokos, Ch., Parsopoulos, K. E., Patsis, P. A., & Vrahatis, M. N. (2005). Particle swarm optimization: an efficient method for tracing periodic orbits in 3D galactic potentials. *Monthly Notices of the Royal Astronomical Society, 359,* 251–260. doi:10.1111/j.1365-2966.2005.08892.x

Skokos, Ch., Patsis, P. A., & Athanassoula, E. (2002a). Orbital dynamics of three-dimensional bars - I. The backbone of three-dimensional bars. A fiducial case. *Monthly Notices of the Royal Astronomical Society, 333*(4), 847–860. doi:10.1046/j.1365-8711.2002.05468.x

Skokos, Ch., Patsis, P. A., & Athanassoula, E. (2002b). Orbital dynamics of three-dimensional bars - II. Investigation of the parameter space. *Monthly Notices of the Royal Astronomical Society, 333*(4), 861–870. doi:10.1046/j.1365-8711.2002.05469.x

Verhulst, F. (1990). *Nonlinear differential equations and dynamical systems.* Berlin: Springer-Verlag.

Vrahatis, M. (1995). An efficient method for locating and computing periodic orbits of nonlinear mappings. *Journal of Computational Physics, 119,* 105–119. doi:10.1006/jcph.1995.1119

Vrahatis, M., & Bountis, T. (1994). An efficient method for computing periodic orbits of conservative dynamical systems. In J. Seimenis (ed.), *Proceedings of the International Conference on Hamiltonian Mechanics, Integrability and Chaotic Behavior* (pp. 261-274).

Vrahatis, M., Bountis, T., & Kollmann, M. (1996). Periodic orbits and invariant surfaces of 4-d nonlinear mappings. *International Journal of Bifurcation and Chaos in Applied Sciences and Engineering, 6,* 1425–1437. doi:10.1142/S0218127496000849

Vrahatis, M. N., Isliker, H., & Bountis, T. (1997). Structure and breakdown of invariant tori in a 4-D mapping model of accelerator dynamics. *International Journal of Bifurcation and Chaos in Applied Sciences and Engineering, 7*(12), 2707–2722. doi:10.1142/S0218127497001825

Vrahatis, M. N., Servizi, G., Turchetti, G., & Bountis, T. C. (1993). *A procedure to compute the fixed points and visualize the orbits of a 2D map* (CERN SL/93-06 (AP)). Geneva, Switzerland: European Center of Nuclear Research (CERN).

Chapter 8
Applications in Operations Research

This chapter is devoted to three representative applications of PSO in operations research. Similarly to the previous chapters, our attention is focused on the presentation of essential aspects rather than reviewing the existing literature. Thus, we present methodologies for formulation of the optimization problem, which is not always trivial, as well as for the efficient treatment of special problem requirements that cannot be handled directly by PSO. Under this prism, we report applications from the fields of scheduling, inventory optimization and game theory. Recent results are also reported per case to provide an idea of the efficiency of PSO.

INTRODUCTION

Operations research (or *operational research*) (OR) is a scientific field that deals with the detection of optimal solutions in complex problems. Optimization itself constitutes a branch of OR, probably the most important one, since the majority of OR problems end up as optimization tasks. However, this is not the only interesting branch of OR. Probability theory, queuing systems, game theory, graph theory, simulation and management sciences, are also intimately related or constitute branches of OR. The procedure of solving an OR problem requires both the selection and application of the most appropriate algorithm for the specific task. Thus, the user interferes crucially in different aspects of the procedure, ranging from solution representation to constraints handling (Hillier & Lieberman, 2005; Winston, 2003).

Algorithms with the plasticity, adaptability and efficiency of evolutionary and swarm intelligence approaches can be more than useful in applications with the aforementioned characteristics. For this reason, OR has always constituted a prosperous field of applications for evolutionary computation and swarm intelligence methods, usually providing problems characterized by peculiarities such as mixed

DOI: 10.4018/978-1-61520-666-7.ch008

Copyright © 2010, IGI Global. Copying or distributing in print or electronic forms without written permission of IGI Global is prohibited.

variable types, varying dimensionality and computationally expensive function evaluations that are obtained through simulation procedures (Hillier & Lieberman, 2005; Winston, 2003).

PSO has been increasingly used in OR applications in the past few years. We present three such applications, from the fields of scheduling, management sciences and game theory, to illustrate the workings of PSO and offer a taster of its performance. For each application, elements that require user interference are analyzed along with the necessary techniques that render PSO applicable to such problems.

SCHEDULING PROBLEMS

In general, *scheduling* refers to the allocation of resources to tasks. This problem type is met very often in real world applications and it has proved to be NP-hard (Johnson & Garey, 1979; Lenstra *et al.*, 1977; Pinedo, 1995). The main objective in scheduling is the assignment of jobs (tasks) to a single or many machines so that several operational criteria are met. These criteria are usually modeled as the minimization of one or more objective functions.

The *single machine total weighted tardiness* (SMTWT) problem is a very challenging scheduling task, with many applications (Pinedo, 1995). Its objective is the sequential processing of n jobs on a single machine, so that its total tardiness is minimized. To put it formally, let $j = 1, 2, \ldots, n$, denote a job; p_j be its processing time; d_j be its due date; and w_j be a weighting factor. A *job sequence* is an n-dimensional ordered vector, $s = (s_1, s_2, \ldots, s_n)^T$, where s_j is an integer denoting the processing order of the j-th job, $j = 1, 2, \ldots, n$. For example, $s_1 = 3$, denotes that job 1 is scheduled as the third job in the processing order, while, $s_2 = 1$ denotes that job 2 is scheduled as the first job in the processing order. All jobs are assumed to be available for processing at time zero.

Let C_j be the completion time of job j in a job sequence s. Since s constitutes a permutation of the vector $(1, 2, \ldots, n)^T$, and time counting starts from zero, C_j can be computed as:

$$C_j = \sum_{\substack{\text{All } i \text{ such} \\ \text{that } s_i \leq s_j}} p_i .$$

Then, the *tardiness*, T_j, of job j is defined as:

$$T_j = \max\{0, C_j - d_j\} .$$

The objective of the SMTWT problem is the determination of a job sequence that minimizes the *total weighted tardiness*, which is defined as:

$$T = \sum_{j=1}^{n} w_j T_j . \tag{1}$$

Branch-and-bound algorithms constitute a standard methodology for tackling SMTWT problems. Unfortunately, they fail to solve large-scale instances of the problem (Abdul-Razaq *et al.*, 1990). In such cases, specially designed heuristic approaches, such as the *earliest due date* and *apparent urgency*, as well as more common optimization algorithms such as simulated annealing, tabu search, genetic algorithms and ant colony optimization have been successfully applied (Crauwels *et al.*, 1998; Den Besten *et al.*, 2000; Potts & Van Wassenhove, 1991). The SMTWT problem has also been tackled within the framework of the *constraint satisfaction problem* (Cesta et al., 2002; Dechter et al., 1991; Schwalb & Vila, 1997).

PSO and its variants have also been applied on task assignment as well as on the SMTWT problem (Liao *et al.*, 2007; Parsopoulos & Vrahatis, 2006; Saldam *et al.*, 2002; Tasgetiren *et al.*, 2004; Zhang *et al.*, 2006). The objective function of SMTWT, as defined in equation (1), allows the straightforward application of PSO. However, this does not hold for the representation of solutions (job sequences). Proper techniques are required to transform a job sequence to a numerical vector that can be handled by PSO. To this end, a simple yet efficient technique, called *smallest positive value* (SPV), was proposed by Tasgetiren *et al.* (2004). SPV was used in combination with the *variable neighborhood search* (VNS) technique (Mladenović & Hansen, 1997), significantly enhancing the performance of PSO on SMTWT problems.

The unified PSO (UPSO) variant, presented in Chapter Four, was also applied in combination with SPV on the SMTWT problem with very promising results. Besides its usefulness on SMTWT, SPV can be the starting point for the development of new appropriate representation techniques for different scheduling problems. Hence, it is a valuable tool that requires special attention. The next section provides a description of SPV, followed by representative experimental results from the application of UPSO with SPV on SMTWT benchmark problems.

The Smallest Positive Value Representation Scheme

Let n be the number of jobs and, $x_i = (x_{i1}, x_{i2}, \ldots, x_{in})^{\mathrm{T}}$, be the i-th particle of the swarm in PSO. Then, a proper representation scheme maps each component of the particle to a corresponding unique job. Thus, in a problem with n jobs, the search space will be n-dimensional. Let also,

$$s_i = (s_{i1}, s_{i2}, \ldots, s_{in})^{\mathrm{T}},$$

be the job sequence that corresponds to x_i, where s_{ij} is the assignment of the j-th job in the processing order and belongs to the set $\{1, 2, \ldots, n\}$.

SPV is based on the assumption that jobs with smaller values of their corresponding components, x_{ij}, are scheduled first. For example, consider the case reported in Table 1. There are $n = 5$ jobs, i.e., $j = 1$, 2,…, 5, and the particle under consideration is, $x_i = (1.45, -3.54, 2.67, -2.29, -4.02)^{\mathrm{T}}$. Each component of x_i corresponds to a job, i.e., the first component, 1.45, corresponds to job 1, the second component, -3.54, corresponds to job 2, etc. According to the SPV technique, the job that will be scheduled first in the processing sequence is the one with the smallest component in x_i. In our example, job 5, whose corresponding component in x_i is equal to -4.02. Then, the second scheduled job will be job 2, which has the second smallest component, -3.54, in x_i. Continuing in the same manner, we obtain the final job sequence, $s_i = (4, 2, 5, 3, 1)^{\mathrm{T}}$, which corresponds to the particle x_i, as reported in Table 1.

Any real-valued vector, regardless of its scaling, can be translated with the SPV scheme to a job sequence. This is a nice property, since the user can restrict the swarm in arbitrarily small search spaces. On the other hand, there is an infinite number of real-valued vectors that have the same ordering in their components, thereby corresponding to the same job sequence. This effect can become detrimental for the exploration capability of the algorithm, since a whole, ostensibly diverse, swarm may actually cor-

Table 1. An example of the SPV technique for the translation of real-valued vectors to job sequences

Job Number	j	1	2	3	4	5
Particle Component	x_{ij}	1.45	-3.54	2.67	-2.29	-4.02
Processing Sequence	s_{ij}	4	2	5	3	1

respond to a unique job sequence in cases where all its particles have the same component ordering.

In order to alleviate the aforementioned deficiencies, the bounding of particles and velocities shall be treated carefully in SMTWT tasks (Parsopoulos & Vrahatis, 2006; Tasgetiren *et al.*, 2004). In the next section, the application of UPSO with SPV on the SMTWT problem serves as a representative example (Parsopoulos & Vrahatis, 2006).

Experimental Results

Parsopoulos and Vrahatis (2006) investigated the performance of three variants of the unified PSO (UPSO) algorithm, presented in Chapter Four, on the SMTWT problem. More specifically, they considered two instances of the main UPSO scheme, with unification factors, $u = 0.2$ and $u = 0.5$, respectively, as well as an instance of UPSO with mutation, with $u = 0.1$ and a multivariate Gaussian random variable, $r_3 \sim N(M, \Sigma)$, with $M = (0, 0, \ldots, 0)^T$, $\Sigma = \sigma^2 I$, where $\sigma = 0.01$ is the standard deviation, and I is the identity matrix. These variants are denoted as $UPSO_1$, $UPSO_2$, and $UPSO_m$, respectively. They also compared their performance with that of the standard global and local PSO variants, denoted as PSO_g and PSO_l, which are obtained by the UPSO scheme for $u = 1$ and $u = 0$, respectively. In all cases, the default parameters, $\chi = 0.729$, $c_1 = c_2 = 2.05$, were used along with the ring neighborhood topology with radius $r = 1$ (Parsopoulos & Vrahatis, 2006).

For the experimentation, they used the established sets of randomly generated benchmark problems of 40 and 50 jobs, each containing 125 instances, which are provided in ORLIB (Den Besten *et al.*, 2000; Tasgetiren *et al.*, 2004). For each test problem, a number of 25 independent experiments were conducted. The swarm size was set equal to $N = 10n$, where n is the number of jobs (recall that under the SPV representation scheme, the dimension of the problem is equal to the number of jobs). In all cases, a maximum number of 2000 iterations were allowed, while the swarm was constrained within the range $[-1,1]^n$ (Parsopoulos & Vrahatis, 2006).

The performance of all approaches was measured in terms of the number of successful experiments (out of 25) and the mean number of iterations required in them, which is a measure of the expected number of iterations. An additional performance measure is the *average relative percent deviation*, which is defined as:

$$\Delta_{avg} = \sum_{i=1}^{R} \left[\frac{1}{R} \left(\frac{100(f_b^i - f^*)}{f^*} \right) \right],$$

where f^* is the value of the true optimal solution; f_b^i is the value of the best solution obtained by the algorithm in the i-th experiment; and R is the total number of experiments for all instances per problem. In the results reported by Parsopoulos and Vrahatis (2006), this number was defined as, $R = 25 \times 125$

Table 2. Experimental results derived from Parsopoulos and Vrahatis (2006) for the application of PSO and UPSO with the SPV representation scheme on the SMTWT problem

Number of jobs	Algorithm	Success (%)	Exp. Iterations	Δ_{avg}
40	PSO_g	56.3%	229.6	1.692
	PSO_l	47.0%	613.0	0.754
	$UPSO_1$	62.5%	158.1	1.765
	$UPSO_2$	54.1%	131.1	2.147
	$UPSO_m$	65.8%	525.6	0.478
50	PSO_g	42.2%	275.6	1.778
	PSO_l	27.0%	338.4	1.292
	$UPSO_1$	43.7%	182.3	1.483
	$UPSO_2$	37.7%	178.4	2.329
	$UPSO_m$	39.6%	472.3	0.720

= 3125. Obviously, if the algorithm detects the true optimal solution in all problem instances, then Δ_{avg} will be equal to zero. Additionally, Δ_{avg} also provides a measure of relative quality for cases where the algorithm failed to detect the optimal solution, since smaller values of Δ_{avg} imply the detection of sub-optimal solutions that lie closer to the optimal one.

Table 2 reports a fraction of results derived by Parsopoulos and Vrahatis (2006). More specifically, the percentage of successful experiments per algorithm and problem are reported, along with the expected number of iterations and the corresponding values of Δ_{avg}. As we see, in the case of 40 jobs, $UPSO_m$ outperformed all other variants, followed closely by $UPSO_1$. We also observe that smaller values of Δ_{avg} are accompanied by a higher expected number of iterations. This holds for both the standard PSO and UPSO, indicating that the obtained accuracy is directly proportional to the required computational burden.

Moreover, PSO_g achieved a better number of expected iterations than its local counterpart, PSO_l, but at reduced robustness as revealed by its higher values of Δ_{avg}. This is a direct consequence of the convergence properties of PSO_g, which renders it faster but also more prone to get stuck in local minima. Similar results were obtained for the 50 jobs case, with $UPSO_1$ outperforming the rest of the approaches. $UPSO_m$ follows closely in performance, although with smaller values of Δ_{avg}, which indicates its tendency for moving closer to the global minimizer in most experiments. For a more thorough presentation of the results, the reader is referred to the original paper by Parsopoulos and Vrahatis (2006).

The reported representative fraction of results is indicative of the potential of PSO and its variants in tackling SMTWT problems. The next section describes and analyzes the application of PSO on another interesting OR problem, namely the continuous inventory optimization problem.

CONTINUOUS REVIEW INVENTORY OPTIMIZATION

Many products undergo deterioration or decay over time. Thus, the proper maintenance of an inventory is of major importance in supply chain business management. Obviously, deterioration and demand rates are positioned in a foremost place in such problems. Deterioration rates have been described with several models in literature (Covert & Philip, 1973; Ghare & Shrader, 1963; Tadikamalla, 1978). Demand rates for a product are usually considered constant in the mature stage of its life cycle, while they can be linearly approximated in the growth and end stage (Donaldson, 1977; Goyal & Giri, 2001; Raafat, 1991; Resh *et al.*, 1976).

Unsatisfied demand is also considered as an important issue of inventory systems. In many cases, the complete backlogging of unsatisfied demand is assumed, although in practice the behavior of customers during the shortage period differs. For this purpose, special attention is paid to recent models, where partial backlogging is considered based on the behavior of the customers (Chang & Dye, 1999; Skouri & Papachristos, 2002; Teng *et al.*, 2002). An additional modeling factor is the inflation effect and time value of money, whose importance increases in practical environments that refer to developing countries.

Generalizing the established developments, Chern *et al.* (2008) studied recently an inventory model for deteriorating items with time-varying demand and partial backlogging, taking into account also the time value of money. However, they used in their study the basic assumption of unlimited storage capacity, which is not valid in practice. Parsopoulos *et al.* (2008) studied the performance of PSO on an extended version of this model with the more realistic property of limited storage capacity. The main goal of this research was the detection of replenishment cycles, instances, and orders, under capacity constraints, which result in a mixed-integer optimization problem with variable-length solutions. The algorithm was applied on three test problems with promising results (Parsopoulos *et al.*, 2008). However, special techniques were needed to render PSO capable of tackling it efficiently. These problem peculiarities are analyzed in the following sections, after a brief description of the review inventory model.

The Review Inventory Model

The considered *continuous review inventory* problem is produced by extending the model of Chern *et al.* (2008) to include limited storage capacity. This model can be considered as a generalized instance that contains different previous models with time-varying demand, product deterioration, and backlogging. The model is based on the following fundamental assumptions:

Assumption 1. The planning horizon is finite and equal to H time units. The initial and final inventory levels during the planning horizon are assumed to be equal to zero.

Assumption 2. Replenishment is instantaneous, i.e., the replenishment rate is infinite.

Assumption 3. The lead-time is zero.

Assumption 4. The on-hand inventory deteriorates with a deterioration rate, $\theta(t)$.

Assumption 5. The demand rate is given with a continuous function, $f(t)$, in $t \in [0, H]$.

Assumption 6. The system admits shortages in all cycles, and each cycle starts with shortages.

Assumption 7. The backlogging rate of shortages is given by a non-negative function, $\beta(x)$, in the waiting time, x, up to the next replenishment, with $0 \leq \beta(x) \leq 1$ and $\beta(0) = 1$.

Henceforth we will use the parameter notation reported in Table 3. If we additionally define as:

$$\delta(t) = \int_0^t \theta(u)du,$$

then, the total cost of the inventory system during the planning horizon, H, is given by (Chern *et al.*, 2008):

Table 3. Notation used for the parameters of the considered model

Parameter	Description
n	Number of replenishment cycles during the planning horizon.
s_i	Starting time of shortage during the i-th cycle, $i = 1, 2,\dots, n$.
t_i	Time instance of the i-th replenishment, $i = 1, 2,\dots, n$.
r	Discount rate.
i_1	Internal inflation rate, varied by the company operational status.
i_2	External inflation rate, varied by the social economical status.
r_1	Difference, r-i_1, between the discount rate and the internal inflation rate.
r_2	Difference, r-i_2, between the discount rate and the external inflation rate.
c_0	Internal fixed purchasing cost per order.
c_p	External variable purchasing cost per unit.
c_{h_1}	Internal inventory holding cost per unit and per unit of time.
c_{h_2}	External inventory holding cost per unit and per unit of time.
c_{b_1}	Internal backlogging cost per unit and per unit of time.
c_{b_2}	External backlogging cost per unit and per unit of time.
c_{l_1}	Internal cost of lost sales per unit and per unit of time.
c_{l_2}	External cost of lost sales per unit and per unit of time.
W	Storage area or volume.

$$
\begin{aligned}
TC(n, s_i, t_i) \;=\; & \sum_{i=1}^{n} c_0 \exp(-r_1 t_i) \\
& + \sum_{i=1}^{n} c_p \exp(-r_2 t_i) \left(\int_{s_{i-1}}^{t_i} \beta(t_i - t) f(t)\,dt + \int_{t_i}^{s_i} \exp(\delta(t) - \delta(t_i)) f(t)\,dt \right) \\
& + \sum_{i=1}^{n} \sum_{j=1}^{2} c_{h_j} \int_{t_i}^{s_i} \exp(-r_j t) \int_{t}^{s_i} \exp(\delta(u) - \delta(t)) f(u)\,du\,dt \\
& + \sum_{i=1}^{n} \sum_{j=1}^{2} \frac{c_{b_j}}{r_j} \int_{s_{i-1}}^{t_i} \left(\exp(-r_j t) - \exp(-r_j t_i) \right) \beta(t_i - t) f(t)\,dt \\
& + \sum_{i=1}^{n} \sum_{j=1}^{2} c_{l_j} \int_{s_{i-1}}^{t_i} \exp(-r_j t)\left(1 - \beta(t_i - t)\right) f(t)\,dt,
\end{aligned}
\tag{2}
$$

with,

$$s_0 = 0, s_{i-1} < t_i \leq s_i, s_n = H.$$

Taking into consideration the capacity constraints, results in the following mixed-integer constrained minimization problem:

$$\min_{n,t_i,s_i} TC(n,t_i,s_i), \quad such \; that \quad \int_{t_i}^{s_i} \exp(\delta(u) - \delta(t_i)) f(u) du \leq W,$$

$$s_0 = 0, \quad s_n = H, \quad s_{i-1} < t_i \leq s_i, \quad i = 1,2,...,n. \tag{3}$$

For a given value of n, and ignoring the constraints, $s_{i-1} < t_i \leq s_i$, $i = 1, 2,..., n$, the problem of equation (3) can be solved using the typical Kuhn-Tucker approach. For this purpose, a number of 2^n nonlinear systems of $2n$ up to $3n$ variables each, have to be solved (Chern *et al.*, 2008; Skouri & Papachristos, 2002). Clearly, using this approach for unknown values of n requires a considerable computational burden.

Parsopoulos *et al.* (2008) applied PSO for solving the problem of equation (3), computing all unknowns, n, t_i, s_i, $i = 1, 2,..., n$, simultaneously. Special treatment was required to render PSO applicable to this problem due to certain peculiarities of the review inventory model. All these necessary manipulations are presented in the following section.

Representation, Manipulations and Assumptions

PSO was applied by Parsopoulos *et al.* (2008) on the continuous review inventory problem to determine both the number of cycles, n, and the corresponding solution, $(s_0, t_1, s_1,..., t_n, s_n)$, concurrently. Based on the time assumptions of the problem, the first and the last component of any candidate solution vector are known a priori to be equal to $s_0 = 0$ and $s_n = H$, respectively. Thus, it is adequate to compute the rest $(2n-1)$ variables, $t_1, s_1,..., t_n$, also called *time components* of the solution vector. These variables, along with n itself, are encoded as components of the particles in the swarm. However, the parameter n varies, requiring the use of particles of variable length; a case that needs special treatment because the update equations of PSO assume particles of equal length.

The aforementioned obstacle was tackled by Parsopoulos *et al.* (2008) by using sufficiently large static vectors to represent the particles. More specifically, a sufficiently large upper bound, n_{max}, on the value of n was determined, i.e.:

$$n_{max} \geq n,$$

based on reasonable estimations of the range of n. Then, the particle dimension was fixed to:

$$D = 2n_{max}. \tag{4}$$

Thus, the i-th particle of the swarm was defined as:

$$x_i = (x_{i1}, x_{i2}, \ldots, x_{iD})^{\mathrm{T}} = \left(n, t_1, s_1, \ldots, t_n, s_n, \ldots, t_{n_{\max}-1}, s_{n_{\max}-1}, t_{n_{\max}} \right)^{\mathrm{T}}.$$

This choice provides a sufficient number of particle components to represent all variables for values of n up to $n_{\max}$. For the evaluation of a particle x_i with $n < n_{\max}$, the cost function, $TC(x)$, defined by equation (3), is computed using only the first $2n$ components of x_i, i.e., $n, t_1, s_1, \ldots, t_n$, along with the fixed s_0 and s_n, while the rest of its components are ignored. Thus, using this representation, solutions with at most $n_{\max}$ cycles can be represented, with only their first $2n$ components being significant, where n is the first component of the vector itself. In other words, the vector yields itself the number of its valid components to be used for its function evaluation, while the rest are simply ignored in the objective function. In this way, Parsopoulos *et al.* (2008) achieved to encapsulate the variable-length problem into a maximal fixed-length problem. Hence, PSO operates on the fixed-length particle vectors, always keeping in mind that the number of their significant components that will be used for their function evaluation is determined by their first component.

Alleviating the obstacle of variable-length particles, another critical issue arises, namely the different type of variables encoded in each particle. More specifically, the number, n, of cycles shall be integer for each candidate solution and lie within the range $[1, n_{\max}]$. Thus, the first component, x_{i1}, of each particle shall be integer, while the rest of the components must be positive real numbers. Under this restriction, special integer-valued operators shall be used in the PSO update equations for the first component of each particle.

In order to avoid the necessity for such operators and retain the simplicity and wide applicability of their approach, Parsopoulos *et al.* (2008) allowed x_{i1} to be a real number within the range $[0.6, n_{\max}]$. Then, for the evaluation of the particle where an integer value of n is required, the real value was rounded to the nearest integer, following the rounding scheme presented in the last section of Chapter Five of the book at hand. For example, a real value, $x_{i1} = 2.34$, would be rounded to the integer value $n = 2$ for the evaluation of the particle, while, $x_{i1} = 3.71$, would correspond to $n = 4$, and $x_{i1} = 6.49$ would imply $n = 6$. Thus, the mixed variable types were tackled without intervening in the operation of PSO. Regarding its initialization, x_{i1} was initialized randomly and uniformly in $[0.6, n_{\max}]$ for each particle.

The rest of the components of a particle, i.e., the time components, shall lie strictly within the range $[0, H]$, and preserve the time ordering:

$$t_1 \leq s_1 \leq \cdots \leq t_{n_{\max}},$$

to avoid meaningless solutions. For this purpose, Parsopoulos *et al.* (2008) imposed the following constraints on the j-th component, x_{ij}, of each particle:

$$x_{i,j-1} \leq x_{ij} \leq x_{i,j+1}, \quad j = 2, 3, \ldots, D\text{-}1.$$

This way, all time instances were enforced to retain a meaningful order. If a component violated a constraint, it was returned to the nearest boundary in the direction of its move, similarly to the restriction of particles within box-shaped search spaces, as described in the first part of the book at hand. However, this approach can bias all components of a particle towards H, if one of the preceding time components assumes a large value close to H. This effect can easily lead PSO to search stagnation, especially if it is encountered in the early stages of a run, e.g., in the initialization phase.

Parsopoulos *et al.* (2008) alleviated the aforementioned deficiency by adopting a special initialization scheme for each time component. More specifically, they initialized it randomly in equidistant intervals within the range [0, *H*], as follows:

$$x_{ij}^{\text{initial}} = (j - 2 + R)\Delta, \quad j = 1, 2, ..., D,$$

where, $\Delta = H/(D-1)$, is a partitioning of the time horizon, and *R* is a random variable uniformly distributed over [0,1]. Thus, each component was able to initially assume feasible values that adhere to all time and ordering constraints.

In cases where constrained warehouse capacity was considered, a penalty function approach was used. The employed penalty function is defined as (Parsopoulos *et al.*, 2008):

$$TC_{\text{pen}}(n, s_i, t_i) = TC(n, s_i, t_i) + \sum_{k=1}^{K} \frac{TC(n, s_i, t_i)}{n}, \tag{5}$$

where $TC(n, s_i, t_i)$ is the cost function defined in equation (2), and *K* is the number of violated constraints, with $0 \leq K \leq n$. Therefore, for each violated constraint, a fixed portion of the cost function is added to the actual cost function value.

Experimental Setting and Results

Parsopoulos *et al.* (2008) applied the constriction coefficient PSO variant on problems $TP_{\text{INV-1}}$, $TP_{\text{INV-2}}$, and $TP_{\text{INV-3}}$, defined in Appendix A of the book at hand. The default parameters of PSO, $\chi = 0.729$, $c_1 = c_2 = 2.05$, as well as a ring neighborhood topology with radius equal to 1 were used. In all problems, the maximum number of replenishments was set equal to $n_{\text{max}} = 20$; therefore, all test problems were 40-dimensional as defined by equation (4) in the previous section. For this dimension, a swarm of 100 particles and a maximum number of 15000 iterations were considered reasonable choices (Parsopoulos *et al.*, 2008).

For each test problem, both the unconstrained and constrained case were investigated, which are defined by equation (3) with or without constraints. The constrained case was addressed using the penalty function of equation (5). A number of 50 experiments were conducted per problem and case. Following the suggestions of the previous section, the first component of each particle, which corresponds to the number of cycles, was initialized randomly and uniformly in [0.6, n_{max}].

PSO succeeded in detecting the optimal solution in all experiments. Table 4 contains an excerpt of the results reported by Parsopoulos *et al.* (2008). The unconstrained cases are indicated with the symbol "∞", which stands for infinite warehouse capacity, under the *W* column. Also, the optimal detected number of cycles, *n**, is reported, along with the corresponding value of the cost function, *TC**, as well as the required computational burden in terms of PSO iterations. Several of the obtained solutions can be found in the original paper by Parsopoulos *et al.* (2008).

As we observe in Table 4, the constrained cases required a significantly increased number of iterations in all test problems. Also, the optimal cost value of the unconstrained case in $TP_{\text{INV-1}}$ differs to that of the constrained one, implying that the unconstrained optimal solution is not even feasible for the constrained case, i.e., it violates the warehouse capacity constraint. Nevertheless, PSO was able to

Table 4. Results derived from Parsopoulos et al. (2008) for the application of PSO on the continuous review inventory problem

Problem	a	W	n^*	TC^*	Success	Mean Iterations
TP_{INV-1}	0.08	∞	3	685.888	100%	223.96
		90	3	688.354	100%	2985.30
	0.05	∞	3	687.686	100%	205.84
		90	3	690.702	100%	1676.90
	0.02	∞	3	689.405	100%	219.66
		90	3	693.010	100%	988.74
TP_{INV-2}		∞	1	21078.04	100%	8.20
		300	1	21078.04	100%	46.48
TP_{INV-3}		∞	1	29990.68	100%	7.64
		300	1	29990.68	100%	28.30

detect the optimal number of cycles, $n^* = 3$, and the corresponding solutions, without user intervention in all cases. Similar observations can be made for the rest of the test problems, although in these cases the unconstrained solutions coincided with the constrained ones. The optimal number of cycles in these cases was equal to $n^* = 1$, with time components positioned on the boundary of the time horizon.

Summarizing, the hard constraints posed on particles did not prevent PSO from detecting the optimal value, although the bounds were changing continuously for each particle and iteration to preserve the ordering of time components. The penalty function defined in equation (5) was adequate to prevent PSO from converging to unfeasible solutions in cases with constrained warehouse capacity, without requiring any further assumption regarding the feasibility of the initial swarm.

Although further experimentation is needed to identify the prospects of PSO on solving inventory optimization problems, the promising preliminary results render it a useful alternative tool that requires limited input from the user. The next section presents the workings of PSO on game theory problems, and specifically on the detection of Nash equilibria, to further reveal its potential in addressing operations research problems.

GAME THEORY PROBLEMS

Game theory is a mathematical discipline that studies socio-economic phenomena that involve a number of interacting decision-makers, also called *players*, whose actions affect each other. The main assumption regarding players is their pursuit of achieving well-defined exogenous goals, based on available knowledge or expectation of the behavior of the rest (Osborne & Rubinstein, 1994). There is a plethora of multidisciplinary applications of game theory, including economics, political sciences, computer science, biology, statistics, sociology, engineering and philosophy (Aumann & Hart, 1992).

The concept of a *game* is the main subject of study in game theory. A game is a model of strategic interactions among players, including all constraints on their actions and interests, but not their actual actions. *Strategic* or *normal form* is a very common game type that has attracted a lot of interest, and will be described later. A *solution* of such a game is a systematic description of its possible outcomes (Osborne & Rubinstein, 1994). *Nash equilibria* constitute the basic solution concept, since they can capture a steady state of the play in which all players have correct expectations regarding the behavior of the rest and they take rational decisions (Nash, 1950, 1951).

The problem of detecting Nash equilibria in a finite strategic game can be given under different formulations. Nevertheless, the computation of its solutions remains a very challenging task until today. Although a plethora of algorithms has been proposed (McKelvey & McLennan, 1996; Von Stengel, 2002, Chapter 45), there are deficiencies that leave room for new developments and applications. For example, an algorithm that computes a single Nash equilibrium can be inadequate in many applications, since, even if a satisfactory equilibrium is computed, more salient equilibria may exist. In such cases, techniques for the detection of more than one equilibria are particularly desirable.

In the following section, our main subject of study will be the formulation of the Nash equilibria problem as a global optimization task. Proper formulations permit the effective application of optimization algorithms, including evolutionary algorithms and PSO, along with techniques such as multistart and deflection for the computation of different solutions. Pavlidis *et al.* (2005) applied PSO, covariance matrix adaptation evolution strategies (CMA-ES) and differential evolution (DE), in combination with the aforementioned techniques for the computation of Nash equilibria, with very promising results. The basic lines of their approach, along with some representative results for PSO, are exposed in the following sections.

Strategic Games and Nash Equilibria

Let us now formally introduce the concepts of strategic games and Nash equilibria. A *strategic game* can be defined as a triplet (Osborne & Rubinstein, 1994):

$$\Gamma = \left\langle (NS), (S_i), (u_i) \right\rangle,$$

where, $NS = \{1, 2, \ldots, N\}$, is a finite set of *players*; $S_i = \{s_{i1}, s_{i2}, \ldots, s_{im_i}\}$ is a set of actions (pure strategies) for the *i*-th player; and, $u_i: S \to \mathbf{R}$, is a *payoff function* for the *i*-th player, where:

$$S = S_1 \times S_2 \times \cdots \times S_N,$$

is the Cartesian product of the sets S_i. Adopting the notation of McKelvey and McLennan (1996), which is also followed by Pavlidis *et al.* (2005), we let P_i be the set of real valued functions on S_i, and $p_{ij} = p_i(s_{ij})$ denote the elements $p_i \in P_i$. Let also:

$$P = \times_{i \in NS} P_i \quad \text{and} \quad m = \sum_{i \in NS} m_i.$$

Then, P is isomorphic to $\mathbf{R}^m$. The elements of P are denoted as vectors, $p = (p_1, p_2, \ldots, p_N)$, where $p_i = (p_{i1}, p_{i2}, \ldots, p_{im_i}) \in P_i$. If $p \in P$ and $p_i' \in P_i$, then (p_i', p_{-i}) denotes the element $q \in P$ that satisfies the relations:

$q_i = p_i'$ and $q_j = p_j$ for $j \neq i$.

If Δ_i is the set of probability measures on S_i, we can define:

$$\Delta = \times_{i \in NS} \Delta_i,$$

so that $\Delta \subseteq \mathbf{R}^m$. Hence, the elements, $p_i \in \Delta_i$, are real valued functions on S_i, i.e., $p_i : S_i \to \mathbf{R}$, and it holds that:

$$\sum_{s_{ij} \in S_i} p_i(s_{ij}) = 1, \quad p_i(s_{ij}) \geq 0, \quad \forall s_{ij} \in S_i.$$

The abusive notation s_{ij} is used to denote the strategy p_i with $p_{ij} = 1$. Thus, (s_{ij}, p_{-i}) constitutes a representation of a strategy where the i-th player adopts the pure strategy, s_{ij}, while the rest players adopt their components of p.

The payoff function, u, is extended to have as its domain the set $\mathbf{R}^m$, as follows:

$$u_i(p) = \sum_{s \in S} p(s) u_i(s),$$

with,

$$p(s) = \prod_{i \in NS} p_i(s_i).$$

Now, we can give the definition of a Nash equilibrium:

Definition 1. A strategy profile, $p^* = (p_1^*, p_2^*, \ldots, p_N^*) \in \Delta$, is a *Nash equilibrium*, if $p^* \in \Delta$ and $u_i(p_i, p_{-i}^*) \leq u_i(p^*)$, for all $i \in NS$ and $p_i \in \Delta_i$.

In simple words, this definition implies that no player can profitably deviate from Nash equilibrium, given the actions of the other players.

The application of PSO for the detection of Nash equilibria requires the existence of a proper objective function, whose global minimizers correspond to the Nash equilibria. In the next section, we describe the transformation of the original game theory problem to an equivalent global minimization task and provide the basic information for its solution with PSO.

Problem Formulation and Solving Issues

The solution of the Nash equilibria problem using optimization algorithms requires a real-valued objective function (McKelvey, 1991). As reported by Pavlidis *et al.* (2005), the following three functions can be defined:

$x: P \to \mathbf{R}^m$, with $x_{ij}(p) = u_i(s_{ij}, p_{-i})$,

$z: P \to \mathbf{R}^m$, with $z_{ij}(p) = x_{ij}(p) - u_i(p)$,

$g: P \to \mathbf{R}^m$, with $g_{ij}(p) = \max\{z_{ij}(p), 0\}$,

$p \in P, i \in NS, s_{ij} \in S_i$.

Then, the following non-negative, real-valued objective function, $v: \Delta \to \mathbf{R}$, is defined:

$$v(p) = \sum_{i \in NS} \sum_{1 \le j \le m_i} g_{ij}^2(p). \tag{6}$$

This function is continuous, differentiable, and it holds that, $v(p) \ge 0$, for all $p \in \Delta$. Obviously, a strategy, p^*, is a Nash equilibrium, if and only if, $v(p^*) = 0$, i.e., p^* is a global minimizer of $v(p)$ (McKelvey, 1991; McKelvey & McLennan, 1996).

PSO can now be applied straightforwardly for the solution of the global optimization problem defined in equation (6). For the detection of several Nash equilibria, the algorithm can be restarted, either applying the deflection procedure or not, on the already detected minimizers (Nash equilibria) as described in Chapter Five. In order for a solution, p, to be a Nash equilibrium, it must hold that $p \in \Delta$, as required by Definition 1. For this purpose, the evaluation of a candidate solution is not performed using its original form but rather a normalized one, which is obtained as follows (Pavlidis *et al.*, 2005):

$$x_{ij}^p = \frac{\| x_{ij} \|}{\sum_{j=1}^{m_i} \| x_{ij} \|},$$

where, $i \in NS$ and $j = 1, 2, \dots, m_i$. This normalization ensures that the evaluated solution lies in Δ. However, the normalized solutions are used only for evaluation purposes and not in the swarm, because, in preliminary experiments, it was shown that their use drastically decreases the population's diversity, thereby reducing the performance of all algorithms studied by Pavlidis *et al.* (2005).

The next section presents representative results from the work of Pavlidis *et al.* (2005) and discusses these to reveal the potential of PSO for solving Nash equilibria problems.

Table 5. The swarm size, available iterations and number of restarts of PSO per test problem

Problem	Swarm Size	Iterations	Restarts
TP_{NM-1}	20	1000	8
TP_{NM-2}	20	1000	10
TP_{NM-3}	50	2000	10
TP_{NM-4}	10	1000	15
TP_{NM-5}	20	1000	18
TP_{NM-6}	10	1000	20

Experimental Setting and Results

PSO was applied by Pavlidis *et al.* (2005) on benchmark problems TP_{NE-1}-TP_{NE-6}, defined in Appendix A. Besides this, DE and CMA-ES were also applied on the same problems and compared to each other.

The main goal was the detection of all Nash equilibria for each problem, with an accuracy equal to 10^{-8}. To achieve this, two different techniques, namely multistart and deflection, were considered. In the simple multistart approach, the algorithm was let to run for a prespecified number of restarts and for a specific number of iterations per restart. In the end, the total number of different Nash equilibria detected was recorded along with the corresponding number of function evaluations. In the case of deflection, the algorithm was restarted after the detection of a Nash equilibrium, applying the deflection transformation described in Chapter Five, on all detected minimizers. The same statistics as for multistart were also recorded in this case.

The global variants of both the inertia weight and the constriction coefficient PSO, denoted as $PSO^{[in]}$ and $PSO^{[co]}$, respectively, were considered. In the first case, the inertia weight was initialized to 1.0 and gradually decreased towards zero for the 75% of the available iterations. In the latter case, the default parameter values, $\chi = 0.729$, $c_1 = c_2 = 2.05$, were used. In both cases, particles were restricted in the range $[-1,1]^n$, where n stands for the corresponding problem dimension. In addition, a maximum velocity equal to 1.0 was used (Pavlidis *et al.*, 2005). For each test problem and algorithm, 30 independent experiments were conducted. The swarm size, allowed number of iterations and number of restarts per experiment are reported in Table 5. Moreover, Table 6 reports the mean number of detected Nash equilibria, as well as the mean number of function evaluations per equilibrium using either deflection or multistart, for all test problems (Pavlidis *et al.*, 2005).

Two main observations can be stated on the results. The first is the superiority of $PSO^{[co]}$ in all cases, in terms of the required number of function evaluations per equilibrium. The second is the superior per-

Table 6. Mean number of detected Nash equilibria and mean number of function evaluations per equilibrium for the constriction coefficient ($PSO^{[co]}$) and the inertia weight ($PSO^{[in]}$) PSO variants, using deflection and multistart as reported in (Pavlidis et al., 2005)

Problem	PSO var.	Deflection		Multistart	
		Nash eq.	Mean F. Eval.	Nash eq.	Mean F. Eval.
TP_{NE-1}	$PSO^{[co]}$	2.97	23219.00	2.67	6265.56
	$PSO^{[in]}$	3.00	30966.67	2.47	23998.56
TP_{NE-2}	$PSO^{[co]}$	4.67	24504.73	3.40	11973.08
	$PSO^{[in]}$	4.90	30276.07	3.23	39025.94
TP_{NE-3}	$PSO^{[co]}$	3.00	255346.25	2.50	78288.75
	$PSO^{[in]}$	3.37	255715.97	2.30	226939.72
TP_{NE-4}	$PSO^{[co]}$	7.03	8569.21	2.93	7455.39
	$PSO^{[in]}$	6.90	15026.70	2.07	40458.78
TP_{NE-5}	$PSO^{[co]}$	10.53	5598.80	1.40	33819.00
	$PSO^{[in]}$	10.50	15116.76	1.23	134998.00
TP_{NE-6}	$PSO^{[co]}$	14.20	3047.69	5.33	6499.35
	$PSO^{[in]}$	14.33	6815.01	4.60	21526.97

formance of PSO[in] against PSO[co], in terms of the mean number of detected Nash equilibria whenever deflection is used; exactly the opposite happens in the case of simple multistart. This indicates that the slower convergence of PSO[in] renders it capable of taking full advantage of the deflection procedure, conducting a more thorough exploration of the search space.

In parallel, we can observe that deflection increases the effectiveness of both PSO approaches significantly, especially for problems with a high number of equilibria (TP_{NE-4}-TP_{NE-6}), contrary to the simple multistart technique. Of course, this gain usually comes at the cost of an increased number of function evaluations, especially in the (simple) cases with small number of equilibria. We must also mention that, as reported by Pavlidis *et al.* (2005), PSO[co] was among the best performing algorithms when deflection was used.

The reported results suggest that PSO with deflection can address Nash equilibria problems satisfactorily. Its additional advantages against gradient-based approaches in optimization problems that lack nice mathematical properties, as well as its potential for easy adaptation to any special requirement imposed by the problem at hand, supports the claim that PSO can serve as a valuable tool in different aspects of game theory.

CHAPTER SYNOPSIS

The application of PSO in OR problems was presented, along with transformations, techniques, representations and manipulations that allow the efficient solution of such problems with PSO variants. Results on different applications from the fields of scheduling, inventory optimization and game theory were reported and discussed. The chapter provided basic knowledge and techniques for the proper construction of objective functions per case, as well as the alleviation of obstacles due to problem peculiarities and requirements. The provided material can be easily adopted for similar problems or constitute the initiating point for further development of more sophisticated approaches.

REFERENCES

Abdul-Razaq, T. S., Potts, C. N., & Van Wassenhove, L. N. (1990). A survey of algorithms for the single machine total weighted tardiness scheduling problem. *Discrete Applied Mathematics*, *26*(2-3), 235–253. doi:10.1016/0166-218X(90)90103-J

Aumann, R., & Hart, S. (Eds.). (1992). *Handbook of game theory with economic applications*. Handbooks in Economics (11), Vol. 1. Amsterdam: North-Holland.

Cesta, A., Oddi, A., & Smith, S. (2002). A constrained-based method for project scheduling with time windows. *Journal of Heuristics*, *8*, 109–136. doi:10.1023/A:1013617802515

Chang, H. J., & Dye, C. Y. (1999). An EOQ model for deteriorating items with time varying demand and partial backlogging. *The Journal of the Operational Research Society*, *50*, 1176–1182.

Chern, M. S., Yang, H. L., Teng, J. T., & Papachristos, S. (2008). Partial backlogging inventory lot-size models for deteriorating items with fluctuating demand under inflation. *European Journal of Operational Research*, *191*(1), 127–141. doi:10.1016/j.ejor.2007.03.053

Covert, R. P., & Philip, G. C. (1973). An EOQ model for items with Weibull distribution deterioration. *American Institute of Industrial Engineers, 5*, 323–326.

Crauwels, H. A. J., Potts, C. N., & Van Wassenhove, L. N. (1998). Local search heuristics for the single machine total weighted tardiness scheduling problem. *INFORMS Journal on Computing, 10*(3), 341–350. doi:10.1287/ijoc.10.3.341

Dechter, R., Meiri, I., & Pearl, J. (1991). Temporal constraint networks. *Artificial Intelligence, 49*, 61–95. doi:10.1016/0004-3702(91)90006-6

Den Besten, M., Stützle, T., & Dorigo, M. (2000). Ant colony optimization for the total weighted tardiness problem. [Berlin: Springer.]. *Lecture Notes in Computer Science, 1917*, 611–620. doi:10.1007/3-540-45356-3_60

Donaldson, W. A. (1977). Inventory replenishment policy for a linear trend in demand: an analytical solution. *Operational Research Quarterly, 28*, 663–670.

Ghare, P. M., & Shrader, G. F. (1963). A model for exponentially decaying inventories. *Journal of Industrial Engineering, 14*, 238–243.

Goyal, S. K., & Giri, B. C. (2001). Recent trends in modeling of deteriorating inventory-an invited review. *European Journal of Operational Research, 134*, 1–16. doi:10.1016/S0377-2217(00)00248-4

Hillier, F. S., & Lieberman, G. J. (2005). *Introduction to operations research*. Boston: McGraw-Hill.

Johnson, D. S., & Garey, M. R. (1979). *Computers and intractability: a guide to the theory of NP-completeness*. Englewood Cliffs, NJ: W.H. Freeman & Co.

Lenstra, J. K., Rinnooy Kan, H. G., & Brucker, P. (1977). Complexity of machine scheduling problem. In *Studies in Integer Programming, Vol. 1 of Annals of Discrete Mathematics* (pp. 343-362). Amsterdam: North-Holland.

Liao, C.-J., Tseng, C.-T., & Luarn, P. (2007). A discrete version of particle swarm optimization for flowshop scheduling problems. *Computers & Operations Research, 34*(10), 3099–3111. doi:10.1016/j.cor.2005.11.017

McKelvey, R. D. (1991). *A Liapunov function for Nash equilibria* (Tech. Rep.). Pasadena: California Institute of Technology.

McKelvey, R. D., & McLennan, A. (1996). Computation of equilibria in finite games. In H.M. Amman, D.A. Kendrick, & J. Rust (Eds.), *Handbook of Computational Economics, Handbooks in Economics (13), Vol. 1* (pp. 87-142). Amsterdam: North-Holland.

Mladenović, N., & Hansen, P. (1997). Variable neighborhood search. *Computers & Operations Research, 24*(11), 563–571. doi:10.1016/S0305-0548(97)00031-2

Nash, J. F. (1950). Equilibrium points in *n*-person games. *Proceedings of the National Academy of Sciences of the United States of America, 36*(1), 48–49. doi:10.1073/pnas.36.1.48

Nash, J. F. (1951). Noncooperative games. *The Annals of Mathematics, 54,* 289–295. doi:10.2307/1969529

Osborne, M. J., & Rubinstein, A. (1994). *A course in game theory.* Cambridge, MA: MIT Press.

Parsopoulos, K. E., Skouri, K., & Vrahatis, M. N. (2008). Particle swarm optimization for tackling continuous review inventory models. [Berlin, Germany: Springer.]. *Lecture Notes in Computer Science, 4974,* 103–112. doi:10.1007/978-3-540-78761-7_11

Parsopoulos, K. E., & Vrahatis, M. N. (2006). Studying the performance of unified particle swarm optimization on the single machine total weighted tardiness problem. In *Lecture Notes in Artificial Intelligence, Vol. 4304* (pp. 760-769). Berlin: Springer.

Pavlidis, N. G., Parsopoulos, K. E., & Vrahatis, M. N. (2005). Computing Nash equilibria through computational intelligence methods. *Journal of Computational and Applied Mathematics, 175*(1), 113–136. doi:10.1016/j.cam.2004.06.005

Pinedo, M. (1995). *Scheduling: theory, algorithms and systems.* Englewood Cliffs, NJ: Prentice Hall.

Potts, C. N., & Van Wassenhove, L. N. (1991). Single machine tardiness sequencing heuristics. *IIE Transactions, 23,* 346–354. doi:10.1080/07408179108963868

Raafat, F. (1991). Survey of literature on continuously deteriorating inventory model. *The Journal of the Operational Research Society, 42,* 27–37.

Resh, M., Friedman, M., & Barbosa, L. C. (1976). On a general solution of the deterministic lot size problem with time-proportional demand. *Operations Research, 24*(4), 718–725. doi:10.1287/opre.24.4.718

Saldam, A., Ahmad, I., & Al-Madani, S. (2002). Particle swarm optimization for task assignment problem. *Microprocessors and Microsystems, 26*(8), 363–371. doi:10.1016/S0141-9331(02)00053-4

Schwalb, E., & Vila, T. (1997). *Temporal constraints: a survey* (Tech. Rep.). Irvine (CA), USA: ICS, University of California at Irvine.

Skouri, K., & Papachristos, S. (2002). A continuous review inventory model, with deteriorating items, time-varying demand, linear replenishment cost, partially time-varying backlogging. *Applied Mathematical Modelling, 26*(5), 603–617. doi:10.1016/S0307-904X(01)00071-3

Tadikamalla, P. R. (1978). An EOQ inventory model for items with Gamma distribution. *American Institute of Industrial Engineers, 5,* 100–103.

Tasgetiren, M. F., Sevkli, M., Liang, Y. C., & Gencyilmaz, G. (2004). Particle swarm optimization algorithm for single machine total weighted tardiness problem. In *Proceedings of the 2004 IEEE Congress on Evolutionary Computation (CEC'04), Portland (OR), USA* (pp. 1412-1419).

Teng, J. T., Chang, H. J., Dye, C. Y., & Hung, C. H. (2002). An optimal replenishment policy for deteriorating items with time-varying demand and partial backlogging. *Operations Research Letters, 30,* 387–393. doi:10.1016/S0167-6377(02)00150-5

Von Stengel, B. (2002). Computing equilibria for two-person games. In R. Aumann & S. Hart (Eds.), *Handbook of Game Theory, Vol. 3* (pp. 1723-1759). Amsterdam: North-Holland.

Winston, W. (2003). *Operations research: applications and algorithms*. Pacific Grove, CA: Duxbury Press.

Zhang, H., Li, H., & Tam, C. M. (2006). Particle swarm optimization for resource-constrained project scheduling. *International Journal of Project Management, 24*(1), 83–92. doi:10.1016/j.ijproman.2005.06.006

Chapter 9
Applications in Bioinformatics and Medical Informatics

This chapter presents two interesting applications of PSO in bioinformatics and medical informatics. The first consists of the adaptation of probabilistic neural network models for medical classification tasks. The second application employs the unified PSO algorithm to tackle magnetoencephalography problems. Our main goal is to clarify crucial points where PSO interferes with the employed computational models and provide details on the formulation of the corresponding optimization problems and experimental settings. Indicative results are reported to illustrate the workings of the algorithms and provide representative samples of their performance.

INTRODUCTION

The research blossoming of the past few years in molecular biology and brain studies has produced a huge amount of data. The need for their processing and assessment gave rise to the field of *bioinformatics*, which deals with the application of information processing methodologies on biological and biomedical tasks. This includes data handling, as well as the development of computational models for the better understanding of the studied physical systems. Modeling, data mining and machine learning occupy a central place in bioinformatic research, which tackles problems such as protein localization, sequence and genome analysis, brain activity monitoring and analysis etc.

Different methodologies have been proposed to address the aforementioned problems. Artificial neural networks, evolutionary algorithms and swarm intelligence are considered among the most popular methodologies today, thanks to their efficiency in analyzing and extracting knowledge from systems with inherent non-deterministic structure and responses. In this chapter, we present two applications

DOI: 10.4018/978-1-61520-666-7.ch009

Copyright © 2010, IGI Global. Copying or distributing in print or electronic forms without written permission of IGI Global is prohibited.

where PSO is used in combination with different modeling techniques to tackle challenging biomedical problems.

The first application refers to the adaptation of probabilistic neural networks (Georgiou *et al.*, 2006). A self-adaptive model is considered, where PSO serves as its basic configuration mechanism. The resulting model is applied on two protein localization problems, as well as on two medical diagnostic tasks, with very promising results. The second application comes from the field of brain studies and refers to the solution of source localization problems in magnetoencephalography (Parsopoulos *et al.*, 2009). In this case, PSO is employed to detect an unknown excitation source, using only a number of sensor measurements. As a second task, we consider the detection of proper parameters in established spherical expansion models. Such models are used in brain studies for the approximation of the brain magnetic potential. A representative portion of published results for both applications are reported and discussed to justify the usefulness of PSO and probe its potential for addressing similar applications.

CALIBRATING PROBABILISTIC NEURAL NETWORKS

Probabilistic neural networks (PNNs) are supervised classification neural network models, closely related to the Bayes classification rule and the Parzen nonparametric probability density function estimation theory (Parzen, 1962; Specht, 1990). Their main advantage against different classifiers is their ability to effectively exploit all the available information on the problem at hand and provide uncertainty measures of the classification accuracy. For example, in a cancer classification task, PNNs can estimate the probability of a tumor being benign or malignant, instead of the yes/no responses provided by most classifiers.

PNNs have been used in a plethora of bioinformatics and medical tasks. Huang (2002) presents a comprehensive study of PNNs combined with a feature extraction method on cancer classification problems. Holmes *et al.* (2001) used PNNs to develop accurate NMR-based metabonomic models for the prediction of xenobiotic-induced toxicity in experimental animals, emphasizing their potential use in accelerated drug discovery programs. Guo *et al.* (2004) considered PNNs for the design of an automatic, reliable, and efficient prediction system for protein subcellular localization in large-scale genome analysis, while Wang *et al.* (1998) used PNNs to identify subtle changes in brain tissue quantities and volumes through magnetic resonance image analysis.

The classification capabilities of PNNs can be attributed to inherent properties stemming from their originating methodologies, namely statistical pattern recognition and artificial NNs. Actually, PNNs constitute a NN implementation of kernel discriminant analysis that employs Bayesian strategies for pattern classification. Each pattern is stored as a separate neuron in the PNN; thus, it can be viewed as an "intelligent" memory (Berthold & Diamond, 1998). This feature results in lower execution time and straightforward training but also in higher storage requirements than the typical feedforward NNs. The next section describes the basics on PNNs.

Probabilistic Neural Networks

The structure of a PNN is similar to that of feedforward NNs, described in Chapter Six. However, they always have only four layers, named as *input*, *pattern*, *summation*, and *output layer*, respectively, as illustrated in Fig. 1. Let the problem at hand consist of K classes with M_k training patterns per class, $k =$

Figure 1. The structure of probabilistic neural networks

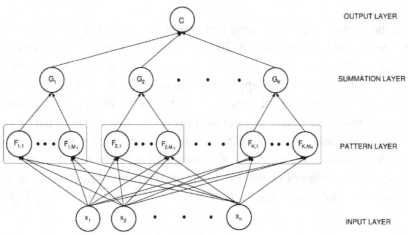

$1, 2,\ldots, K$, and, $X_{i,k} \in \mathbf{R}^n$ be the i-th training pattern of the k-th class, $i = 1, 2,\ldots, M_k$. Then, for each pattern vector, $X_{i,k}$, a neuron in the pattern layer of the PNN is created. This neuron represents a Gaussian kernel activation function with center $X_{i,k}$. Moreover, each class has a neuron in the summation layer, which is used to sum the responses of all neurons belonging to this class.

The classification of a new input vector, $X \in \mathbf{R}^n$, starts with its presentation to the input layer. This layer simply forwards X to all neurons of the pattern layer. Each neuron produces an output using a Gaussian kernel of the following form:

$$f_{i,k}(X) = \frac{1}{(2\pi)^{n/2} \det(\Sigma)^{1/2}} \exp\left(-\frac{1}{2}(X - X_{i,k})^{\mathrm{T}} \Sigma^{-1} (X - X_{i,k}) \right), \tag{1}$$

$i = 1, 2,\ldots, M_k, \quad k = 1, 2,\ldots, K,$

where $X_{i,k}$ is the center of the kernel; Σ is the matrix of *spread parameters* (also called *smoothing parameters*); and $\det(\Sigma)$ denotes its determinant. The spread parameters determine the size and shape of the receptive field of the kernel. Specht (1990) initially proposed the use of a single global spread parameter, σ^2. PNNs that follow this approach are called *homoscedastic*. On the other hand, *heteroscedastic* PNNs use separate uncorrelated spread parameters.

The summation layer computes the approximation of the conditional class probability functions by combining the previously computed densities as follows:

$$G_k(X) = w_k \sum_{i=1}^{M_k} f_{i,k}(X), \tag{2}$$

for $k = 1, 2,\ldots, K$, where w_k is a positive weight standing for the prior probability of the k-th class. Thus, it must hold that:

$$\sum_{k=1}^{K} w_k = 1.$$

The input vector, X, is classified to the class with the highest summation unit value. Thus, if $C(X)$ denotes the class where X is classified, it holds that:

$$C(X) = \arg\max_{1 \leq k \leq K} \{G_k\}. \tag{3}$$

The prior probabilities, w_k, are always problem-dependent; hence, their estimation is nontrivial, since the training set may not contain useful information regarding the priors. A common setting in such cases is the assumption of equiprobable classes with:

$$w_k = 1/K, \quad 1 \leq k \leq K.$$

PNNs have a unique feature that distinguishes them from all different classification techniques. This is the property of its outputs to correspond to conditional posterior probabilities of class membership for any input vector. This property is particularly desirable, since it gives the opportunity to classify a pattern vector as unknown if all posterior probabilities fall below a predefined threshold. Additionally, Bayesian confidence intervals can be easily derived for the classification, rendering PNNs a valuable tool.

In addition to the aforementioned properties, heteroscedastic PNNs can also infer information related to feature selection in problems of large dimensionality. This is achieved by monitoring the spread parameters. Indeed, if the i-th variable (feature) assumes large spread parameter, σ_i^2, it can be considered as less important and hence omitted. On the other hand, the "curse of dimensionality" remains the main drawback of PNNs (Hämäläinen & Holmström, 1996). Besides, dependent variables in the data set can also reduce performance. These problems can be addressed by applying principal components analysis (PCA) on the data, although the resulting variables can become even harder to interpret.

In the next section, we describe and analyze a recently proposed self-adaptive PNN model, which requires the contribution of PSO to adapt its spread parameters.

A Self-Adaptive Probabilistic Neural Network Model

As already mentioned, the performance of PNNs depends on a multitude of factors. The spread matrix, Σ, is included among the most important factors, as it has a direct effect on the main classification element of PNNs, i.e., the kernels. In general, Σ can be a diagonal matrix with dimension equal to that of the pattern vectors. Obviously, in one-dimensional cases, the matrix is reduced to a single scalar value, σ^2.

The detection of the optimal spread matrix for a given problem can significantly enhance the classification quality of PNNs. Moreover, since any optimal spread matrix refers solely to a specific data set, it is desirable to equip the PNN model with a technique that allows the computation of new spread parameters under different conditions (data sets, problems etc.) For this purpose, Georgiou *et al.* (2006) proposed a *self-adaptive* PNN model (denoted as SA-PNN). This model incorporates an optimization module that employs optimization algorithms to detect the optimal spread parameters, and it applies both to homoscedastic and heteroscedastic PNNs. The original version of Georgiou *et al.* (2006) employed PSO in both high- and single-dimensional problems. In the latter, the golden section algorithm was also applied alternatively to PSO.

Specifically for PSO, which is of our main interest, a swarm of spread matrices was randomly initialized and adapted through the typical PSO update equations. The spread matrix is diagonal; thus, it can be encoded as a particle by considering only its main diagonal as a vector. Then, for a specific sample

Table 1. Pseudocode of the SA-PNN model

Input:	Training set, $S_{train} = \{X_1^{train}, X_2^{train}, ..., X_M^{train}\}$; test set, S_{test}; swarm size, N; PSO parameters, χ, c_1, c_2; bounding box, B.
Step 1.	**Construct** the PNN by passing all training pattern vectors.
Step 2.	**Set** $t \leftarrow 0$.
Step 3.	**Initialize** particles, $x_i(t) = \text{diag}(\Sigma_i(t)) \in B$, and velocities, $v_i(t) \in B$, $i = 1, 2, ..., N$.
Step 4.	**Initialize** best positions, $p_i(t)$, and neighborhood best indices g_i, $i = 1, 2, ..., N$.
Step 5.	**While** (maximum number of iterations is not reached) **Do**
Step 6.	**Update** velocities, $v_i(t+1)$, and particles, $x_i(t+1)$, $i = 1, 2, ..., N$.
Step 7.	**Constrain** particles within B.
Step 8.	**Evaluate** particles using the objective function, $f(x_i(t+1))$.
Step 9.	**Update** best positions, $p_i(t)$, and g_i, $i = 1, 2, ..., N$.
Step 10.	**Set** $t \leftarrow t+1$.
Step 11.	**End While**
Step 12.	**Report** optimal spread matrix and performance of the PNN for the test set S_{test}.

from the data set, the pseudocode reported in Table 1 is applied to detect the best spread matrix with respect to the specific data sample.

The objective function, $f(x)$, constitutes an essential element of the SA-PNN model. Georgiou *et al.* (2006) used the *leave-one-out* misclassification proportion of the training set as the objective value of a particle (spread matrix). According to this, the PNN is trained using all but one of the patterns from the training set. Then, the excluded pattern is used to assess the classification accuracy of the PNN. This procedure is repeated excluding a different pattern each time, until all patterns have been excluded from the training set once.

This procedure would be very expensive computationally for feedforward NNs, since it requires the retraining of the NN for each excluded pattern. However, it is efficient for the PNN, as it requires simply the omission of the output, $f_{i,k}$, of the excluded pattern, $X_{i,k}$, from the summation unit, G_k. The adaptation of the spread parameter terminates after a prescribed number of PSO iterations.

In the next paragraph, we present and discuss the application of the SA-PNN model with PSO on four widely used classification tasks.

Experimental Setting and Results

The SA-PNN model with PSO was applied on the *E.coli*, Yeast, Breast Cancer and Pima Indian Diabetes data sets, using three different sampling techniques, namely stratified random sampling, λ-fold cross-validation and train-validation-test partitioning of the data set. Brief descriptions of the data sets are provided in Appendix A of the book at hand, while the sampling techniques are described below:

1. **Stratified random sampling (SRS):** According to this technique, a data set of size N with K classes is divided in K non-overlapping subsets, called *strata*, each containing N_k pattern vectors from the k-th class, $k \in \{1, 2, ..., K\}$, with:

Table 2. PSO parameters for the SA-PNN used by Georgiou et al. (2006)

Parameter	Value
Swarm size	5 for homoscedastic and 10 for heteroscedastic PNNs.
Maximum iterations	50
Bounds of spread parameters	[0,5]
Constriction PSO parameters	$\chi = 0.729$, $c_1 = c_2 = 2.05$.
Inertia PSO parameters	$w = 1.0 \rightarrow 0.1$ for 37 iterations (75% of maximum) and fixed to 0.1 for the rest.
Maximum velocity	0.3
Neighborhood topology	Ring with radius equal to 1.

$$\sum_{k=1}^{K} N_k = N.$$

Then, a random sample of size a_k is selected independently from each stratum. All the selected samples are put together to form the stratified random sample, which is used as the training set and has size:

$$a = \sum_{k=1}^{K} a_k,$$

while the rest of the data are used as the test set. Proportionate allocation uses a constant sampling fraction, a_k / N_k, per stratum.

2. **λ-fold cross-validation (λ-CV):** According to this technique, the data set is divided into λ parts of approximately equal size. Then, each one of the λ subsets is used as the test set, while the rest subsets serve as training sets. Finally, the average error over all λ trials is computed.

3. **Train-validation-test partitioning (TVT):** According to this technique, the data set is divided into three components; a procedure which is widely used in NN training. In the SA-PNN framework, the network is constructed using the train set. Then, PSO determines the spread matrix, Σ, by minimizing the leave-one-out misclassification error on the training set, while the error on the validation set is monitored at each iteration. In order to avoid over-training of the network, the algorithm stops if the validation error increases for a number, μ, of subsequent iterations. Finally, the spread matrix that yielded the lowest error on the validation set is reported.

Georgiou *et al.* (2006) applied homoscedastic and heteroscedastic SA-PNNs, optimized by the constriction coefficient and inertia weight PSO variants. In addition, they compared their performance with that of different methodologies, namely the standard PNNs, adaptive PNNs optimized with golden section, feedforward NNs, and C-support vector machines. In all approaches, they used the SRS, TVT and λ-CV sampling techniques. For the SRS approach, the random sample size was set equal to 50% of the corresponding class size. In TVT, the value $\mu = 5$ was used, while $\lambda = 4$ and $\lambda = 10$ were considered for cross-validation, depending on the data set at hand. The homoscedastic cases with constriction coefficient and inertia weight PSO will be henceforth denoted as SA-PNN$_c$[HOM] and SA-PNN$_i$[HOM], respectively, while the corresponding heteroscedastic cases will be denoted as SA-PNN$_c$[HET] and SA-PNN$_i$[HET]. All PSO parameters used by Georgiou *et al.* (2006) are summarized in Table 2.

Tables 3-6 contain representative results derived from Georgiou *et al.* (2006) to provide intuition regarding the performance of different SA-PNNs models. More specifically, the mean number and standard deviation of the achieved classification accuracy per problem, sampling technique and SA-PNN model is reported. In all cases, SA-PNNs with PSO were among the best performing models as reported by Georgiou *et al.* (2006). Regarding their combination with classification techniques, λ-CV was distinguished in the *E.coli*, Yeast and Pima Indian Diabetes data sets, with TVT and SRS exhibiting no statistically significant differences in Kolmogorov-Smirnov significance tests (Georgiou *et al.*, 2006). In the Breast Cancer data set, all three approaches had statistically equivalent performance.

Table 3. Classification accuracy of SA-PNN models for the E.coli data set problem under different sampling techniques

Sampling	Model	Mean	StD
SRS	SA-PNN$_c^{[HOM]}$	86.39	2.10
	SA-PNN$_i^{[HOM]}$	85.72	3.33
	SA-PNN$_c^{[HET]}$	81.42	5.27
	SA-PNN$_i^{[HET]}$	81.78	4.65
4-CV	SA-PNN$_c^{[HOM]}$	86.79	0.58
	SA-PNN$_i^{[HOM]}$	86.77	0.67
	SA-PNN$_c^{[HET]}$	83.44	1.62
	SA-PNN$_i^{[HET]}$	83.07	2.48
TVT	SA-PNN$_c^{[HOM]}$	86.33	3.21
	SA-PNN$_i^{[HOM]}$	86.08	3.92
	SA-PNN$_c^{[HET]}$	81.69	5.33
	SA-PNN$_i^{[HET]}$	82.41	5.48

Table 4. Classification accuracy of SA-PNN models for the Yeast data set problem under different sampling techniques

Sampling	Model	Mean	StD
SRS	SA-PNN$_c^{[HOM]}$	57.73	1.59
	SA-PNN$_i^{[HOM]}$	57.73	1.60
	SA-PNN$_c^{[HET]}$	58.01	1.71
	SA-PNN$_i^{[HET]}$	57.66	1.73
10-CV	SA-PNN$_c^{[HOM]}$	58.84	0.39
	SA-PNN$_i^{[HOM]}$	58.87	0.33
	SA-PNN$_c^{[HET]}$	58.86	0.69
	SA-PNN$_i^{[HET]}$	58.92	0.77
TVT	SA-PNN$_c^{[HOM]}$	57.70	2.10
	SA-PNN$_i^{[HOM]}$	57.76	2.25
	SA-PNN$_c^{[HET]}$	57.89	2.62
	SA-PNN$_i^{[HET]}$	57.50	2.38

Moreover, the most efficient combination of SA-PNNs with PSO seems to depend on the problem at hand and the sampling technique. Indeed, as we observe in Tables 3-6, SRS and TVT perform better with the constriction coefficient PSO variants combined with either SA-PNN model. On the other hand, for the (overall best-performing) CV cases, constriction coefficient PSO variants performed better in the *E.coli* problem combined with homoscedastic SA-PNNs, as well as in the Pima Indian Diabetes problem combined with heteroscedastic SA-PNNs. Contrary to this, inertia weight PSO variants performed better for the Yeast and Breast Cancer data sets combined with heteroscedastic and homoscedastic models, respectively. Finally, we can observe that homoscedastic SA-PNNs were overall more efficient in the

Table 5. Classification accuracy of SA-PNN models for the Breast Cancer data set problem under different sampling techniques

Sampling	Model	Mean	StD
SRS	SA-PNN$_c$[HOM]	96.20	0.82
	SA-PNN$_i$[HOM]	96.20	0.78
	SA-PNN$_c$[HET]	94.71	1.44
	SA-PNN$_i$[HET]	95.11	1.23
10-CV	SA-PNN$_c$[HOM]	95.83	0.26
	SA-PNN$_i$[HOM]	95.85	0.23
	SA-PNN$_c$[HET]	95.13	0.46
	SA-PNN$_i$[HET]	94.96	0.61
TVT	SA-PNN$_c$[HOM]	95.49	1.47
	SA-PNN$_i$[HOM]	95.49	1.41
	SA-PNN$_c$[HET]	95.72	1.36
	SA-PNN$_i$[HET]	95.55	1.34

Table 6. Classification accuracy of SA-PNN models for the Pima Indians Diabetes data set problem under different sampling techniques

Sampling	Model	Mean	StD
SRS	SA-PNN$_c$[HOM]	73.59	2.08
	SA-PNN$_i$[HOM]	73.40	2.01
	SA-PNN$_c$[HET]	73.96	1.87
	SA-PNN$_i$[HET]	74.05	2.10
10-CV	SA-PNN$_c$[HOM]	74.13	0.82
	SA-PNN$_i$[HOM]	74.12	0.77
	SA-PNN$_c$[HET]	75.29	0.72
	SA-PNN$_i$[HET]	75.11	0.87
TVT	SA-PNN$_c$[HOM]	72.93	1.76
	SA-PNN$_i$[HOM]	72.72	1.92
	SA-PNN$_c$[HET]	74.32	2.89
	SA-PNN$_i$[HET]	74.21	3.19

E.coli and Breast Cancer cases, while heteroscedastic models suited better the Yeast and Pima Indian Diabetes cases.

The reported results suggest that PSO can provide a means of improvement for PNN models, resulting in approaches that exhibit competitive performance against other important bioinformatic methodologies. Recent work (Georgiou *et al.*, 2009) incorporates PSO in combination with Bayesian estimation models to calibrate further aspects of the PNN structure, opening new research horizons regarding the interaction of these methodologies in the development of more efficient adaptive classification models. The next section presents the application of PSO on a different yet important medical informatics task.

TACKLING MAGNETOENCEPHALOGRAPHY PROBLEMS

Magnetoencephalography (MEG) is a methodology for studying the functional human brain. It is based on capturing and analyzing the magnetic fields produced by excitations of small regions of the human brain by using sensor measurements. The obtained data can be used to formulate the problem as a challenging optimization task. To date, MEG is considered one of the most important biomedical methodologies, accompanied by an extensive literature (Bronzan, 1971; Dassios, 2006, 2007, 2008a, 2008b; Dassios & Fokas, in press; Dassios *et al.*, 2005; 2007a, 2007b; Dassios & Kariotou, 2003a, 2003b, 2004, 2005; Fokas et al., 1996, 2004; Geselowitz, 1970; Grynspan & Geselowitz, 1973; Ilmoniemi *et al.*, 2005; Nolte & Curio, 1997).

Parsopoulos *et al.* (2009) studied PSO in two different MEG problems. The first is the well-known *inverse problem*, where an unknown excitation source is detected using a number of (accurate or noisy) sensor measurements. The second problem is called *forward task for inverse use* and refers to the computation of proper coefficients that optimize the approximation of the magnetic potential through spherical expansions as well as their tolerance under variations of the number of sensors. These two tasks are formally described in the following sections.

The Considered MEG Problems

The workings of MEG are based on the fact that the brain is activated via an electrochemical excitation of a small region in the cerebral tissue, which produces a very weak yet measurable magnetic field outside the head. These biomagnetic signals are received by a set of sensors distributed almost uniformly upon a helmet, which covers the whole head except the frontal face and neck. SQUID is the most sensitive equipment ever built for such measurements, as it can measure magnetic flux down to 50-500 fT (Hamalainen *et al.*, 1993).

Biomagnetic fields live in the realm of the quasistatic Maxwell equations (Plonsey & Hepner, 1967). Thus, in the source-free exterior space, the magnetic field is provided through the gradient of a harmonic function, called *scalar magnetic potential*, and denoted as $U(r)$. For a spherical conductor, which commonly models the human head, and for the most widely used source model of a dipolar current with moment Q, located at an interior point, r_0, the magnetic potential at an exterior point, r, is defined as follows (Sarvas, 1987):

$$U(r) = \frac{(Q \times r_0)^{\mathrm{T}} r}{F(r; r_0)},$$

(4)

where,

$$F(r; r_0) = \|r\| \|r\text{-}r_0\|^2 + \|r\text{-}r_0\| r^{\mathrm{T}} (r\text{-}r_0),$$

with $\|r\|$ denoting the measure of vector r. The two MEG problems addressed with PSO by Parsopoulos *et al.* (2009) are defined below.

The Inverse Problem

This problem deals with the identification of a source position and moment based in data obtained through sensor measurements. More specifically, let:

$$r_0 = (r_{01}, r_{02}, r_{03})^{\mathrm{T}} \text{ and } Q = (q_1, q_2, q_3)^{\mathrm{T}},$$

be the position and moment vector of a given source, respectively. These two vectors are assumed to be normal to each other, and they can be used through equation (4) to compute the magnetic potential, $U(r)$, for a set of sensor positions, r_i, $i = 1, 2, \ldots, K$. Then, the essential *inverse problem* is defined as the identification of the original source (position and moment vector) with a desirable accuracy using solely equation (4) and the provided measurements.

The Forward Task for Inverse Use

The magnetic potential, $U(r)$, is a harmonic function; hence, it enjoys a spherical expansion, as follows:

$$U(r) = \sum_{l=1}^{\infty} A_l \frac{P_l(\cos(\theta))}{\rho^{l+1}} + \sum_{l=1}^{\infty} \sum_{k=1}^{l} \left[B_l^k \cos(k\varphi) + C_l^k \sin(k\varphi) \right] \frac{P_l^k(\cos(\theta))}{\rho^{l+1}},$$

(5)

with respect to the Legendre functions, $P_l(\cos(\theta))$ and $P_l^k(\cos(\theta))$, where, A_l, B_l^k, and C_l^k, are coefficients under determination, depending on the source defining vectors, r_0 and Q; and (ρ, φ, θ) are the spherical coordinates of the sensor position, r.

As reported in Parsopoulos *et al.* (2009), the first eight coefficients,

$$A_1, A_2, B_1^1, B_2^1, B_2^2, C_1^1, C_2^1, C_2^2,$$

which correspond to $l = 1, 2$, in equation (5), provide more than adequate information for the analytical acquisition of the source. However, large fluctuations in their values impose problems when the number of sensor positions varies. Thus, it is essential to compute their values and study their behavior under a variable number of sensor measurements.

This task can be achieved by explicitly varying the number of sensors from very small up to reasonably large values with fixed increment, detecting for each instance the coefficients that produce the smallest error in the approximation of $U(r)$ with equation (5). Then, crucial information regarding the robustness of the coefficients can be gained by analyzing the absolute relative error between consecutive values of each coefficient for the different measurements. This is the main goal of the *forward task for inverse use* problem.

The methodology for tackling the presented MEG problems is given in the following paragraphs, along with representative results from Parsopoulos *et al.* (2009).

Experimental Setting and Results for the Inverse Problem

In their experiments, Parsopoulos *et al.* (2009) considered three different source points, defined by their position and moment vectors. The three sources were generated randomly inside a 3-dimensional sphere of radius, $r_{sphere} = 9$, simulating stimulation sources of the human brain. Also, three different numbers of sensors were considered, namely $K_{sensor} = 50, 100, 200$, and the corresponding equidistant sensor positions were generated on a sphere of radius, $r_{kask} = 10$, which simulates the helmet used for the measurement of the magnetic potential. Then, for each source point, the corresponding K_{sensor} measurements for all sensor positions were computed using equation (4) for all different numbers of sensors.

Parsopoulos *et al.* (2009) considered both the case of exact measurements and (the more realistic) case of noisy measurements. For this purpose, multiplicative noise was added on the measurements, U_k, $k = 1, 2,\ldots, K_{sensor}$, of the magnetic potential as follows:

$$U_k' = U_k (1+\eta), \quad k = 1, 2,\ldots, K_{sensor}, \tag{6}$$

where η is a normally distributed random variable, $\eta \sim N(0, \sigma_M^2)$. A noise value, η, corresponds to a percentage, $100\eta\%$, of measurement alteration. The only restriction of this noise addition scheme is the requirement for ensuring that $\eta > -1$ holds, to avoid alteration of the measurement sign. Values of U_k', produced by equation (6) using $\sigma_M = 0.01, 0.05, 0.10$, were used instead of the original, U_k, in the noisy experiments of Parsopoulos *et al.* (2009).

PSO was employed to detect the three sources for the different sensors' numbers and noise levels. Each particle of the swarm was considered to be a candidate source position, randomly initialized in a sphere with $r_{sphere} = 9$. The components of the defining position and moment vectors, $r_0^{(i)}$ and $Q^{(i)}$, respectively, of a candidate source were encoded in a particle, x_i, as a 5-dimensional vector:

$$x_i = \left(q_1^{(i)}, q_2^{(i)}, r_{01}^{(i)}, r_{02}^{(i)}, r_{03}^{(i)} \right)^T,$$

resulting in a 5-dimensional highly nonlinear optimization problem. Since the moment and position of a candidate source shall be normal to each other, the third component, $q_3^{(i)}$, of $Q^{(i)}$ was analytically derived, as follows:

$$q_3^{(i)} = \frac{q_1^{(i)} r_{01}^{(i)} + q_2^{(i)} r_{02}^{(i)}}{r_{03}^{(i)}},$$

Table 7. Parameters of the inverse problem and UPSO for the experiments reported in Parsopoulos et al. (2009)

Parameter type	Description	Values
Problem-related	Number of sensors	$K_{sensor} = 50, 100, 200$
	Sphere radius (head)	$r_{sphere} = 9$
	Kask radius	$r_{kask} = 10$
	Measurement noise	$\sigma_M = 0.0$ (noiseless), $0.01, 0.05, 0.1$
	Solution bounds	$[-9, 9]^5$
	Solution accuracy	10^{-16}
UPSO-related	Unification factor	$u = 0.0, 0.1, 0.2,\ldots, 1.0$
	Mutation strength	$\sigma = 1.0$
	PSO parameters	$\chi = 0.729, c_1 = c_2 = 2.05$
	Swarm size	$N = 50$
	Maximum iterations	$t_{max} = 3000$

in order to preserve orthogonality, while special care was taken when $r_{03}^{(i)} = 0$ (Parsopoulos *et al.*, 2009).

The objective function used for particle (candidate sources) evaluation was defined as (Parsopoulos *et al.*, 2009):

$$f(x_i) = \sum_{k=1}^{K_{sensor}} \left(U_k - U_k^{(i)} \right)^2, \tag{7}$$

where $U_k^{(i)}, k = 1, 2,\ldots, K_{sensor}$, are the measurements obtained by equation (4) for the specific candidate source encoded in particle x_i, and U_k are the actual measurements for the true source.

Parsopoulos *et al.* (2009) performed experiments on the inverse problem with the constriction coefficient variant of the unified PSO (UPSO), described in Chapter Four, and for unification factor values, $u = 0.0, 0.1, 0.2,\ldots, 1.0$. Recall that $u = 0.0$ and $u = 1.0$ correspond to the standard local and global PSO variants, respectively, while all intermediate values result in UPSO variants with different exploration/ exploitation properties. Additionally, they also considered the mutated UPSO variant with $\mu = 0.0$ and $\sigma = 1.0$. The whole parameter set of their experiments is reported in Table 7. 100 experiments were conducted for each source and algorithm, recording the number of successes in achieving the desirable accuracy, 10^{-16}, within the required number of iterations. Results were analyzed statistically and their significance was investigated through Wilcoxon rank sum hypothesis testing.

The results reported in Parsopoulos *et al.* (2009) are divided in two cases: without and with noise. In the first case, only the local PSO variant (i.e., UPSO with $u = 0.0$) and UPSO with $u = 0.1$ achieved success rates of 100% in all cases and for all sources. The rest of the algorithms suffered efficiency loss, especially as unification factor approaches $u = 1.0$. This is a strong indication of the high nonlinearity of the underlying objective function, which renders exploitation-promoting variants more efficient.

An indicative portion of results for the first source point considered in Parsopoulos *et al.* (2009) is reported in Table 8. As we observe, there is an increase of the required number of iterations with the number of sensors for the specific source. Statistical significance tests reported by Parsopoulos *et al.*

Table 8. Results of the most successful UPSO variants for the inverse problem without noise, in terms of the number of successes and mean number of required iteration, as reported in Parsopoulos et al. (2009)

Unification factor	K_{sensor}	Mutation	Success	Mean iterations
$u = 0.0$	50	No	100%	1136.79
	100	No	100%	1213.59
	200	No	100%	1284.04
$u = 0.1$	50	No	100%	347.67
		Yes	100%	674.82
	100	No	100%	351.60
		Yes	100%	716.66
	200	No	100%	361.59
		Yes	100%	778.44

(2009) reveal that there is a significant difference in the performance of the algorithm when the number of sensors increases. However, it was shown that this does not hold for all sources. Thus, it must be expected that increasing the number of sensors (and therefore the computational burden) does not always produce performance differences. In addition, we observe that both UPSO variants outperformed the standard local PSO, exhibiting significantly superior performance. This was also observed for the rest of the sources, justifying the usefulness of UPSO in the noiseless inverse problem.

In the case of noisy measurements, a solution cannot be identified by the measured function value, since it is contaminated by noise. Because of this, the algorithms were let to exhaust the available 3000 iterations, recording the distance (in terms of the ℓ_2-norm) of the final best solution from the actual one. This distance was used for assessing and comparing their performance under the noise levels 1%, 5% and 10% (Parsopoulos *et al.*, 2009).

Representative results for the noisy case are reported in Table 9 for the two most efficient algorithms of the noiseless case, i.e., UPSO with $u = 0.0$ (local PSO) and $u = 0.1$, without mutation. The results offer some interesting observations. First, keeping the number of sensors fixed while increasing noise levels results in worst solutions with higher distance from the actual source. This was expected, since higher noise can easily mislead the algorithm.

Secondly, keeping the noise level constant while increasing the number of sensors lent an obscure behavior to the algorithm; in some cases, it was improved, while in others it had the exactly opposite effect. Thus, it is not guaranteed that the addition of sensors will produce better results under noisy measurements (which is the most realistic case). A possible explanation for this effect can be the accumulated additional noise that accompanies higher sensor numbers when evaluating particles with equation (7).

Finally, we observe that UPSO and local PSO achieve almost the same mean values, differing only in their last few decimal digits, as indicated by the magnitude of the reported standard deviations. Thus, the performance of UPSO degrades to the levels of standard local PSO because noise imposes perturbations in the search vectors, neutralizing the effect of the combined exploration and exploitation oriented search directions of UPSO. This is also verified for the rest sources as reported in Parsopoulos *et al.* (2009).

Summarizing the conclusions for the inverse problem, UPSO was shown to be significantly better than the standard PSO in cases without noise; however, in the presence of noise, their performance was

Table 9. Results of the most successful UPSO variants for the inverse problem with noise, in terms of the distance of the obtained solution from the actual source (Parsopoulos et al., 2009)

K_{sensor}	Noise level	Local PSO ($u = 0.0$)		UPSO with $u = 0.1$	
		Mean	StD	Mean	StD
50	1%	3.540907e-02	6.230506e-09	3.540907e-02	8.304670e-09
	5%	2.585456e-01	9.259604e-09	2.585456e-01	1.037415e-08
	10%	3.411434e-01	1.486605e-08	3.411434e-01	2.083517e-08
100	1%	3.507655e-02	6.427307e-09	3.507655e-02	8.130485e-09
	5%	1.304847e-01	1.195694e-08	1.304847e-01	1.471345e-08
	10%	2.825110e-01	2.165706e-08	2.825110e-01	3.071261e-08
200	1%	4.380892e-02	6.263326e-09	4.380892e-02	6.494797e-09
	5%	1.865135e-01	1.268799e-08	1.865135e-01	1.683970e-08
	10%	5.076858e-01	2.315223e-08	5.076858e-01	3.024798e-08

competitive. Also, UPSO was shown to be more robust under small changes in the number of sensors. Nevertheless, in all cases the final outcome seems to be highly related to the corresponding source.

Experimental Setting and Results for the Forward Task for Inverse Use

In this task, Parsopoulos *et al.* (2009) considered only the overall best-performing algorithm of the inverse problem, namely UPSO with $u = 0.1$. Due to the increased dimension of the problem (8-dimensional) the maximum number of iterations was increased to 5000, while the same source points as for the inverse problem were used. The number of sensors varied from $K_{sensor} = 10$ up to 1000 with fixed increments of 10. For each value of K_{sensor} and given source, the potential function was approximated by equation (5). Each particle of the swarm consisted a candidate set of the eight coefficients bounded within $[-100,100]^8$. Each experiment was repeated 5 times and the final values of the coefficients were averaged to avoid possible inaccuracies due to the stochasticity of UPSO.

For two consecutive values, K_1 and K_2, of K_{sensor} (e.g., $K_1 = 10$ and $K_2 = 20$), the absolute relative error between the two corresponding (averaged) values of each coefficient was recorded. This error is defined as (Parsopoulos *et al.*, 2009):

$$\varepsilon^Y_{K_1 \to K_2} = \frac{\left| Y_{K_2} - Y_{K_1} \right|}{\left| Y_{K_1} \right|},$$

where Y_{K_i} stands for the averaged value of the coefficient $Y \in \{A_1, A_2, B_2^1, B_2^1, B_2^2, C_1^1, C_2^1, C_2^2\}$, when computed using K_i sensors. The statistical analysis of these errors offered useful information regarding the behavior of the coefficients. Thus, for the 100 different number of sensors, $K_{sensor} = 10, 20, 30,...,$ 1000, there were 99 error values received per coefficient:

$$\varepsilon_{10 \to 20}, \varepsilon_{20 \to 30}, \varepsilon_{30 \to 40},..., \varepsilon_{900 \to 1000}.$$

These values were analyzed statistically per source and they are depicted in boxplots in Fig. 2.

As we observe, different coefficients exhibit different fluctuations in their values. For the specific source, coefficient B_2^1, which is denoted with the index number 4 in Fig. 2, is the most sensitive, as it assumes values in a wide range, also having the highest number of outliers (denoted as crosses outside of the boxes in Fig. 2). However, this is not always the case, since different sources have a different effect on the range of all coefficients, as reported by Parsopoulos *et al.* (2009). Thus, there are strong indications that the source at hand plays again a crucial role on the behavior of coefficients. Nevertheless, for the three sources studied by Parsopoulos *et al.* (2009), C_1^1 and C_2^1 appear to be the most robust coefficients overall. The interested reader is referred to the original paper for a more thorough analysis of the forward task for inverse use and a complete exposition of the results.

Summarizing our experience from the application of PSO on MEG problems, we can infer that performance depends more or less on the source under investigation. It would be a frail conclusion to consider all PSO variants adequately efficient regardless of the problem instance. Thus, a per-case investigation, which takes into consideration the specific requirements and adjustments implied by the source at hand, constitutes the most suitable approach in tackling MEG problems with PSO.

CHAPTER SYNOPSIS

We presented and discussed two applications of PSO from the field of bioinformatics and medical informatics. The first refers to the adaptation of probabilistic neural network models in four medical classification tasks, while the second deals with two different magnetoencephalography tasks. Application details, such as the objective function formulation, which are directly related to the transformation of the original problem to the corresponding optimization problem, were presented along with the experimental settings provided in the original works. Interesting conclusions were derived, bringing deficiencies and general performance trends of the algorithm to the reader's notice. There is a lot more to be done in

Figure 2. The 99 values of the absolute relative error per coefficient in boxplot representation. Coefficient indices are as follows: $1\text{-}A_1$, $2\text{-}A_2$, $3\text{-}B_1^1$, $4\text{-}B_2^1$, $5\text{-}B_2^2$, $6\text{-}C_1^1$, $7\text{-}C_2^1$, $8\text{-}C_2^2$

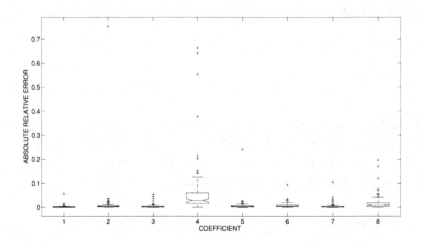

this wide research area to extrapolate sound conclusions regarding the potential and limitations of PSO in bioinformatics and medical tasks. The presented results can constitute a good starting point for the further development and thorough examination of PSO in such important applications.

REFERENCES

Berthold, M., & Diamond, J. (1998). Constructive training of probabilistic neural networks. *Neurocomputing, 19*, 167–183. doi:10.1016/S0925-2312(97)00063-5

Bronzan, J. B. (1971). The magnetic scalar potential. *American Journal of Physics, 39*, 1357–1359. doi:10.1119/1.1976655

Dassios, G. (2006). What is recoverable in the inverse magnetoencephalography problem? *Contemporary Mathematics, 408*, 181–200.

Dassios, G. (2007). The magnetic potential for the ellipsoidal MEG problem. *Journal of Computational Mathematics, 25*(2), 145–156.

Dassios, G. (2008a). Electric and magnetic activity of the brain in spherical and ellipsoidal geometry. In H. Ammari (Ed.), *Lecture Notes from the Mini-course on Mathematics of Emerging Biomedical Imaging*. Paris: Springer.

Dassios, G. (2008b). Neuronal currents and EEG-MEG fields. *Mathematical Medicine and Biology, 25*, 133–139. doi:10.1093/imammb/dqn007

Dassios, G., & Fokas, A. S. (in press). Electro-magnetoencephalography and fundamental solutions. *Quarterly of Applied Mathematics*.

Dassios, G., Fokas, A. S., & Hadjiloizi, D. (2007a). On the complementarity of electroencephalography and magnetoencephalography. *Inverse Problems, 23*, 2541–2549. doi:10.1088/0266-5611/23/6/016

Dassios, G., Fokas, A. S., & Kariotou, F. (2005). On the non-uniqueness of the inverse MEG problem. *Inverse Problems, 21*, L1–L5. doi:10.1088/0266-5611/21/2/L01

Dassios, G., Giapalaki, S., Kandili, A. N., & Kariotou, F. (2007b). The exterior magnetic field for the multilayer ellipsoidal model of the brain. *The Quarterly Journal of Mechanics and Applied Mathematics, 60*(1), 1–25. doi:10.1093/qjmam/hbl022

Dassios, G., & Kariotou, F. (2003a). Magnetoencephalography in ellipsoidal geometry. *Journal of Mathematical Physics, 44*(1), 220–241. doi:10.1063/1.1522135

Dassios, G., & Kariotou, F. (2003b). On the Geselowitz formula in biomagnetics. *Quarterly of Applied Mathematics, LXI*(2), 387–400.

Dassios, G., & Kariotou, F. (2004). On the exterior magnetic field and silent sources in magnetoencephalography. *Abstract and Applied Analysis, 4*, 307–314. doi:10.1155/S1085337504306032

Dassios, G., & Kariotou, F. (2005). The direct MEG problem in the presence of an ellipsoidal shell inhomogeneity. *Quarterly of Applied Mathematics, LXIII*(4), 601–618.

Fokas, A. S., Gelfand, I. M., & Kurylev, Y. (1996). Inversion method for magnetoence-phalography. *Inverse Problems*, *12*, L9–L11. doi:10.1088/0266-5611/12/3/001

Fokas, A. S., Kurylev, Y., & Marinakis, V. (2004). The unique determination of neuronal currents in the brain via magnetoencephalography. *Inverse Problems*, *20*, 1067–1082. doi:10.1088/0266-5611/20/4/005

Georgiou, V. L., Malefaki, S., Parsopoulos, K. E., Alevizos, Ph. D., & Vrahatis, M. N. (2009). Expeditive extensions of evolutionary Bayesian probabilistic neural networks. In [LNCS], *Lecture Notes in Computer Science, 5851*, 30-44.

Georgiou, V. L., Pavlidis, N. G., Parsopoulos, K. E., Alevizos, Ph. D., & Vrahatis, M. N. (2006). New self-adaptive probabilistic neural networks in bioinformatic and medical tasks. *International Journal of Artificial Intelligence Tools*, *15*(3), 371–396. doi:10.1142/S0218213006002722

Geselowitz, D. B. (1970). On the magnetic field generated outside an inhomogeneous volume conductor by internal current sources. *IEEE Transactions on Magnetics*, *6*, 346–347. doi:10.1109/TMAG.1970.1066765

Grynspan, F., & Geselowitz, D. B. (1973). Model studies of the magnetocardiogram. *Biophysical Journal*, *13*, 911–925. doi:10.1016/S0006-3495(73)86034-5

Guo, J., Lin, Y., & Sun, Z. (2004). A novel method for protein subcellular localization based on boosting and probabilistic neural network. In *Proceedings of the 2nd Asia-Pacific Bioinformatics Conference (APBC 2004), Dunedin, New Zealand* (pp. 20-27).

Hämäläinen, A., & Holmström, L. (1996). Complexity reduction in probabilistic neural networks. In [LNCS]. *Lecture Notes in Computer Science, 1112*, 65–70.

Hämäläinen, M. S., Hari, R., Ilmoniemi, R. J., Knuutila, J., & Lounasmaa, O. (1993). Magnetoencephalography–theory, instrumentation, and applications to noninvasive studies of the working human brain. *Reviews of Modern Physics*, *65*, 413–497. doi:10.1103/RevModPhys.65.413

Holmes, E., Nicholson, J. K., & Tranter, G. (2001). Metabonomic characterization of genetic variations in toxicological and metabolic responses using probabilistic neural networks. *Chemical Research in Toxicology*, *14*, 182–191. doi:10.1021/tx000158x

Huang, C. J. (2002). A performance analysis of cancer classification using feature extraction and probabilistic neural networks. In *Proceedings of the 7th Conference on Artificial Intelligence and Applications (TAAI 2002), Taiwan* (pp. 374-378).

Ilmoniemi, R. J., Hämäläinen, M. S., & Knuutila, J. (2005). The forward and inverse problems in the spherical model. In H. Weinberg, G. Stroink, & T. Katila (Eds.), *Biomagnetism: Applications and Theory* (pp. 278-282). Oxford, UK: Pergamon Press.

Nolte, G., & Curio, G. (1997). On the calculation of magnetic fields based on multipole modeling of focal biological current sources. *Biophysical Journal*, *73*, 1253–1262. doi:10.1016/S0006-3495(97)78158-X

Parsopoulos, K. E., Kariotou, F., Dassios, G., & Vrahatis, M. N. (2009). Tackling magnetoencephalography with particle swarm optimization . *International Journal of Bio-Inspired Computation*, *1*(1/2), 32–49. doi:10.1504/IJBIC.2009.022772

Parzen, E. (1962). On the estimation of a probability density function and mode. *Annals of Mathematical Statistics*, *33*, 1065–1076. doi:10.1214/aoms/1177704472

Plonsey, R., & Hepner, D. B. (1967). Considerations of quasistationarity in electrophysiological systems. *The Bulletin of Mathematical Biophysics*, *29*, 657–664. doi:10.1007/BF02476917

Sarvas, J. (1987). Basic mathematical and electromagnetic concepts of the biomagnetic problem. *Physics in Medicine and Biology*, *32*, 11–22. doi:10.1088/0031-9155/32/1/004

Specht, D. F. (1990). Probabilistic neural networks. *Neural Networks*, *3*(1), 109–118. doi:10.1016/0893-6080(90)90049-Q

Wang, Y., Adali, T., Kung, S., & Szabo, Z. (1998). Quantification and segmentation of brain tissues from MR images: a probabilistic neural network approach. *IEEE Transactions on Image Processing*, *7*(8), 1165–1181. doi:10.1109/83.704309

Chapter 10
Applications in Noisy and Dynamic Environments

This chapter discusses the workings of PSO in two research fields with special importance in real-world applications, namely noisy and dynamic environments. Noise simulation schemes are presented and experimental results on benchmark problems are reported. In addition, we present the application of PSO on a simulated real world problem, namely the particle identification by light scattering. Moreover, a hybrid scheme that incorporates PSO in particle filtering methods to estimate system states online is analyzed, and representative experimental results are reported. Finally, the combination of noisy and continuously changing environments is shortly discussed, providing illustrative graphical representations of performance for different PSO variants. The text focuses on providing the basic concepts and problem formulations, and suggesting experimental settings reported in literature, rather than on the bibliographical presentation of the (prohibitively extensive) literature.

OPTIMIZATION IN THE PRESENCE OF NOISE

In many real-world applications, function evaluations are either the outcome of complex simulation procedures or come directly from measurement equipment. In such cases, the obtained values are almost surely contaminated by noise, which is induced by accumulated computational errors or equipment measurement tolerances. Noise limits the usefulness of classical gradient-based algorithms, rendering function and gradient values misleading for the algorithm. Thus, there is an increasing need for robust algorithms, capable of addressing noise and providing satisfactory solutions with the least possible computational burden.

A classical optimization algorithm for solving noisy problems is the nonlinear simplex method of Nelder and Mead (1965), presented for PSO initialization purposes in Chapter Three of the book at hand.

DOI: 10.4018/978-1-61520-666-7.ch010

Copyright © 2010, IGI Global. Copying or distributing in print or electronic forms without written permission of IGI Global is prohibited.

Its simplex update scheme, which considers only relative differences among search points instead of their actual values, allows it to work with imprecise functions. Thus, possible inaccuracies in measurement have a less critical effect on its operation than for algorithms that require precise values. Torczon (1991) further improved this scheme by incorporating a more sophisticated approach that works equally well in noiseless problems. Of course, evolutionary and swarm intelligence algorithms could not be ignored in noisy problems. The use of populations and their inherent tolerance of imprecise information renders them a very appealing alternative.

Arnold (2001) investigated several optimization methods in the presence of noise. Specifically, he considered the following algorithms:

1. Direct pattern search of Hooke and Jeeves (1961)
2. Nonlinear simplex method of Nelder and Mead (1965)
3. Multidirectional search algorithm of Torczon (1991)
4. Implicit filtering algorithm of Gilmore and Kelley (1995)
5. Simultaneous perturbation stochastic approximation algorithm of Spall (1992)
6. Evolutionary gradient search algorithm of Salomon (1998). Evolution strategy with cumulative mutation strength adaptation of Hansen and Ostermeier (2001)

The behavior of these algorithms was studied on a simple multi-dimensional sphere model under Gaussian additive noise. The obtained results suggested the existence of a threshold on noise strength that, if surpassed, leads to the unreliable convergence of any algorithm, often resulting in search stagnation. In addition, the following remarks were reported:

1. The algorithm of Hooke and Jeeves declined rapidly in noisy problems even for small dimensions.
2. The nonlinear simplex method failed even on the sphere model, unless dimension was kept low, while it got easily stuck in suboptimal points.
3. The algorithm of Torczon alleviated stagnation but diverged for high noise levels.
4. The implicit filtering approach approximated the gradient, thus exhibiting poor performance especially in highly noisy cases.
5. Similar behavior was observed for the simultaneous perturbation stochastic approximation method, although it appeared to be more sensitive even for small dimensions, as well as for the evolutionary gradient search approach.
6. The evolution strategy with cumulative mutation strength adaptation of Hansen and Ostermeier was shown to be the most robust algorithm with respect to the effects of noise.

The adequacy of evolutionary algorithms in noisy problems was also verified and analyzed in other works (see Arnold, 2002; Beyer, 2000; Beyer *et al.*, 2004; Darwen & Pollack, 1999; Eskandari & Geiger, 2009). PSO was also studied in the presence of noise. The first study of Parsopoulos and Vrahatis (2001) was followed by more sophisticated approaches for different PSO variants (Bartz-Beilstein *et al.*, 2007; Han & He, 2007; Janson & Middendorf, 2006; Pan *et al.*, 2006; Parsopoulos & Vrahatis, 2002a, 200b), suggesting the ability of PSO to tackle noisy problems.

In most studies, additive Gaussian noise was used to simulate noise in nature. However, multiplicative noise is also an interesting alternative, perhaps closer to reality, since it considers noise strength proportional to the actual function value. The two noise types are formally described in the following paragraphs.

Types of Noise

Let $f(x)$ be the objective function, producing accurate values for any x in its domain. Then, *additive noise* assumes that function values are obtained through the approximation:

$$\hat{f}(x) = f(x) + \eta, \tag{1}$$

where $\eta \sim N(0,\sigma^2)$ is a Gaussian random variable with mean, $\mu = 0$, and standard deviation σ. Its probability density function is then defined as follows:

$$P(\eta) = \frac{1}{\sigma\sqrt{2\pi}}\exp\left[-\frac{1}{2}\left(\frac{\eta}{\sigma}\right)^2\right].$$

Obviously, additive noise does not take into consideration the magnitude of the original function values. Thus, a given noise value, η, may have tremendous impact on one function while being almost inconspicuous in another.

For example, due to the Gaussian selection of η, we know that more than 99% of its values will lie within the range $(-3\sigma, 3\sigma)$. Thus, if the actual $f(x)$ assumes comparable values with η, the noise will dominate the actual function and completely corrupt it. On the other hand, if $f(x)$ takes values several orders of magnitude larger than η, the addition of noise becomes faint and the problem closely approximates its noiseless counterpart. Thus, cautious selection is required when simulating noise in benchmark problems, in order to avoid possible deterioration of the objective function.

Alternatively to the additive case, *multiplicative noise* can be used. According to it, the original function $f(x)$ is approximated by:

$$\hat{f}(x) = f(x)(1 + \eta), \tag{2}$$

where $\eta \sim N(0,\sigma^2)$ is again a Gaussian random variable with mean, $\mu = 0$, and standard deviation σ, with the additional constrain:

$$1 + \eta > 0. \tag{3}$$

This type of noise accords a relative stochastic error to the actual function values as a percentage of their magnitude. For example, a noise value, $\eta = 0.01$, introduces an increase of 1% of the magnitude of $f(x)$ in its value, while, $\eta = -0.1$, reduces the value of $f(x)$ by 10% of its magnitude. In both cases, the sign of $f(x)$ is retained; this is a desirable property, attributed to the constraint of equation (3). Multiplicative noise simulates equipment measurement errors more plausibly, since they usually come as a percentage of the measured quantity. The user can tune the percentage of alteration by selecting a proper standard deviation of the noise. The usual machinery measurement tolerances do not exceed ±1%, but, for simulation purposes, percentages up to ±10%, i.e., $\eta \in (-0.1, 0.1)$, have been used. Naturally, as the value of

Table 1. Experimental setting per problem. The dimension, swarm initialization interval and desirable accuracy of solution per problem are reported

Problem	Dimension	Initial interval	Accuracy
TP_{UO-1}	2	$[-5,5]^2$	10^{-6}
TP_{UO-2}	2	$[-3,3]^2$	10^{-6}
TP_{UO-3}	2	$[-1,1]^2$	10^{-3}
TP_{UO-11}	2	$[-1,1]^2$	10^{-3}
TP_{UO-12}	2	$[0,10]^2$	10^{-6}
TP_{UO-13}	2	$[-1,1]^2$	10^{-3}

η approaches the ends of the interval $(-1,1)$, the obtained values of the noise become comparable to the actual function values, rendering optimization quality questionable.

In both the additive and multiplicative noise types, Gaussian errors were considered. Although this choice is not mandatory, it is adopted in almost all relevant works, since measurement errors in nature and technology are very often modeled with this distribution. Moreover, most stochastic processes have mean values that follow this distribution (central limit theorem of statistics) (Beyer, 2001). For these reasons, it is used in the vast majority of noisy optimization works, as well as in all cases presented in the rest of this chapter.

Many applications, especially from the fields of electronics and circuit design, involve the concept of *white noise*. This is also met in noisy optimization simulations of relative models. This concept refers to the case of time-uncorrelated noise values. Gaussian distribution can produce white noise, but not all Gaussian noise is white. The two properties are not interrelated, as the one refers to the type of distribution, while the other refers to its correlation with time. In most cases, noisy benchmark problems in literature assume white noise. Contrary to the white noise is *colored noise*, where time correlations exist in its values. An example for simulating colored noise is provided in Fox *et al.* (1988).

The following sections report applications of PSO on benchmark problems as well as on a simulated real world noisy problem.

Application of PSO on Noisy Benchmark Problems

Parsopoulos and Vrahatis (2001, 2002a, 2002b) reported preliminary results of PSO on noisy benchmark problems. They considered the test problems, TP_{UO-1}, TP_{UO-2}, TP_{UO-3}, TP_{UO-11}, TP_{UO-12} and TP_{UO-13}, described in Appendix A of the book at hand, with multiplicative noise, as defined in equations (2) and (3) (Parsopoulos & Vrahatis, 2002b). The experimental setting per problem is reported in Table 1. More specifically, the dimension, swarm initialization interval and the desirable accuracy for considering an experiment as successful are reported for each problem (Parsopoulos & Vrahatis, 2002b). The parameters of the employed inertia weight PSO variant, as well as the noise strength (variance), σ^2, of η in equation (2) are reported in Table 2.

Although in practice it is common to re-evaluate a particle for a number of times and average these values in order to balance the influence of noise, Parsopoulos and Vrahatis (2002b) did not apply re-evaluation on PSO, as their primary goal was to investigate whether it is capable of detecting funnels that lead towards the region of the global minimizer or if it succumbs to the misleading information induced by noise.

Table 2. PSO and experimental setting for the test problems reported in Table 1

Parameter	Values
Swarm size	20
Inertia weight	$w = 1.2 \rightarrow 0.4$ (linearly decreasing)
Cognitive/social	$c_1 = c_2 = 0.5$
Maximum iterations	5000
Noise strength (σ^2)	0.01, 0.02, 0.04, 0.07, 0.09
Experiments per case	30

A portion of the results reported by Parsopoulos and Vrahatis (2002b) are reported in Table 3. More specifically, for each problem and variance value, the percentages of success in 30 experiments, as well as the required number of iterations to attain a solution with the desirable accuracy, are reported. As we can see, PSO exhibited nice performance in all cases with its success rate remaining high even for high noise strength. In addition, the required number of iterations did not fluctuate significantly with the increase of noise. Therefore, it seems that noise did not cause crucial instability in the operation of PSO on these widely used test problems.

The promising results triggered further research in noisy applications of PSO, especially in problems closely related to the real world. The next section presents such an application on inverse scattering.

Particle Identification by Light Scattering with PSO

Astronomy, meteorology, medicine and bioengineering are just some of the fields where the size and optical characteristics of particles are studied through laser light scattering. This is generally called the *particle identification by light scattering* (PILS) problem, and it comprises a very challenging task since it is always characterized by noisy measurements and imprecise functions.

Table 3. Results derived from Parsopoulos and Vrahatis (2002b) for the application of PSO in noisy problems. The success percentages over 30 experiments, as well as the mean number of iterations per problem and value of variance are reported

Variance	Statistic	TP_{UO-1}	TP_{UO-2}	TP_{UO-3}	TP_{UO-11}	TP_{UO-12}	TP_{UO-13}
0.01	Success	100%	100%	100%	100%	100%	100%
	Mean It.	1694.0	1950.5	1650.2	1668.9	1729.8	1628.5
0.02	Success	100%	96%	100%	100%	100%	100%
	Mean It.	1687.6	2057.7	1654.9	1719.6	1741.0	1637.7
0.04	Success	100%	92%	100%	100%	100%	100%
	Mean It.	1683.8	2108.3	1672.4	1757.1	1742.1	1637.8
0.07	Success	100%	92%	100%	100%	100%	100%
	Mean It.	1690.6	2111.6	1691.4	1875.7	1988.2	1661.4
0.09	Success	100%	88%	100%	100%	100%	100%
	Mean It.	1691.5	2272.6	1697.2	1839.7	1723.7	1658.1

Two techniques are mainly used in PILS: *dynamic light scattering* (DLS) and *static light scattering* (SLS) (Hodgson, 2000). In DLS, laser light is focused onto a small volume of solution containing the colloidal particles and the scattered light is collected over a small solid angle. The phase and polarization of the light scattered by a molecule is determined by its size, shape and composition. Random Brownian motion causes the total intensity at the detector to fluctuate in time. Autocorrelation functions are generated from these fluctuations and then inverted to obtain the distribution of particle sizes (Hodgson, 2000).

On the other hand, in SLS, the total intensity of the scattered light is measured as a function of angle and this information is used to determine the particle size distributions. For spherical particles, the angular dependence of the scattered light is described by the *Mie scattering function* (Bohren & Huffman, 1983), which depends on the size and refractive indices of the particles, as well as on their surrounding medium.

A fundamental problem underlying in inverse light scattering is the determination of the *refractive index*, n, and *radius*, r, of homogeneous spherical particles suspended in a known medium. This problem can be addressed by applying polarized light of known wavelength on the particles and measuring the intensity, I, of the scattered light for different angles. The standard *Lorenz-Mie theory* (Bohren & Huffman, 1983) can be used to describe this procedure, while the observed scattering pattern can characterize both single particles and particle distributions. Thus, for unknown particles, we can assume hypothetical values of n and r, and experimentally measure the intensities of the scattered light in different angles. Then, the actual values of these intensities, as they are computed theoretically for the specific refractive index, radius, and for the same angles are derived and compared to the experimentally measured ones. If our hypothesis is correct, the intensity values must coincide. Otherwise, there will be an error per angle. The summed square-error for all angles determines the underlying optimization problem in n and r. The solutions of this problem are exactly the wanted refractive index and radius of the particles.

Putting it formally, let, $I_s(\theta_1), I_s(\theta_2), \ldots, I_s(\theta_m)$, be the experimentally measured intensities of the scattered light for the angles, $\theta_1, \theta_2, \ldots, \theta_m$. The goal is to determine the refractive index, n, and the radius, r, of the particles under consideration. The value of n can be either real or complex, while r can assume, by definition, only real values. Since the intensity varies widely with respect to the angle, it is preferable to work with its logarithm rather than the actual values. Thus, the corresponding optimization problem is defined by the following objective function:

$$E(r,n) = \sum_{i=1}^{m} \left(LI_s(\theta_i) - LI_{TH}(\theta_i, r, n) \right)^2,$$

(4)

where,

$$LI_s(\theta_i) = \log(I_s(\theta_i)), \quad i = 1, 2, \ldots, m,$$

and $LI_{TH}(\theta_i, r, n)$ is the logarithm of the theoretically computed intensities for the specific values of n and r at angles $\theta_i, i = 1, 2, \ldots, m$. In the case of real values of n, the objective function is 2-dimensional. On the other hand, complex values of n can be described as pairs of real numbers, thus, the problem becomes 3-dimensional. In practice, the objective function is always contaminated by noise, due to inaccuracies in intensity measurement.

Several techniques have been applied on the problem defined by equation (4). For real values of n, random search and multilevel single-linkage clustering have been used (Hodgson, 2000; Zakovic *et al.*,

1998). However, the objective function has numerous local minima, rendering a good initial guess of the solution very difficult. Evolutionary algorithms are, in general, less dependent on the initial guess. Hence, they have been considered as a promising alternative to address this problem (Hodgson, 2000).

Parsopoulos *et al.* (2003) applied PSO and differential evolution (DE) on the PILS problem. In their experiments, they simulated the intensity functions using the BHMIE routine of Bohren and Huffman (1983), contaminating its values with multiplicative noise by using equation (2). Using the BHMIE routine, the user is able to set the global minimizers, i.e., refractive index and radius of the particles, measure the corresponding intensities in several angles through simulation, and then use the measurements to solve the inverse problem, i.e., identify the originating refractive index and radius that produced these intensities, by using the error function of equation (4).

For this purpose, they considered the following cases, previously studied by Hodgson (2000) with genetic algorithms:

Case 1: Refractive index, $n^* = 1.40$, and radius, $r^* = 1.55 \mu m$

Case 2: Refractive index, $n^* = 1.65$, and radius, $r^* = 4 \mu m$

Figure 1 illustrates the objective function for Case 1, where its high nonlinearity becomes apparent. In all experiments, the surrounding medium was water with refractive index 1.336 and radius $0.5145 \mu m$. The angular range was set between 0 and 180 degrees with increments of 9 degrees, resulting in $m = 20$ different angles. The value of n was bounded within [1,2] in both cases, while $r \in [1,2]$ and $r \in [3,5]$ were considered for Case 1 and 2, respectively (Parsopoulos *et al.*, 2003).

The parameter setting of Table 4 was used for the constriction coefficient and inertia weight PSO variant, denoted as PSO[co] and PSO[in], respectively. Again, no re-evaluation of particles was used. PSO was let to exhaust the available number of iterations, and the final best solution was recorded along with its actual (i.e., noiseless) function value and its distance from the true global minimizer under the ℓ_2-norm. These values were statistically analyzed, in terms of their median, standard deviation and minimum value (Parsopoulos *et al.*, 2003).

Figure 1. The objective function for refractive index, $n^* = 1.40$, and radius, $r^* = 1.55 \mu m$

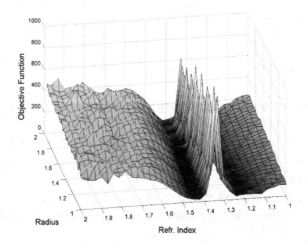

Table 5 reports a portion of the results for both PSO variants. More specifically, the median of the function value, as well as the median of the measured distance from the global minimizer are reported for Case 1 and 2, for different noise levels. As we observe, PSO performed efficiently in both problems, with PSO[in] exhibiting better performance in Case 1, while PSO[co] was shown to be better in Case 2 with respect to both the reported medians. Additional results provided in Parsopoulos *et al.* (2003) suggest that PSO also outperformed the DE algorithm, while PSO[in] was, overall, the most robust algorithm. For further details, the reader is referred to Parsopoulos *et al.* (2003).

Table 4. Parameter setting used in Parsopoulos et al. (2003) for the PILS problem

Parameter	Values
Swarm size	30
Inertia weight	$w = 1.2 \rightarrow 0.1$ (linearly decreasing)
Cognitive/social	$c_1 = c_2 = 2.05$
Constriction coefficient	$\chi = 0.729$
Maximum iterations	200
Noise strength (σ)	0.0 (noiseless), 0.01, 0.05, 0.1
Experiments per case	20

Table 5. Results for the two PILS problems considered by Parsopoulos et al. (2003). The median of the obtained function values, as well as of the measured distance from the global minimizer, are reported for different noise levels.

Case	Noise strength	Method	Median Function value	Median Distance from global
Case 1	0.00	PSO[co]	1.69e-13	2.33e-09
$n^* = 1.40$		PSO[in]	2.71e-21	5.20e-13
$r^* = 1.55\mu m$	0.01	PSO[co]	3.51e-15	3.75e-10
		PSO[in]	5.29e-20	1.99e-12
	0.05	PSO[co]	3.29e-15	5.92e-10
		PSO[in]	2.69e-20	1.63e-12
	0.10	PSO[co]	9.61e-15	4.44e-10
		PSO[in]	5.48e-21	4.24e-13
Case 2	0.00	PSO[co]	3.80e-13	1.74e-08
$n^* = 1.65$		PSO[in]	2.68e-12	3.68e-08
$r^* = 4\mu m$	0.01	PSO[co]	1.16e-13	1.08e-08
		PSO[in]	5.85e-13	1.78e-08
	0.05	PSO[co]	4.13e-13	1.97e-08
		PSO[in]	1.01e-11	6.62e-08
	0.10	PSO[co]	1.47e-12	3.47e-08
		PSO[in]	3.55e-12	5.98e-08

An equally interesting application of PSO in the particle filtering problem in systems with multiplicative noise was recently proposed, and it is described in the next section.

Particle Filtering in Systems with Multiplicative Noise Using PSO

Particle filtering (PF) is a modeling methodology for the online state estimation of a dynamical system as observations become available (Arulampalam *et al.*, 2002; Gordon *et al.*, 1993). This is possible by generating a set of random samples, which is propagated and updated recursively to approximate the state probability density function (PDF) of the system. Their inherent stochastic properties render PFs capable of addressing nonlinearities and noise of the estimated distribution.

PFs rely on importance sampling, i.e., they use proposal distributions to approximate the posterior one. The most common proposal is the probabilistic model of state evolution, i.e., the transition prior. This choice, however, fails if the likelihood is highly peaked compared to the prior, or if it lies in its tail. Several methods have been proposed to address this problem. A popular approach is the *unscented particle filter* (UPF) (Van der Merwe *et al.*, 2000), which employs the *unscentedKalman filter* (Julier & Uhlmann, 1997) as proposal distribution. This combination outperforms other existing filters, although at the cost of heavy computational burden.

Different approaches are based on optimization methods to avoid resampling by biasing the prior sample towards regions of the state space with high likelihood (Krohling, 2005; Kwok *et al.*, 2005; Uosaki & Hatanaka, 2005; Wang *et al.*, 2006). Recently, Tong *et al.* (2006) proposed a similar approach that employs PSO. The produced filter outperformed the generic PF and UPF approaches in terms of the computational load; however, their experiments were restricted solely to a single system with very small noise in observations. Thus, although preliminary indications were promising, questions regarding the efficiency of the PSO-based approach remained unanswered. In all the aforementioned studies only additive noise was considered.

Klamargias *et al.* (2008) extended the PSO-based approaches by considering a PF model with PSO for systems with multiplicative noise. The proposed model employs a conventional multiobjective optimization approach to address the problem of sample biasing towards the prior or the likelihood. This approach appears to be very appealing, especially due to the combination of noisy with multiobjective optimization; for this reason, we will consider it in more detail.

The next paragraph briefly presents the necessary mathematical background, followed by the formulation of the corresponding optimization problem. Finally, the experimental setting and indicative results from (Klamargias *et al.*, 2008) are reported to illustrate the approach on a widely used test problem.

Basic Concepts of Particle Filtering

In the following paragraphs, we briefly present the mathematical background of PFs, following closely the presentations of Gordon *et al.* (1993) and Klamargias *et al.* (2008). The general discrete-time, non-linear, non-Gaussian state estimation problem is at the core of interest in PF. A signal (state) vector:

$$x_t \in \mathbf{R}^{n_x}, \quad t \geq 0,$$

is considered to evolve according to a system model:

$$x_{t+1} = f(x_t, w_t),$$

where,

$$f : \mathbf{R}^{n_x} \times \mathbf{R}^{n_w} \rightarrow \mathbf{R}^{n_x},$$

is the system's *transition function*, and $w_t \in \mathbf{R}^{n_w}$ is a zero-mean, white-noise sequence with known state-independent PDF. At discrete time moments, measurements (observations):

$$y_t \in \mathbf{R}^{n_y}, \quad t \geq 1,$$

become available. There is a *measurement function* defined as:

$$h : \mathbf{R}^{n_x} \times \mathbf{R}^{n_e} \rightarrow \mathbf{R}^{n_y},$$

which connects measurements with the state vector of the system as follows:

$$y_t = h(x_t, e_t),$$

where $e_t \in \mathbf{R}^{n_e}$ is a zero-mean, white-noise sequence with known PDF, which is independent of the current and past system states and noise. Therefore, the signal is a hidden Markov process with initial distribution, $p(x_0)$, and prior distribution, $p(x_{t+1} \mid x_t)$, while measurements are conditionally independent for the given system process, with marginal distribution, $p(y_t \mid x_t)$. These three distributions are adequate to describe the hidden Markov model.

The main goal of PF is the recursive estimation of the filtering distribution, $p(x_t \mid y_{1:t})$, in time. The PDF of the current system state, x_t, is approximated by taking into consideration all measurements from the beginning up to time t. The approximation consists of the *prediction phase* and the *update phase* described below.

If $p(x_t \mid y_{1:t})$ is available at time t, then the prior PDF of the system state at time $(t+1)$ can be predicted as follows:

$$p(x_{t+1} \mid y_{1:t}) = \int_{\mathbf{R}^{n_x}} p(x_{t+1} \mid x_t) p(x_t \mid y_{1:t}) dx_t, \tag{5}$$

where the transition PDF, $p(x_{t+1} \mid x_t)$, is defined as:

$$p(x_{t+1} \mid x_t) = p_{w_t}\left(f^{-1}(x_t, x_{t+1})\right) \times \left|\det\left(J_{f^{-1}}\right)\right|, \tag{6}$$

with J denoting the Jacobian. At time $(t+1)$, a measurement, y_{t+1}, becomes available. This measurement can be used to update the predicted prior via the Bayes rule, as follows:

$$p(x_{t+1} \mid y_{1:t+1}) = \frac{p(y_{t+1} \mid x_{t+1}) p(x_{t+1} \mid y_{1:t})}{p(y_{t+1} \mid y_{1:t})}, \tag{7}$$

where the likelihood function, $p(y_{t+1} \mid x_{t+1})$, is defined as:

$$p(y_{t+1} \mid x_{t+1}) = p_{e_{t+1}}\left(h^{-1}(x_{t+1}, y_{t+1})\right) \times \left|\det\left(J_{h^{-1}}\right)\right|, \tag{8}$$

and

$$p(y_{t+1} \mid y_{1:t}) = \int_{\mathbf{R}^{n_x}} p(y_{t+1} \mid x_{t+1}) p(x_{t+1} \mid y_{1:t}) \mathrm{d}x_{t+1}.$$

Although $p(y_{t+1} \mid y_{1:t})$ is usually unknown, it is sufficient to evaluate (Klamargias *et al.*, 2008):

$$p(x_{t+1} \mid y_{1:t+1}) \propto p(y_{t+1} \mid x_{t+1}) p(x_{t+1} \mid y_{1:t}).$$

Let $\{x_t^{(i)}\}_{i=1,2,\ldots,N}$ be a set of random samples from the PDF $p(x_t \mid y_{1:t})$. Then the PF propagates and updates these samples to obtain the values $\{x_{t+1}^{(i)}\}_{i=1,2,\ldots,N}$, which are approximately distributed as $p(x_{t+1} \mid y_{1:t+1})$. This is accomplished in the following two phases:

Prediction Phase

Each sample, $x_t^{(i)}$, is passed through the system model, and a new sample:

$$\overline{x}_{t+1}^{(i)} = f(x_t^{(i)}, w_t^{(i)}), \quad i = 1, 2, \ldots, N,$$

is obtained for a sample, $w_t^{(i)}$, drawn from the PDF, $p(w_t)$, of system noise.

Update Phase

As soon as the measurement y_{t+1} becomes available, the likelihood $p(y_{t+1} \mid \overline{x}_{t+1}^{(i)})$ is evaluated and a normalized weight is computed for each prior sample, as follows:

$$q_{t+1}^{(i)} = \frac{p(y_{t+1} \mid \overline{x}_{t+1}^{(i)})}{\sum_{j=1}^{N} p(y_{t+1} \mid \overline{x}_{t+1}^{(j)})}, \quad i = 1, 2, \ldots, N,$$

defining a discrete distribution over the predicted samples, $\{\overline{x}_{t+1}^{(i)}\}_{i=1,2,\ldots,N}$. Then, resampling is performed N times from this discrete distribution, generating new samples, $\{x_{t+1}^{(i)}\}_{i=1,2,\ldots,N}$, such that:

$$\Pr\left(x_{t+1}^{(j)} = \overline{x}_{t+1}^{(i)}\right) = q_{t+1}^{(i)}, \quad i = 1, 2, \ldots, N,$$

for any $j = 1, 2, \ldots, N$. The produced new samples are approximately distributed as $p(x_{t+1} \mid y_{1:t+1})$ (Gordon *et al.*, 1993).

The resampling stage of PFs is the source of their most important deficiency, namely the *particle impoverishment* problem. According to this, many of the samples, $\{\overline{x}_{t+1}^{(i)}\}_{i=1,2,\ldots,N}$, assume very small weights, $q_{t+1}^{(i)}$, failing to be selected during resampling. The problem arises in cases where the region of the state space with significant likelihood, $p(y_{t+1} \mid \overline{x}_{t+1}^{(i)})$, is small compared to the region where the prior, $p(x_{t+1} \mid y_{1:t})$, is significant, and it becomes more intense when the narrow likelihood falls in a tail of the prior.

The aforementioned problem can be addressed by intervening between the prediction and update phase of the PF operation. Klamargias *et al.* (2008) proposed an approach that utilizes PSO for this purpose. Their approach requires the determination of a proper optimization problem, which is defined in the next section.

The Optimization Problem Formulation

The main goal is the biasing of the prior samples towards regions of the state space with significant likelihood, without abandoning regions with significant prior. This dual problem constitutes a multiobjective task with two competing objectives (Klamargias *et al.*, 2008). The first objective can be modeled as a function, $F_1(x)$, which takes its maximum values at regions of high likelihood. On the other hand, the second objective can be modeled by another function, $F_2(x)$, which assumes its maximum values at regions of high prior.

Assuming that the two objective functions are available, the simplest approach to tackle the multiobjective problem is the *conventional weighted aggregation* (CWA) approach. According to this, the two tasks are aggregated into a single objective function and the problem can be treated as a standard optimization problem, defined as follows (Klamargias *et al.*, 2008):

$$\max_{x \in S} \ F(x) = a_1 F_1(x) + a_2 F_2(x), \tag{9}$$

where S is the state space of the prior samples, and a_1, a_2, are non-negative weights, such that $a_1 + a_2 = 1$. Thus, the generated samples, $\{\overline{x}_{t+1}^{(i)}\}_{i=1,2,\dots,N}$, are considered as initial conditions and they are adapted to maximize $F(x)$. In the framework of PSO, these samples would constitute the initial swarm. The resulting sample, obtained after the execution of PSO, constitutes the sample that will undergo resampling. The rest of the PF operation is left unchanged (Klamargias *et al.*, 2008).

The CWA approach was selected by Klamargias *et al.* (2008) despite the existence of a plethora of more sophisticated multiobjective approaches. The reason was the combination of simplicity with satisfactory performance offered by CWA in benchmark problems, as reported in previous studies (Parsopoulos & Vrahatis, 2002c). In addition, it allows the user to control the influence of each objective by modifying the weights of equation (9) properly. Thus, if it is desirable, more attention can be paid to either the system model (prior) or to the observation (likelihood). Nevertheless, different multiobjective approaches can be considered. Such approaches are presented in another chapter of the book at hand.

So far, we assumed that the two objective functions, $F_1(x)$ and $F_2(x)$, are available. However, their determination is nontrivial, as they are intimately related to the likelihood and prior of the system at hand. Therefore, they can only be determined on a per problem base. In the following paragraphs, we use a simple example that also served as the benchmark problem of Klamargias *et al.* (2008) to illustrate the problem formulation procedure.

Consider the following nonlinear model (Gordon *et al.*, 1993) with multiplicative noise:

$$x_{t+1} = s(x_t, w_t) = f(x_t)(1+w_t), \tag{10}$$

$$y_t = m(x_t, e_t) = h(x_t)(1+e_t), \tag{11}$$

where,

$$f(x_t) = \frac{1}{2}x_t + \frac{25x_t}{1+x_t^2} + 8\cos(1.2t),$$

and,

$$h(x_t) = \frac{x_t^2}{20},$$

and w_t, e_t, are zero-mean Gaussian white noises with variances Q and R, respectively. In order to avoid computational singularities, the functions $f(x_t)$, $h(x_t)$, as well as a sample point, x, are prohibited from taking an exact zero value. From equation (10), it is derived that:

$$w_t = \frac{x_{t+1} - f(x_t)}{f(x_t)} = s^{-1}(x_t, x_{t+1}),$$

while equation (11) gives:

$$e_t = \frac{y_t - h(x_t)}{h(x_t)} = m^{-1}(x_t, x_{t+1}).$$

Thus, the transition PDF is given by the following equation:

$$p(x_{t+1} \mid x_t) = p_{w_t}\left(s^{-1}(x_t, x_{t+1})\right)\left|\frac{d}{dx_{t+1}}s^{-1}\right| = \left|\frac{1}{f(x_t)}\right| p_{w_t}\left(\frac{x_{t+1} - f(x_t)}{f(x_t)}\right),$$

and the likelihood is given by:

$$p(y_t \mid x_t) = p_{e_t}\left(m^{-1}(x_t, y_t)\right)\left|\frac{d}{dy_t}m^{-1}\right| = \left|\frac{1}{h(x_t)}\right| p_{e_t}\left(\frac{y_t - h(x_t)}{h(x_t)}\right),$$

where p_{w_t}, p_{e_t}, are Gaussian PDFs. Then, if $x_{t+1}^{(i),\text{PSO}}$ denotes the i-th particle in the swarm of PSO, $\overline{x}_{t+1}^{(i)}$ is the corresponding sample point where it was initialized, and y_{t+1} is the observation, then the two objective functions of equation (9) are defined as follows:

$$F_1\left(x_{t+1}^{(i),\text{PSO}}\right) = \left|\frac{1}{h\left(\overline{x}_{t+1}^{(i)}\right)}\right|\frac{1}{\sqrt{2\pi R}}\exp\left\{-\frac{1}{2R}\left[\frac{y_{t+1} - h\left(\overline{x}_{t+1}^{(i)}\right)}{h\left(\overline{x}_{t+1}^{(i)}\right)}\right]^2\right\},$$

(12)

$$F_2\left(x_{t+1}^{(i),\text{PSO}}\right) = \frac{1}{\sqrt{2\pi Q}}\exp\left\{-\frac{1}{2Q}\left[\frac{x_{t+1}^{(i),\text{PSO}} - \overline{x}_{t+1}^{(i)}}{\overline{x}_{t+1}^{(i)}}\right]^2\right\}.$$

(13)

Note that the time index, t, in equations (12) and (13) refers to the PF system iteration, not the PSO one. In the next section, we present the experimental setting and portion of the results reported by Klamargias *et al.* (2008) for the problem presented above.

Experimental Setting and Results

Klamargias *et al.* (2008) considered the system model defined by equations (10) and (11) with multiplicative noise, under different levels of noise strength:

$$Q, R \in \{0.01, 0.05, 0.10, 0.30, 0.50\}.$$

The PF was let to operate for 60 time steps with a PSO-based optimization phase taking place at each iteration, as described in the previous section. All parameter values used in the experiments are reported in Table 6.

Three different levels of the weight a_1 in equation (9) were considered, namely $a_1 = 0.2$ (prior more important), 0.5 (equal importance) and 0.8 (likelihood more important), with $a_2 = 1\text{-}a_1$. The prior sample was passed as initial swarm in PSO, which returned its best positions as the final sample, after 50 iterations (Klamargias *et al.*, 2008). At each experiment, the root mean square-error (RMS) was computed for the PSO-based PF, denoted hereafter as PSO-PF, and compared with the standard PF approach. The experiments were accompanied by Wilcoxon rank sum hypothesis testing to justify the statistical significance of the results.

A portion of the results reported by Klamargias *et al.* (2008) is reported in Table 7. More specifically, for each different level of Q and R, the RMS of the standard PF as well as the best performing PSO-PF case, in terms of its RMS, are reported for the case of the sample (and, consequently, swarm) size, $N = 20$. Regarding PSO-PF, the value of a_1, the RMS and the p-value produced by the Wilcoxon test regarding the significance of the observed performance difference between the best performing PSO-PF and the standard PF, are reported.

Two significant observations can be stated regarding the reported results. First, we can clearly see that PSO-PF outperformed the standard PF for higher system noise levels. Indeed, for $Q = 0.01$ and 0.05, PF outperforms or has marginal difference from PSO-PF. However, for $Q > 0.05$, PSO-PF outperforms PF significantly. This is also verified by the Wilcoxon tests in most cases (a p-value is considered to be significant if it is smaller than 0.05). Additionally, this observation holds for all PSO-PF variants, as it is reported in (Klamargias *et al.*, 2008), and not only for the best one reported in Table 7. The second observation is that, in all cases, smaller values of a_1 exhibited superior performance, with $a_1 = 0.2$ being the overall most efficient value. This suggests that the prior shall be considered as more important than likelihood in the optimization procedure. Finally, an expected increase in the observed RMS values for all approaches was observed for higher values of the measurement noise level.

Klamargias *et al.* (2008) reported similar results for the case of $N = 50$, as well. Thus, incorporation of PSO in PF has shown the potential of providing nice results, enhancing the performance of the PF methodology. Thus, it can be considered as a promising alternative, especially in cases with high system noise levels. Further experimentation using different multiobjective optimization approaches will unveil the interaction between the noisy and multiple criteria nature of the underlying optimization problem, providing a better understanding of the problem as well as the potential for further improvements.

In the next section, PSO is illustrated on an application field that is intimately related to noisy optimization, namely optimization in changing and imprecise environments.

Table 6. Parameter setting used in Klamargias et al. (2008) for the PF problem

Parameter	Values
Swarm size	20, 50
PSO parameters	$\chi = 0.729$, $c_1 = c_2 = 2.05$
Maximum PSO iterations	50
PF time steps	60
Noise strength Q and R	0.01, 0.05, 0.10, 0.30, 0.50
Experiments per case	100

OPTIMIZATION IN CONTINUOUSLY CHANGING ENVIRONMENTS

The property of noise is often accompanied in real-world applications by another interesting property, namely continuously changing landscapes. In contrast to the static objective functions considered so far, the position of the global minimizer (and probably the whole landscape) in dynamic environments changes during the optimization procedure. In this case, the ability of an algorithm to track and follow the trajectory of the minimizer is at the core of our interest. Noise can be absent, resulting in problems with accurate function evaluations but moving minimizer(s).

Excellent introductions to the application of evolutionary algorithms in dynamic environments exist in the relevant literature (Blackwell & Branke, 2006; Branke, 2002; Jin & Branke, 2005). In addition, several attempts have been made to apply PSO variants on such problems (Blackwell & Bentley, 2002; Carlisle & Dozier, 2000; Eberhart & Shi, 2001; Hu & Eberhart, 2002; Li, 2004; Parrott & Li, 2004; Parsopoulos & Vrahatis, 2001, 2005). In most cases, only the dynamic movement of the minimizers is considered, without the addition of noise.

Parsopoulos and Vrahatis (2001) were among the first to consider both non-static minimizers and noisy function evaluations. They performed a set of preliminary experiments on the 2-dimensional instances of test problems TP_{UO-2}, TP_{UO-15} and TP_{UO-22}, described in Appendix A of the book at hand, where the whole search space was rotated at each iteration, while Gaussian multiplicative noise was added to the function values. The rotation angle was taken randomly between 0 and 360 degrees, while the standard deviation of the Gaussian noise term assumed random values from 0 up to 0.9, to simulate a whole range of system fluctuations, from the milder to the radical. Their experiments used the inertia weight PSO variant with linearly decreasing inertia weight, w, from 1.0 towards 0.4, and $c_1 = c_2 = 0.5$. In addition, a fixed offset was added to the global minimizer at each iteration, to assess the tracking capabilities of PSO in the presence of noise (Parsopoulos & Vrahatis, 2001).

Figures 2, 3 and 4 illustrate the mean fitness value of the swarm, averaged over 100 experiments, for different offset values. As we observe in all figures, there is a preliminary exploration period of nearly 100 iterations, where PSO assumes higher function values. After this period, the distance from the global minimum, in terms of the swarm's mean function value, is stabilized and fluctuates within a narrow problem-dependent range for the rest of the run. This trend is observed for all offsets and test problems, although different offsets may result in different magnitudes of fluctuation.

This behavior is in line with the behavior of the inertia weight PSO variant, as illustrated in Fig. 4 of Chapter Two. Thus, it seems that noise does not modify the dynamics of PSO. Results are also reported

Table 7. Results for the PSO-PF and PF reported in Klamargias et al. (2008) for sample (and swarm) size, N = 20

System noise (Q)	Measurement noise (R)	RMS for PF	Best PSO-PF		
			a_1	RMS	p-value
0.01	0.01	1.720	0.2	2.164	3.5e-04
	0.0	2.739	0.2	2.560	3.8e-01
	0.1	2.191	0.2	2.917	1.7e-04
	0.3	2.590	0.5	3.663	1.2e-07
	0.5	2.598	0.2	3.715	1.3e-06
0.05	0.01	2.438	0.5	2.826	6.3e-03
	0.0	3.831	0.2	2.797	1.0e-02
	0.1	3.811	0.2	3.000	7.7e-02
	0.3	4.128	0.5	3.984	9.4e-01
	0.5	3.712	0.5	4.403	5.9e-04
0.10	0.01	2.674	0.2	3.179	3.5e-03
	0.0	4.033	0.2	3.249	2.3e-01
	0.1	4.447	0.2	3.498	1.3e-02
	0.3	5.159	0.5	4.577	6.2e-02
	0.5	5.294	0.5	4.889	2.5e-01
0.30	0.01	7.350	0.5	5.219	1.4e-06
	0.0	7.297	0.5	4.934	4.4e-08
	0.1	6.933	0.2	4.987	2.8e-07
	0.3	7.967	0.5	6.078	3.5e-08
	0.5	7.628	0.2	6.604	9.4e-03
0.50	0.01	9.537	0.2	6.532	2.3e-11
	0.0	8.434	0.2	6.866	8.4e-05
	0.1	8.628	0.5	7.112	7.7e-04
	0.3	8.923	0.5	8.038	3.1e-02
	0.5	9.135	0.2	8.255	1.5e-02

in Parsopoulos and Vrahatis (2001) for the detection of the global minimizer under different noise levels with a desired accuracy. The derived results agree with the remarks already stated in the previous sections. Thus, the reader is referred to the original paper for further details.

In a more recent work (Parsopoulos & Vrahatis, 2005), the unified PSO (UPSO) approach, described in Chapter Four, was assessed in purely dynamic environments without noisy function evaluations. However, this time, the experimental setting was completely different. More specifically, the global minimizer was let to perform a random unrestricted movement in the search space. This movement was simulated by adding a Gaussian offset with zero mean and standard deviations equal to 0.01, 0.10 and 0.50 to its position. Moreover, the movement was asynchronous, i.e., it was performed with a probability equal to 0.5 at each iteration. No bounds were imposed on particles and velocities, while the best positions were re-evaluated after each movement of the minimizer. For this purpose, a technique for the detection of environment changes, such as the one presented by Carlisle (2002), can be used.

Several values of the unification factor were considered for the 30-dimensional instances of test problems TP_{UO-1}-TP_{UO-4}, as well as for the 2-dimensional problem TP_{UO-5}, described in Appendix A of the book at hand. Their performance, in terms of the mean best function value averaged over 100 experiments, was recorded for the different offset noise levels. Figure 5 illustrates the performance of different unification factor values. The graphical representations are accompanied by a plethora of numerical results, for which the user is referred to the original paper of Parsopoulos and Vrahatis (2005). According to their analysis, $u = 0.5$, was shown to be overall the most efficient and robust value. The study was further extended to also include UPSO with mutation. Extensive experimentation revealed that mutated UPSO can achieve even better performance, especially in the case of large offset noise strength. Among the mutation settings, the one with mean equal to 1 and mutation on the global velocity component was shown to be the most promising in most cases.

Nevertheless, preliminary results suggested that UPSO is capable of producing satisfactory results under different settings in dynamic environments. This opens a challenging research field, where many aspects remain to be investigated, especially regarding the interaction of PSO with different established methodologies for identifying and tracking continuously changing minimizers.

CHAPTER SYNOPSIS

Challenging noisy problems were presented and tackled with PSO. The basic concepts of noisy optimization were reported along with two interesting applications of PSO, namely a simulated real-world problem on particle identification by light scattering, as well as the improvement of particle filtering methods for system states prediction. Emphasis was placed on problem formulation, while promising experimental settings from the relative literature were reported. Moreover, the case of dynamic envi-

Figure 2. Mean fitness value of the swarm for TP_{UO-2} and different offset values

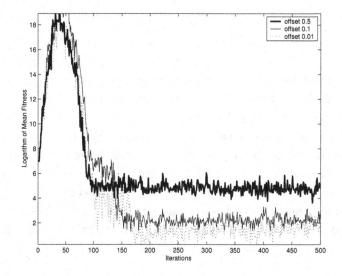

ronments was considered in the presence of noise and under accurate function evaluations. Graphical presentations of results reported in literature were given, along with conclusions derived for different PSO variants. Although there is a significant number of works in noisy and dynamically changing environments to date, there is an ongoing interest on such problems. This is mostly driven by real-world applications that involve such problems, opening a wide research horizon where many new developments are expected.

Figure 3. Mean fitness value of the swarm for TP$_{UO\text{-}15}$ and different offset values

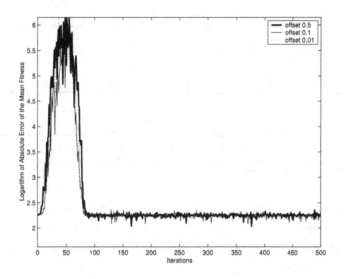

Figure 4. Mean fitness value of the swarm for TP$_{UO\text{-}22}$ and different offset values

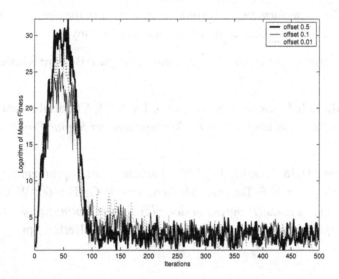

Figure 5. Performance of UPSO for (A) u = 0.0, (B) u = 0.2, (C) u = 0.5 and (D) u = 1.0. Test problems are indexed as follows: 1-TP$_{UO-1}$, 2-TP$_{UO-2}$, 3-TP$_{UO-3}$, 4-TP$_{UO-4}$ and 5-TP$_{UO-5}$

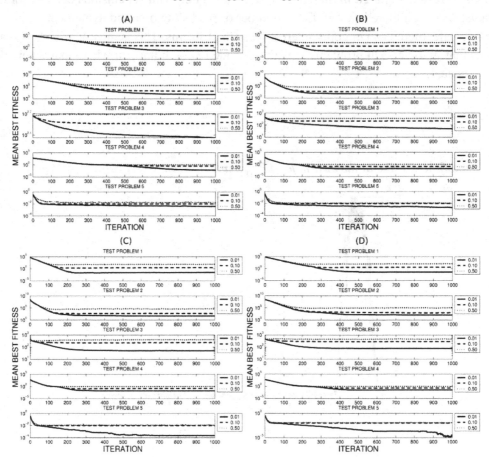

REFERENCES

Arnold, D. V. (2001). *Local performance of evolution strategies in the presence of noise*. Ph.D. thesis, Department of Computer Science, University of Dortmund, Germany.

Arnold, D. V. (2002). *Noisy optimization with evolution strategies*. Boston: Kluwer Academic Publishers.

Arulampalam, M. S., Maskell, S., Gordon, N., & Clapp, T. (2002). A tutorial on particle filters for online nonlinear/non-gaussian Bayesian tracking. *IEEE Transactions on Signal Processing, 50*(2), 174–188. doi:10.1109/78.978374

Bartz-Beilstein, T., Blum, D., & Branke, J. (2007). Particle swarm optimization and sequential sampling in noisy environments. In K.F. Doerner, M. Gendreau, P. Greistorfer, W. Gutjahr, R.F. Hartl, & M. Reimann (Eds.), *Metaheuristics (Progress in Complex System Optimization), Operations Research/Computer Science Interfaces series, Vol. 39, Part V* (pp. 261-273). Berlin: Springer.

Beyer, H.-G. (2000). Evolutionary algorithms in noisy environments: theoretical issues and guidelines for practice. *Computer Methods in Applied Mechanics and Engineering, 186*(2-4), 239–267. doi:10.1016/S0045-7825(99)00386-2

Beyer, H.-G. (2001). *The theory of evolution strategies.* Berlin: Springer.

Beyer, H.-G., Olhofer, M., & Sendhoff, B. (2004). On the impact of systematic noise on the evolutionary optimization performance - a sphere model analysis. *Genetic Programming and Evolvable Machines, 5*(4), 327–360. doi:10.1023/B:GENP.0000036020.79188.a0

Blackwell, T., & Branke, J. (2006). Multi-swarms, exclusion, and anti-convergence in dynamic environments. *IEEE Transactions on Evolutionary Computation, 10*(4), 459–472. doi:10.1109/TEVC.2005.857074

Blackwell, T. M., & Bentley, P. J. (2002). Dynamic search with charged swarms. In *Proceedings of the 2002 Genetic and Evolutionary Computation Conference (GECCO'02), New York (NY), USA* (pp. 19-26).

Bohren, C. F., & Huffman, D. R. (1983). *Absorption and scattering of light by small particles.* New York: Wiley.

Branke, J. (2002). *Evolutionary optimization in dynamic environments.* Amsterdam: Kluwer Academic Publishers.

Carlisle, A. (2002). *Applying the particle swarm optimizer to non-stationary environments.* PhD thesis, Auburn University, Alabama, USA.

Carlisle, A., & Dozier, G. (2000). Adapting particle swarm optimization to dynamic environments. In *Proceedings of the International Conference on Artificial Intelligence (IC-AI 2000), Las Vegas (NV), USA* (pp. 429-434).

Darwen, P. J., & Pollack, J. B. (1999). Co-evolutionary learning on noisy tasks. In *Proceedings of the 1999 IEEE Congress on Evolutionary Computation (CEC'99), Washington (DC), USA* (pp. 1724-1731).

Eberhart, R. C., & Shi, Y. (2001). Tracking and optimizing dynamic systems with particle swarms. In *Proceedings of the 2001 IEEE Congress on Evolutionary Computation (CEC'01), Seoul, Korea* (pp. 94-100).

Eskandari, H., & Geiger, C. D. (2009). Evolutionary multiobjective optimization in noisy problem environments. *Journal of Heuristics, 15*(6), 559-595.

Fox, R. F., Gatland, I. R., Roy, R., & Vemuri, G. (1988). Fast, accurate algorithm for numerical simulation of exponentially correlated colored noise. *Physical Review A., 38*, 5938–5940. doi:10.1103/PhysRevA.38.5938

Gilmore, T., & Kelley, C. T. (1995). An implicit filtering algorithm for optimization of functions with many local minima. *SIAM Journal on Optimization, 5*, 269–285. doi:10.1137/0805015

Gordon, N. J., Salmond, D. J., & Smith, A. F. M. (1993). Novel approach to nonlinear/non-gaussian Bayesian state estimation. *IEE–Proceedings–F, 140*(2), 107–113.

Han, L., & He, X. (2007). A novel opposition-based particle swarm optimization for noisys problems. In *Proceedings of the 3ʳᵈ International Conference on Natural Computation (ICNC'07), Haikou, China* (pp. 624-629).

Hansen, N., & Ostermeier, A. (2001). Completely derandomized self-adaptation in evolution strategies. *Evolutionary Computation, 9*, 159–195. doi:10.1162/106365601750190398

Hodgson, R. J. W. (2000). Genetic algorithm approach to particle identification by light scattering. *Journal of Colloid and Interface Science, 229*, 399–406. doi:10.1006/jcis.2000.6989

Hooke, R., & Jeeves, T. A. (1961). Direct search solution of numerical and statistical problems. *Journal of the ACM, 8*, 212–229. doi:10.1145/321062.321069

Hu, X., & Eberhart, R. (2002). Adaptive particle swarm optimisation: detection and response to dynamic systems. In *Proceedings of the 2002 IEEE Congress on Evolutionary Computation (CEC'02), Honolulu (HI), USA* (pp. 1666-1670).

Janson, S., & Middendorf, M. (2006). A hierarchical particle swarm optimizer for noisy and dynamic environments. *Genetic Programming and Evolvable Machines, 7*(4), 329–354. doi:10.1007/s10710-006-9014-6

Jin, Y., & Branke, J. (2005). Evolutionary optimization in uncertain environments - a survey. *IEEE Transactions on Evolutionary Computation, 9*(3), 303–317. doi:10.1109/TEVC.2005.846356

Julier, S. J., & Uhlmann, J. K. (1997). A new extension of the Kalman filter to nonlinear systems. In *Proceedings of the Aerosense: 11ᵗʰ International Symposium on Aerospace/Defence Sensing, Simulation and Controls, Orlando (FL), USA* (Vol. "Multi Sensor Fusion", Tracking and Resource Management II).

Klamargias, A. D., Parsopoulos, K. E., Alevizos, Ph. D., & Vrahatis, M. N. (2008). Particle filtering with particle swarm optimization in systems with multiplicative noise. In *Proceedings of the 2008 Genetic and Evolutionary Computation Conference (GECCO'08), Atlanta (GA), USA* (pp. 57-62).

Krohling, R. A. (2005). Gaussian particle swarm and particle filter for nonlinear state estimation. In A.P. del Pobil (Ed.), *Proceeding of the 9th IASTED Conference on Artificial Intelligence and Soft Computing (ASC 2005), Benidorm, Spain*.

Kwok, N. M., Zhou, W., Dissanayke, G., & Fang, G. (2005). Evolutionary particle filter: re-sampling from the genetic algorithm perspective. In *Proceedings of the 2005 IEEE International Conference on Intelligent Robots and Systems (IROS 2005), Edmonton, Canada* (pp. 2935-2940).

Li, X. (2004). Adaptively choosing neighbourhood bests using species in a particle swarm optimizer for multimodal function optimization. In [LNCS]. *Lecture Notes in Computer Science, 3102*, 105–116.

Nelder, J. A., & Mead, R. (1965). A simplex method for function minimization. *The Computer Journal, 7*, 308–313.

Pan, H., Wang, L., & Liu, B. (2006). Particle swarm optimization for function optimization in noisy environment. *Applied Mathematics and Computation, 181*(2), 908–919. doi:10.1016/j.amc.2006.01.066

Parrott, D., & Li, X. (2004). A particle swarm model for tacking multiple peaks in a dynamic environment using speciation. In *Proceedings of the 2004 IEEE Congress on Evolutionary Computation (CEC'04), Portland (OR), USA* (pp. 98-103).

Parsopoulos, K. E., Laskari, E. C., & Vrahatis, M. N. (2003). Particle identification by light scattering through evolutionary algorithms. In *Proceedings of the 1ˢᵗ International Conference for Mathematics and Informatics for Industry (MII 2003), Thessaloniki, Greece* (pp. 97-108).

Parsopoulos, K. E., & Vrahatis, M. N. (2001). Particle swarm optimizer in noisy and continuously changing environments. In M.H. Hamza (Ed.), *Proceedings of the 2001 IASTED Artificial Intelligence and Soft Computing Conference, Cancun, Mexico* (pp. 289-294).

Parsopoulos, K. E., & Vrahatis, M. N. (2002a). Recent approaches to global optimization problems through particle swarm optimization. *Natural Computing, 1*(2-3), 235–306. doi:10.1023/A:1016568309421

Parsopoulos, K. E., & Vrahatis, M. N. (2002b). Particle swarm optimization for imprecise problems. In D. Fotiadis, & C. Massalas (Eds.), *Scattering and Biomedical Engineering, Modeling and Applications* (pp. 254-264). Singapore: World Scientific.

Parsopoulos, K. E., & Vrahatis, M. N. (2002c). Particle swarm optimization method in multiobjective problems. In *Proceedings of the 2002 ACM Symposium on Applied Computing (SAC 2002), Madrid, Spain* (pp. 603-607).

Parsopoulos, K. E., & Vrahatis, M. N. (2005). Unified particle swarm optimization in dynamic environments. In [LNCS]. *Lecture Notes in Computer Science, 3449*, 590–599.

Salomon, R. (1998). Evolutionary search and gradient search: Similarities and differences. *IEEE Transactions on Evolutionary Computation, 2*, 45–55. doi:10.1109/4235.728207

Spall, J. C. (1992). Multivariate stochastic approximation using a simultaneous perturbation gradient approximation. *IEEE Transactions on Automatic Control, 37*, 332–341. doi:10.1109/9.119632

Tong, G., Fang, Z., & Xu, X. (2006). A particle swarm optimized particle filter for nonlinear system state estimation. In *Proceedings of the 2006 IEEE Congress on Evolutionary Computation (CEC'06), Vancouver (BC), Canada* (pp. 438-442).

Torczon, V. (1991). On the convergence of the multidirectional search algorithm. *SIAM Journal on Optimization, 1*, 123–145. doi:10.1137/0801010

Uosaki, K., & Hatanaka, T. (2005). Nonlinear state estimation by evolution strategies based Gaussian sum particle filter. In [LNCS]. *Lecture Notes in Computer Science, 3681*, 635–642.

Van der Merwe, R., Doucet, A., De Freitas, N., & Wan, E. (2000). *The unscented particle filter* (Tech. Rep.). Department of Engineering, Cambridge University, UK.

Wang, Q., Xie, L., Liu, J., & Xiang, Z. (2006). Enhancing particle swarm optimization based particle filter tracker. In [LNCS]. *Lecture Notes in Computer Science, 4114*, 1216–1221. doi:10.1007/11816171_151

Zakovic, S., Ulanowski, Z., & Bartholmew-Biggs, M. C. (1998). Application of global optimisation to particle identification using light scattering. *Inverse Problems, 4*(4), 1053–1067. doi:10.1088/0266-5611/14/4/019

Chapter 11
Applications in Multiobjective, Constrained and Minimax Problems

This chapter is devoted to the application of PSO and its variants on three very interesting problem types, namely (a) multiobjective, (b) constrained, and (c) minimax optimization problems. The biggest part of the chapter refers to the multiobjective case, since there is a huge bibliography with a rich assortment of PSO approaches developed to date. Different algorithm types are presented and briefly discussed, focusing on the most influential approaches.

APPLICATION IN MULTIOBJECTIVE OPTIMIZATION

Multiobjective optimization (MO) is the optimization branch that deals with problems where different objectives have to be handled simultaneously. In such problems, two or more competing and/or incommensurable objective functions have to be minimized concurrently. Due to the special nature of MO problems, *Pareto optimality* substitutes the standard optimality concept of single-objective optimization. Each Pareto optimal solution corresponds to an acceptable trade-off of the objective functions. Thus, two solutions may have the same fitness value, although possessing different inherent properties. For example, in shape optimization, two Pareto optimal solutions may correspond to structure configurations of equal fitness but different properties. Therefore, it is highly desirable to find the largest possible number of solutions with different inherent properties for a given MO problem.

Evolutionary algorithms are particularly suited to MO problems, due to their ability to evolve multiple Pareto optimal solutions concurrently (Coello Coello *et al.*, 2002; Deb, 1999; Schaffer, 1984). To date, a plethora of MO evolutionary approaches have been proposed, based on different concepts such as fitness sharing, niching and elitism (Deb *et al.*, 2002; Erickson *et al.*, 2001; Fonseca & Fleming, 1993; Horn *et al.*, 1994; Srinivas & Deb, 1994; Zitzler & Thiele, 1999). The detected Pareto optimal solutions are stored

DOI: 10.4018/978-1-61520-666-7.ch011

Copyright © 2010, IGI Global. Copying or distributing in print or electronic forms without written permission of IGI Global is prohibited.

in memory structures, called *external archives*. The use of archives significantly enhances the performance of MO approaches, although it requires special procedures for its efficient maintenance.

PSO has also been intensively studied in MO problems, giving rise to new, specialized variants (Reyes-Sierra & Coello Coello, 2006a). In Chapter Four of the book at hand we already presented such a variant, namely vector evaluated PSO (VEPSO), while a conventional MO approach was used with PSO in the previous chapter for the optimization of particle filtering models. The extensive work with PSO on MO problems means that any attempt to present it thoroughly within the strict confines of a single chapter is condemned to failure. For this reason, the following sections briefly describe only the most significant developments, with emphasis on methodologies and concepts that can trigger the development of new approaches. In addition, we underline the most active research directions along with future trends.

The next section provides concise descriptions of the necessary background material and fundamental concepts of MO problems.

Fundamental Concepts

Let $A \subset \mathbf{R}^n$ be an n-dimensional search space, and, $f_i(x)$, $i = 1, 2,\ldots, k$, be k objective functions defined over A. Also, let $\mathbf{f}(x)$ be a vector function defined as:

$$\mathbf{f}(x) = (f_1(x), f_2(x),\ldots,f_k(x))^{\mathrm{T}}, \tag{1}$$

and,

$$g_i(x) \leq 0, \quad i = 1, 2,\ldots, m, \tag{2}$$

be m inequality constraints. The objective functions, $f_i(x)$, may be conflicting with each other; thus, the detection of a single global minimum for all is impossible. For this purpose, optimality of solution in MO problems needs to be redefined properly.

Let $u = (u_1, u_2,\ldots, u_k)^{\mathrm{T}}$ and $v = (v_1, v_2,\ldots, v_k)^{\mathrm{T}}$ be two k-dimensional vectors. Then, we can state the following definitions:

Definition 1 (Pareto dominance): The vector u is said to *dominate* vector v, if and only if it holds that:

$u_i \leq v_i$, for all $i = 1, 2,\ldots, k$, and, $u_i < v_i$, for at least one component i.

This property is known as *Pareto dominance*.

Definition 2 (Pareto optimality): A solution, $x \in A$, of the MO problem is said to be *Pareto optimal*, if and only if there is no other solution, $y \in A$, such that $\mathbf{f}(y)$ dominates $\mathbf{f}(x)$. Alternatively, we can say that x is *nondominated* with respect to A. The set of all Pareto optimal solutions is called *Pareto optimal set* and will be henceforth denoted as P^*.

Definition 3 (Pareto front): The set of vector function values of all Pareto optimal solutions,

$$PF^* = \{\mathbf{f}(x): x \in P^*\}, \tag{3}$$

is called *Pareto front*. A Pareto front is *convex* if and only if, for all u, $v \in PF^*$, and all $\lambda \in (0,1)$, there exists a vector, $w \in PF^*$, such that:

$$\lambda \|u\| + (1-\lambda) \|v\| \geq \|w\|,$$

while it is called *concave*, if and only if:

$$\lambda \|u\| + (1-\lambda) \|v\| \leq \|w\|.$$

A Pareto front can be also partially convex and/or concave as well as discontinuous. These cases are considered the most difficult for most multiobjective optimization algorithms.

The main goal of MO is the detection of (hopefully) all Pareto optimal solutions, $x^* = (x_1^*, x_2^*,\ldots, x_n^*)$, of $\mathbf{f}(x)$. However, the Pareto optimal set can be infinite, while computation is usually restricted within strict time and space limitations, dictating more realistic expectations. Thus, the main goal of MO is reduced to the detection of the highest possible number of Pareto optimal solutions, with the smallest possible deviation from the Pareto front and adequate spread.

Important Application Issues

Multiobjective PSO approaches can be divided in two categories (Reyes-Sierra & Coello Coello, 2006a). The first consists of PSO variants that consider each objective function separately. In these approaches, each particle is evaluated only with one objective function at a time, and the best positions are determined following the standard single-objective PSO rules, using the corresponding objective function. The main challenge in these PSO variants is the proper manipulation of information from each objective function, in order to guide particles towards Pareto optimal solutions. The second category consists of approaches that evaluate all objective functions for each particle, and, based on the concept of Pareto optimality, produce nondominated best positions (often called *leaders*) to guide the particles. The determination of leaders is nontrivial, since they have to be selected among a plethora of nondominated solutions in the neighborhood of a particle. This is the main challenge related to the second category.

In all aforementioned approaches, a large number of Pareto optimal solutions can be detected during a single run of the algorithm, giving rise to a very important issue: the problem of solution maintenance. Storing nondominated solutions as best positions of the particles is the most trivial solution to this problem. However, it suffers two major drawbacks. First, it is not valid in cases where the desirable size of the Pareto front exceeds swarm size. Second, the selection of one between two equally good nondominated solutions to serve as the best position of a particle requires carefully selected problem-dependent criteria.

The first problem can be addressed by using an external archive for storing the nondominated solutions discovered during search. In this case, an update procedure shall take place at each iteration of the algorithm. During this procedure, the archive is scanned and new solutions are introduced into it, while the dominated ones are removed so that its size adheres to storage restrictions. The second problem is related to inherent properties of the problem at hand and can only be addressed per case. A general multiobjective PSO scheme can be described with the following pseudocode:

```
Begin
    Initialize swarm, velocities and best positions.
    Set an empty set as the external archive.
    While (stopping criteria not satisfied) Do
        For each particle of the swarm
        Select a leader from the external archive (if applicable).
        Update particle velocity and position.
        Evaluate new particle position.
        Update best position and external archive.
        End For
    End While
End
```

Evidently, the selection of leaders, as well as archive and best positions update, constitute key concepts of multiobjective PSO approaches, albeit not the only ones. Different factors, such as swarm diversity, also have a crucial impact on performance.

Leader selection requires the determination of quality measures for archive members. Density estimators, which promote archive diversity, have been proposed for this purpose. The most commonly used approaches are the *nearest neighbor density estimator* (Deb *et al.*, 2002) and *kernel density estimator* (Deb & Goldberg, 1989), which both provide estimations of the number and proximity of neighbors for a given point.

Archive update is a more complex problem. A new solution shall be included in the archive if it is nondominated by all its members. On the other hand, if some of the archive members are dominated by a new solution, they shall be removed. Domination tests can be computationally very expensive as the archive size increases rapidly with time. In practice, a small number of iterations is usually enough to fill the archive fully. Thus, archive size shall be carefully considered to avoid long delays due to dominance tests at each iteration of the algorithm.

The special case of a candidate new solution that is nondominated by all archive members has also to be addressed. This solution must compete with all archive members to replace an existing one. In such cases, archive diversity serves as the acceptance criterion, i.e., the decision to replace an existing archive member with a new solution is based on whether it retains the maximum possible archive diversity. For this purpose, clustering techniques have been proposed (Knowles & Corne, 2000; Zitzler & Thiele, 1999). Alternatively, the concept of ε-dominance has been used to separate the Pareto front in boxes and retain one solution per box (Mostaghim & Teich, 2003b).

Updating the best positions of particles depends on the PSO variant. In variants with distinct evaluation of each objective function, it is performed as in standard PSO for single-objective optimization. In Pareto-based approaches, a best position is replaced only if the new one dominates it. If they are nondominated, the new one is preferred if it promotes swarm diversity. At this stage, the employed neighborhood topology and PSO variant have also crucial impact on performance. To date, there is no sufficient experimental evidence suggesting superiority of specific variants and topologies in general MO problems.

In the following sections, we briefly present the state-of-the-art multiobjective PSO approaches, following the aforementioned categorization in approaches that exploit each objective function separately and Pareto-based schemes. Although this distinction is helpful for presentation purposes, it is not strict, as various approaches combine features from both categories.

PSO Variants that Exploit Each Objective Function Separately

This category consists of approaches that either combine all objective functions in a single one or consider each objective function in turn for the evaluation of particles. In these approaches, minor modifications are required, as PSO retains the main update procedures for particles and best positions. External archives are usually employed for storing nondominated solutions. Their main drawback is the difficulty of defining a scheme for the proper manipulation of the objective functions to achieve convergence to the actual Pareto front (Jin *et al.*, 2001). This category includes weighted aggregation, objective function ordering, and non-Pareto vector evaluated approaches, described in the following sections.

Weighted Aggregation Approaches

These approaches aggregate all objective functions in a weighted combination, producing a single one:

$$F(x) = \sum_{i=1}^{k} w_i f_i(x),$$

where w_i are non-negative weights, such that:

$$\sum_{i=1}^{k} w_i = 1.$$

Thus, the MO problem is transformed to a single-objective one that can be addressed straightforwardly with PSO. If weights remain fixed during a run, we have the case of *conventional weighted aggregation* (CWA), which is characterized by simplicity but suffers some crucial deficiencies. For instance, the algorithm has to be applied repeatedly with different weight settings to detect a desirable number of nondominated solutions, as only a single solution can be attained per run. Moreover, CWA is unable to detect solutions in concave regions of the Pareto front (Jin *et al.*, 2001).

The aforementioned limitations can be addressed by using approaches with dynamically adjusted weights, such as *bang-bang weighted aggregation* (BWA). For bi-objective problems, the weights in BWA are adapted as follows (Jin *et al.*, 2001):

$$w_1(t) = \text{sign}(\sin(2\pi t/a)), \quad w_2(t) = 1 - w_1(t),$$

where a is a user-defined adaptation frequency, and t stands for the iteration number. Another popular approach is the *dynamic weighted aggregation* (DWA), where weights are adapted as follows:

Table 1. Parameter setting for the experiments reported in Parsopoulos and Vrahatis (2002a, 2002b)

Parameter	Value
Swarm size	Problem dependent
PSO parameters	$w = 1.0 \rightarrow 0.4$ (linearly decreasing) $c_1 = c_2 = 0.5$
Particle boundaries	$[0,1]^2$
Maximum iterations	150
Size of archive	20 for TP_{MO-1}-TP_{MO-3}, 40 for TP_{MO-4} and TP_{MO-5}
Weight adaptation frequency	100 for BWA, 200 for DWA

$$w_1(t) = |\sin(2\pi t/a)|, \quad w_2(t) = 1 - w_1(t).$$

The use of the sign function in BWA results in abrupt changes of weights, enforcing the algorithm to keep moving towards the Pareto front. The same effect is achieved with DWA, although with milder weight changes than BWA. Experiments with genetic algorithms revealed the superiority of DWA in convex Pareto fronts, while their performance was almost identical in concave Pareto fronts.

Parsopoulos and Vrahatis (2002a, 2002b) proposed a multiobjective PSO weighted aggregation approach and studied its performance in the bi-objective problems TP_{MO-1}-TP_{MO-5}, defined in Appendix A of the book at hand, using CWA, BWA and DWA. In their experiments, the weight adaptation frequencies, $a = 100$ for BWA and $a = 200$ for DWA, were used. They employed the inertia weight PSO variant, with the experimental setting of Table 1.

Figure 1 illustrates the reported Pareto fronts for the CWA approach. We must note that for TP_{MO-3}, only the two end points, $(0,1)^T$ and $(1,0)^T$, of the Pareto front were detected, due to the already mentioned disability of CWA to detect solutions in concave regions of the Pareto front. The same hold for TP_{MO-4}, where only convex parts of the (partially convex/concave) Pareto front were detected. The corresponding results for BWA and DWA are illustrated in Fig. 2, except of TP_{MO-2}, which is omitted due to space limitations, as it has the same (convex) Pareto front type with TP_{MO-1}. We must note that under CWA, one Pareto optimal point can be detected per run. Thus, PSO was repeated 20 times in TP_{MO-1}-TP_{MO-3}, and 40 times in TP_{MO-4} and TP_{MO-5} to detect the desirable number of Pareto optimal points (20 and 40, respectively). On the other hand, for the dynamic weight cases, the archiving technique of Jin *et al.* (2001) was employed and the illustrated Pareto fronts consist of the corresponding archived points. As expected, dynamic schemes outperformed fixed weights, as reported in Parsopoulos and Vrahatis (2002a, 2002b). Although simplicity and straightforward applicability render weighted aggregation schemes very attractive, their efficiency on problems with more than two objectives has not been investigated extensively.

A similar approach was proposed by Baumgartner *et al.* (2004), where subswarms that use a different weight setting each are combined with a gradient-based scheme for the detection of Pareto optimal solutions. More specifically, the swarm is divided in subswarms, each using a specific weight setting. The best particle of each subswarm serves as the leader for itself. In addition, a preliminary Pareto decision is made in order to investigate candidate Pareto optimal solutions more thoroughly. This decision is made for each particle, x, based on the relation:

Figure 1. Pareto fronts reported in Parsopoulos and Vrahatis (2002a, 2002b) for CWA with PSO, for the test problems TP$_{MO-1}$-TP$_{MO-5}$, defined in Appendix A of the book at hand

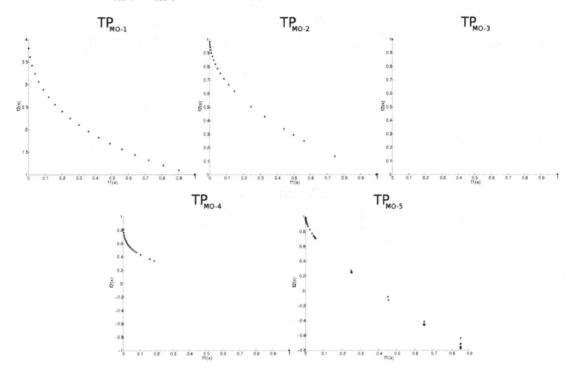

$$\frac{1}{k}\left|\sum_{j=1}^{k}\mathrm{sgn}\Big(f_{j}(x(t+1))-f_{j}(x(t))\Big)\right|\neq 1,$$

where t stands for the iteration counter. If this relation holds, x could be a Pareto optimal point and the gradients of the objective functions, f_i, $i = 1, 2,\ldots, k$, are computed on a perturbed point, $x + \Delta x$. If no objective function improves at the perturbed point, then x is considered to be Pareto optimal and it is removed from the swarm. Although results on a limited set of test problems are promising, the algorithm has not been fully evaluated and compared with other PSO approaches.

Mahfouf *et al.* (2004) proposed a dynamically modified weights approach, where the PSO variant with linearly decreasing inertia weight was modified by incorporating a mutation operator to alleviate swarm stagnation. In addition, an acceleration term was used to accelerate convergence at the later stages of the algorithm. More specifically, the standard velocity update of PSO becomes:

$$v_{ij}(t+1) = w\, v_{ij}(t) + a\, [r_1\, (p_{ij}(t) - x_{ij}(t)) + r_2\, (p_{gj}(t) - x_{ij}(t))],$$

where a is an acceleration factor depending on the current iteration number, defined as:

$$a = a_0 + \frac{t}{t_{\max}},$$

Figure 2. Pareto fronts reported in Parsopoulos and Vrahatis (2002a, 2002b) for BWA and DWA with PSO, for the test problems TP_{MO-1}-TP_{MO-5}, defined in Appendix A of the book at hand

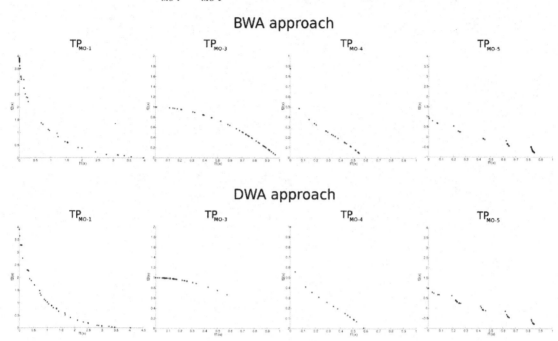

where t_{max} is the maximum number of iterations, and a_0 lies within the range [0.5,1.0]. After the computation of new particle positions, both new and old positions are entered in a list. The *non-dominated sorting technique* (Li, 2003) is applied on this list, and the non-dominated particles (that approximate the Pareto front) are selected. These particles suffer a mutation procedure in an attempt to further improve them. The resulting particles constitute the swarm in the next iteration of the algorithm.

This scheme was applied on a problem from the steel industry, with promising results (Mahfouf *et al.*, 2004). The algorithm combines characteristics of different approaches that were shown to enhance the performance of multiobjective methods. Its competitive performance to PSO and different evolutionary approaches, such as NSGA-II and SPEA2, can be attributed to the mutation operator that preserves swarm diversity, as well as to the sorting technique that allows the direct exploitation and evolution of points approximating the Pareto front, instead of using an external archive.

Ordering Approaches

These approaches are based on ranking schemes that determine the importance of each objective function. Minimization takes place for each function separately, starting from the most important one. Hu and Eberhart (2002) proposed such an ordering scheme. Since Pareto front constitutes a boundary of fitness values, their algorithm keeps the simplest objective function fixed and minimizes the rest. Experiments used a special local PSO variant with dynamic neighborhoods defined in the fitness value space. Non-dominated solutions were stored as the best positions of particles; thus, no external archive was needed.

Figure 3. Pareto fronts reported in Parsopoulos and Vrahatis (2002a, 2002b) for VEPSO

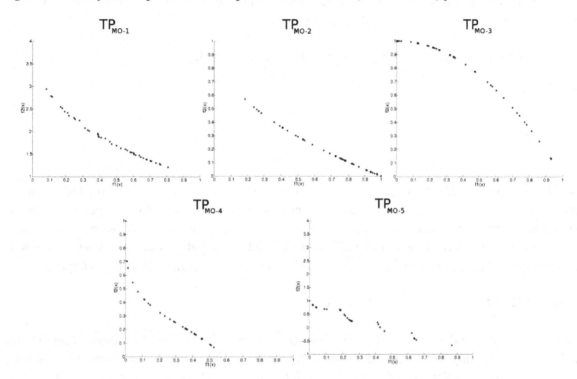

Although promising results were received for problems with two objective functions, there was limited justification on the ordering procedure, which can be crucial for its performance.

The previous approach was extended by Hu *et al.* (2003a) by incorporating an external archive in the form of external memory for storing nondominated solutions and reducing computational cost. Albeit promising results were reported in problems with two objective functions, further investigation is needed to reveal its potential, since the authors reported the inability of this approach to tackle the binary string problem.

Non-Pareto Vector Evaluated Approaches

Parsopoulos and Vrahatis (2002a, 2002b) also proposed the vector evaluated PSO (VEPSO) scheme, described in Chapter Four, and investigated its performance on test problems TP_{MO-1}-TP_{MO-5}, previously used for the weighted aggregation approaches. We remind that VEPSO employs one swarm per objective function and evaluates it only with this function, while best positions of one swarm are used to update velocities of another swarm that corresponds to a different objective function. Figure 3 illustrates the Pareto fronts received by VEPSO under the same experimental setting with the weighted aggregation approaches described in a previous section, by communicating the overall best positions between the two swarms (Parsopoulos & Vrahatis, 2002b).

An approach similar to VEPSO, called *multi-species PSO*, was proposed by Chow and Tsui (2004) within a generalized autonomous agent response-learning framework related to robotics. It uses a set of subswarms, each evaluated with one objective function, and information of best particles is communicated to neighboring subswarms, in terms of an augmented velocity update rule. Thus, the velocity of the *i*-th particle in the *k*-th swarm is updated as follows:

$$v_{ij}^{[k]}(t+1) = v_{ij}^{[k]}(t) + a_1^{[k]} (p_{ij}^{[k]}(t) - x_{ij}^{[k]}(t)) + a_2^{[k]} (p_{gj}^{[k]}(t) - x_{ij}^{[k]}(t))) + A,$$

where,

$$A = \sum_{l=1}^{H_k} \left(p_{gj}^{[l]}(t) - x_{ij}^{[k]}(t) \right),$$

with H_k being the number of swarms that communicate with the *k*-th one and $p_g^{[l]}$ being the best position of the *l*-th swarm, $l = 1, 2,..., H_s$. The algorithm was shown to be competitive to other established multiobjective PSO approaches, although in limited number of experiments. At the same time, there are several unanswered questions, since velocity update did not include any constriction coefficient or inertia weight, while the scheme for defining neighboring swarms was not adequately analyzed.

Variants Based on Pareto Dominance

These approaches use the concept of Pareto dominance to determine the best positions (leaders) that guide the swarm during the search. Additional criteria that take into consideration further issues, such as swarm diversity and Pareto front spread, are needed, rendering Pareto-based PSO methods a blossoming research area with significant developments. In the following paragraphs, we briefly describe the most significant ones.

Coello Coello and Salazar Lechuga (2002) proposed the Multiobjective PSO (MOPSO), one of the first Pareto-based PSO approaches (Coello Coello *et al.*, 2004). In MOPSO, the nondominated solutions are stored in an archive (also called *repository*). In addition, the search space is divided in hypercubes. Each hypercube is assigned a fitness value, inversely proportional to the number of particles it contains. Then, the classical roulette wheel selection is used to select a hypercube and a leader from it. Thus, the velocity update for the *i*-th particle of the employed inertia weight PSO variant becomes:

$$v_{ij}(t+1) = w \, v_{ij}(t) + c_1 r_1 (p_{ij}(t) - x_{ij}(t)) + c_2 r_2 (R_h(t) - x_{ij}(t)),$$

where p_i is its best position, and R_h is the selected leader from the repository. The best position, p_i, is updated at each iteration, based on its domination by the new one.

In addition, a *retentioncriterion* is used to address the limited repository size. This criterion allows new solutions to enter a full repository, giving priority to solutions located in less crowded areas of the objective space. MOPSO was shown to be competitive against NSGA-II and PAES on widely used test problems, and it is currently considered as one of the most typical multiobjective PSO approaches.

Fieldsend and Singh (2002) proposed a multiobjective PSO scheme that addresses problems due to the truncation of limited archives. This is achieved by using a tree-like structure, called *dominated tree*, for unconstrained archiving maintenance (Fieldsend *et al.*, 2003). The algorithm works similarly to MOPSO except the repository, which is maintained through the aforementioned structure. An additional feature that works beneficially for the algorithm is the use of mutation, called *craziness*, on particle velocity to preserve swarm diversity. The algorithm was shown to be competitive with PAES, although it suffers a general deficiency of such approaches in cases where closeness in objective space is loosely related to closeness in parameter space.

Bartz-Beielstein *et al.* (2003) proposed DOPS, a method based on elitist archiving. Their analysis considered different schemes for updating the archive and selecting the most proper solutions using functions that assess performance and contribution of each particle to Pareto front spreading. More specifically, two functions, F_{sel} and F_{del}, are used to assign a *selection* and a *deletion* fitness value to each particle, respectively. The selection value is a measure of particle influence on the Pareto front spreading and increases with the distance from the nearest neighbors. Thus, every time a personal or a globally best position is needed, a member is chosen from the archive based on a roulette wheel selection over F_{sel}. If the number of available non-dominated solutions surpasses archive size, then a member of the archive is selected for deletion based on F_{del}. Different selection and deletion functions are proposed and evaluated. The method was supported by a parameter sensitivity analysis, providing useful hints on the archiving impact on multiobjective PSO variants.

Mostaghim and Teich (2003a, 2003b, 2004) proposed several algorithms based on MOPSO, incorporating special schemes for selecting archive members for particle position update. In 2003, they proposed a MOPSO approach in combination with the *sigma method* that assigns a numerical value to each particle and archive member (Mostaghim & Teich, 2003a). For example, in a bi-objective problem, if a particle has objective values, (f_1, f_2), it is assigned a sigma value:

$$\sigma = \frac{(K_2 f_1)^2 - (K_1 f_2)^2}{(K_2 f_1)^2 + (K_1 f_2)^2},$$

where K_1, K_2, are the maximum objective values of particles for f_1 and f_2, respectively. Then, the archive member with the closest sigma value can be used as the leader of this particle. Also, a turbulence (mutation) factor is used in particle update to maintain swarm diversity. The algorithm outperformed SPEA2 in widely used bi-objective problems but the opposite happened for problems with three objectives. In addition, the authors underline the necessity for large swarm sizes, as an adequate number of distributed solutions are required in the objective space.

Mostaghim and Teich (2003b) also studied the performance of MOPSO using the concept of ε-dominance and compared it to clustering-based approaches with promising results. Their work focused mainly on the archiving methodology rather than the algorithm, indicating the superiority of MOPSO with ε-dominance in terms of the obtained Pareto fronts.

Furthermore, they proposed an algorithm for covering the Pareto front by using subswarms and an unbounded external archive (Mostaghim & Teich, 2004). In this approach, an initial approximation of the Pareto front is detected through MOPSO and subswarms are initialized around each nondominated solution to perform a more refined search around it. The algorithm outperformed an evolutionary approach (Hybrid MOEA) that incorporates a space subdivision scheme on an antenna design problem. The supported applicability of the proposed scheme on problems of any dimension and number of objectives is also counted in its advantages against similar approaches.

Li (2004) proposed the *maximin PSO* approach, which exploits the maximin fitness function (Balling, 2003). According to it, the fitness function of a decision vector, x, is defined as:

$$\max_{\substack{j=1,2,\ldots,N \\ x\neq y}} \min_{i=1,2,\ldots,k} \{f_i(x) - f_i(y)\},$$

where k is the number of objective functions, and N is the swarm size. Obviously, only decision vectors with a maximin function value less than zero can be nondominated solutions with respect to the current population. The maximin function promotes swarm diversity, since it penalizes particles that form clusters. It also favors middle solutions in convex Pareto fronts and extreme solutions in concave fronts (Balling, 2003). Li (2004) has addressed this effect by using adequately large swarms, while nondominated solutions are stored in an archive to serve as leaders (randomly selected by the particles). Maximin PSO outperformed NSGA-II on typical benchmark problems. However, experiments were restricted in bi-objective unconstrained problems; thus, no sound conclusions can be derived regarding its efficiency in more demanding cases.

Toscano Pulido and Coello Coello (2004) proposed *another MOPSO* (AMOPSO), an approach similar to VEPSO, where subswarms are used to probe different regions of the search space. Each subswarm has its own group of leaders, formed from a large set of nondominated solutions through clustering. Each subswarm is assigned a group of leaders and selects randomly those that will serve as its guides towards the Pareto front. This approach can alleviate problems related to disconnected search spaces, where a particle may be assigned leaders that lie in a disconnected region, wasting a lot of search effort. At some points, information exchange is also permitted among subswarms. The authors showed that AMOPSO is competitive to NSGA-II and can be a viable alternative.

AMOPSO does not use an external archive (nondominated solutions are stored as best positions of the particles), in contrast to *OMOPSO*, proposed by Reyes-Sierra and Coello Coello (2005), which employs two external archives. This approach uses the nearest neighbor estimator and stores the selected best positions for the current iteration of PSO in the one archive and the overall nondominated solutions (final solutions) in the other. Established concepts such as turbulence (mutation) and ε-dominance are also used for diversity and archive maintenance respectively, increasing complexity compared to AMOPSO. The special feature of OMOPSO is a mechanism for removing leaders when their number exceeds a threshold, lending it increased efficiency and effectiveness. OMOPSO was shown to outperform previously presented MOPSO approaches, as well as NSGA-II and SPEA2, rendering it a highly efficient method.

Raquel and Naval (2005) proposed MOPSO-CD, an approach that incorporates *crowding distance* for selecting the global best particle, as well as for the deletion of nondominated solutions from the external archive. Mutation is also employed to maintain diversity of nondominated solutions in the archive. Crowding distance is computed for each nondominated solution separately. If $f_1, f_2, \ldots, f_k$, are the objective functions and AR is the external archive, then for the computation of the crowding distance of $p \in AR$ with respect to f_j, $j = 1, 2, \ldots, k$, denoted as $CD_p(f_j)$, all points in AR are sorted with respect to their f_j value, and:

$$CD_p(f_j) = f_j(q) - f_j(r),$$

where q is the member of AR that follows immediately after p in the sorting order with respect to the f_j values, and r is the point that precedes p in the same ordering. Thus, the total crowding distance of p is given by:

$$\sum_{j=1}^{k} CD_p(f_j).$$

A proportion of the nondominated points of *AR* with the highest crowding distances (selected randomly) serve as swarm leaders. In addition, swarm diversity is retained by using mutation on particles at randomly selected iterations. Typical constraint-handling techniques adopted from the NSGA-II algorithm (Deb *et al.*, 2002) are incorporated for addressing constrained problems. MOPSO-CD was compared to MOPSO with promising results.

Alvarez-Benitez *et al.* (2005) proposed *Rounds*, *Random*, and *Prob*, three techniques for selecting leaders from the external archive based solely on the concept of Pareto dominance. Each technique equips the algorithm with different properties. For a given particle, x_i, Rounds promotes as its global guide the nondominated solution that dominates the fewest particles of the swarm, including x_i. This solution is then excluded from selection for the remaining particles. The procedure can be computationally expensive for large archives; however, it was shown to promote diversity. On the other hand, Random uses a probabilistically selected nondominated solution that dominates x_i as its guide. For selection, each nondominated solution is assigned the same probability. Prob constitutes an extension of Random that favors archive members that dominate the smallest number of particles. Mutation was also employed, along with constraint-handling techniques for constrained problems.

Salazar Lechuga and Rowe (2005) introduced MOPSO-*fs*, a MOPSO variant with explicit fitness sharing. According to this approach, each particle, p_i, in the archive is assigned a fitness value:

$$F_i^{\text{sh}} = \frac{10}{\sum_{j=1}^{n} s_i^j},$$

where,

$$s_i^j = \begin{cases} 1 - \left(\dfrac{d_i^j}{\sigma_{share}} \right)^2, & \text{if } d_i^j < \sigma_{share}, \\ 0, & \text{otherwise,} \end{cases}$$

with σ_{share} being a user-defined distance, and d_i^j be a distance measure between the nondominated solutions, p_i and p_j. This fitness-sharing scheme assigns higher fitness values to solutions with small number of other solutions surrounding them. Leaders of the swarm are selected through a roulette wheel selection technique based on the assigned fitness values. MOPSO-*fs* was shown to be competitive with MOPSO, as well as with NSGA-II and PAES, although the analysis of choosing fitness sharing parameters remains under investigation.

Reyes-Sierra and Coello Coello (2006b) reported an interesting study on the online parameter adaptation of MOPSO. More specifically, the inertia weight, w, acceleration coefficients, c_1 and c_2, and selection method (dominance or crowding) probability, P_s, were statistically investigated using analysis of variance (ANOVA). The analysis has shown that large values of P_s, w, and c_2 provide better results, while c_1 seems to have a milder effect on performance. After identifying the most crucial parameters, they proposed different adaptation techniques based on a reward system. According to their analysis, the parameter level selection could be *proportional*, *greedy*, or based on the *soft max strategy* that employs a Gibbs distribution. Their results were very promising, opening the way towards more efficient self-adaptive multiobjective approaches.

In our opinion, the aforementioned algorithms constitute the most promising and influential multi-objective PSO approaches. The relative research has produced a remarkable number of new approaches and improvements. Of course, it is impossible to include them all in this chapter. For this reason, we refer the interested reader to the review works of Reyes-Sierra and Coello Coello (2006a) and Parsopoulos and Vrahatis (2008) for further acquisition. The next section presents another interesting application field of PSO, constrained optimization.

APPLICATION IN CONSTRAINED OPTIMIZATION

The general constrained optimization (CO) problem was defined by equation (1) in Chapter One of the book at hand, as follows:

$$\min_{x \in A} f(x), \quad \text{subject to} \quad C_i(x) \le 0, \; i = 1, 2, ..., k.$$

Several authors consider even the bounding boxes of the problem variables, x_i, that define the search space, A, as constraints. We will consider as CO problems only those with possible additional inequality relations, besides the ones that define A. Such problems appear very often in engineering applications, such as structural optimization, engineering design, VLSI design and allocation problems (Floudas & Pardalos, 1990; Himmelblau, 1972).

CO problems can be solved using either deterministic or stochastic algorithms. However, deterministic approaches, such as *feasible direction* and *generalized gradient descent*, require strong mathematical properties of the objective function, such as continuity and differentiability (Floudas & Pardalos, 1990; Himmelblau, 1972; Hock & Schittkowski, 1981). In cases where these properties are absent, evolutionary computation offers reliable alternative methods. This has been verified in several studies in the past few years (Coello Coello, 1999; Homaifar *et al.*, 1994; Joines & Houck, 1994; Parsopoulos & Vrahatis, 2002c; Yeniay, 2005).

Since most evolutionary approaches were primarily designed to address unconstrained problems, constraint-handling techniques are usually required to detect only feasible solutions. The most common constraint handling technique is the use of penalty functions, described in Chapter Five of the book at hand. According to this approach, the original objective function is transformed to another function that penalizes infeasible solutions by adding penalties to their values based on the number and magnitude of violated constraints. The reader is referred back to Chapter Five for a formal presentation of the penalty function approach.

Parsopoulos and Vrahatis (2002c) applied both the constriction coefficient and inertia weight PSO variants, denoted as $PSO^{[co]}$ and $PSO^{[in]}$, respectively, on the CO problems TP_{CO-1}-TP_{CO-6}, defined in Appendix A of the book at hand, using the penalty function defined by equations (11) and (12) of Chapter Five. Following the experimental setting of Yang *et al.* (1997) they used the following penalty function parameters:

$$\theta(q_i(x)) = \begin{cases} 10, & q_i(x) < 0.001, \\ 20, & 0.001 \le q_i(x) \le 0.1, \\ 100, & 0.1 \le q_i(x) \le 1.0, \\ 300, & \text{otherwise}, \end{cases} \qquad \gamma(q_i(x)) = \begin{cases} 1, & q_i(x) < 1.0, \\ 2, & \text{otherwise}, \end{cases}$$

and

$$h(t) = \begin{cases} \sqrt{t}, & \text{for test problem } TP_{CO-1}, \\ t\sqrt{t}, & \text{for the rest test problems}. \end{cases}$$

The tolerance to consider a constraint as violated was equal to 10^{-5} in all cases. The parameter setting of the PSO variants is reported in Table 2, while the best detected solutions over 10 experiments, in terms of their function value and sum of violated constraints, per problem and PSO variant are reported in Table 3.

As reported in Parsopoulos and Vrahatis (2002c), the results were better in most cases than the corresponding in Yang *et al.* (1997) for different evolutionary algorithms. In addition, we can observe that the inertia weight variant usually provided solutions with a smaller sum of violated constraints than the constriction coefficient variant. This can be attributed to its better exploration properties that permit a more thorough search around the boundaries by the constraints.

Another interesting variant, namely unified PSO (UPSO), described in Chapter Four of the book at hand, was investigated on CO problems in Parsopoulos and Vrahatis (2005). However, this study focused on widely used engineering design problems and employed the penalty function defined in equations (13) and (14) of Chapter Five. Also, a simple feasibility preserving scheme was used for the best positions. More specifically, following the approach of Hu *et al.* (2003b), a best position was updated only to feasible new best positions. Thus, the feasibility of solutions was explicitly preserved, attracting particles towards feasible regions of the search space.

The UPSO variants with $u = 0.2$ and $u = 0.5$, as well as the one with $u = 0.1$ and mutation, denoted as $UPSO_1$, $UPSO_2$ and $UPSO_m$, respectively, were applied on the test problems TP_{ED-1}-TP_{ED-4}, defined in Appendix A, and compared with the standard global and local PSO variant with constriction coefficient. In all test problems, the weights of the penalty function were set to the values, $w_1 = w_2 = 100$, and 100 independent experiments were performed. The parameter setting is summarized in Table 4, while Table 5 reports a portion of the results reported in Parsopoulos and Vrahatis (2005), namely the best

Table 2. Parameter setting of the PSO experiments reported in Parsopoulos and Vrahatis (2002c) for constrained optimization problems

Parameter	Value
PSO parameters	$\chi = 0.73$, $c_1 = c_2 = 2.0$ $w = 1.2 \rightarrow 0.1$ (linearly decreasing)
Maximum velocity	$v_{max} = 4$
Swarm size	100
Violation tolerance	10^{-5}

Table 3. Best solutions detected by the two PSO variants, in terms of their function value and sum of violated constraints, as reported in Parsopoulos and Vrahatis (2002c)

Problem	PSO variant	Best solution	
		Function value	Sum of viol. con.
TP_{CO-1}	PSO[in]	1.393431	0.000020
	PSO[co]	1.393432	0.000020
TP_{CO-2}	PSO[in]	-6961.798	0.0000087
	PSO[co]	-6961.837	0.000019
TP_{CO-3}	PSO[in]	680.639	0.000019
	PSO[co]	680.635	0.00130
TP_{CO-4}	PSO[in]	-31543.484	1.311
	PSO[co]	-31542.578	1.311
TP_{CO-5}	PSO[in]	-31544.036	0.997
	PSO[co]	-31543.312	0.996
TP_{CO-6}	PSO[in]	-213.0	0.0
	PSO[co]	-213.0	0.0

performing variant along with the mean function value of the solutions averaged over all experiments per problem.

The mutated UPSO variant, as well as the balanced one with $u = 0.5$, were shown to be the most promising, outperforming both the standard local and global PSO variants. In addition, as reported in Parsopoulos and Vrahatis (2005), the results conform to different evolutionary approaches, indicating that PSO and its variants can work satisfactorily in CO problems with widely used constraint-handling techniques. Recently, promising approaches were proposed by Muñoz Zavala *et al.* (2005), where perturbation operators and simple feasibility rules were incorporated in PSO, as well as by He and Wang (2007), where a coevolutionary PSO approach co-evolves solutions and penalty factors using two independent swarms. Both approaches exhibit efficient behavior, compared favorably to established approaches.

Another efficient coevolutionary approach was proposed by Krohling and Santos Coelho (2006), where Gaussian distributions were also incorporated in a coevolutionary PSO scheme. This approach transforms the CO problem to a minimax optimization one. In the next section, we report on the application of PSO on this interesting problem type.

APPLICATION IN MINIMAX OPTIMIZATION

Numerous problems encountered in optimal control, engineering design, discrete optimization, Chebyshev approximation, and game theory involve *minimax problems* (Demyanov & Molozemov, 1974; Du & Pardalos, 1995; Zuhe *et al.*, 1990). In general, the minimax problem can be defined in the following explicit form:

$$\min_x f(x), \tag{4}$$

Table 4. Parameter setting of the UPSO experiments reported in Parsopoulos and Vrahatis (2005) for engineering design problems

Parameter	Value
PSO parameters	$\chi = 0.729$, $c_1 = c_2 = 2.05$
Maximum iterations	5000
Swarm size	20
Neighborhood topology	Ring with radius equal to 1
Mutation parameters (UPSO$_m$)	$\mu = (0,0,...,0)^T$, $\sigma = 0.01$

Table 5. The best performing approach reported in Parsopoulos and Vrahatis (2005), along with the mean function value of the solutions, averaged over 100 experiments

Test problem	Best performing approach	Mean function value
TP$_{ED-1}$	UPSO$_m$	2.29478×10^{-2}
TP$_{ED-2}$	UPSO$_2$	1.96820×10^{0}
TP$_{ED-3}$	UPSO$_m$	3.80562×10^{-8}
TP$_{ED-4}$	UPSO$_2$	8.01637×10^{3}

where,

$$f(x) = \max_{1 \leq i \leq m} f_i(x),$$ (5)

with $f_i(x)$: $S \subset \mathbf{R}^n \to \mathbf{R}$, $i = 1, 2,..., m$. This problem type is closely related to constrained optimization. Indeed, a constrained problem of the form:

$$\min_x f(x), \quad \text{subject to the constraints } C_i(x) \geq 0, \quad i = 2,3,...,m,$$ (6)

can be given in an equivalent implicit minimax form, as follows:

$$\min_x \max_{1 \leq i \leq m} f_i(x),$$ (7)

where,

$$f_1(x) = f(x),$$
$$f_j(x) = f(x) - a_j C_j(x),$$
$$a_j > 0, \quad j = 2,3,...,m.$$ (8)

Note that constraints in equation (6) are of the type, $C_i(x) \geq 0$. If the form $C_i(x) \leq 0$ is used instead, then the sign in equation (8) must change as follows: $f_j(x) = f(x) + a_j C_j(x)$. It has been proved that for suf-

Table 6. Parameter setting of the PSO experiments reported in Laskari et al. (2002) for minimax problems

Parameter	Value
PSO parameters	$\chi = 0.729$, $c_1 = c_2 = 2.0$ $w = 1.0 \rightarrow 0.1$ (linearly decreasing)
Maximum velocity	$v_{max} = 4$
Swarm size	20 (TP_{MX-1}-TP_{MX-3} and TP_{MX-5}), 50 (TP_{MX-4} and TP_{MX-6})
Accuracy	10^{-4}
Maximum Func. Ev.	20000

ficiently large values of a_j, the optimum of the minimax problem coincides with that of the constrained problem (Bandler & Charalambous, 1974).

Due to its peculiar form, the minimax problem is addressed with classical optimization algorithms by using smoothing techniques. Such techniques produce a penalty function that combines all functions in a single one, which can be solved through gradient-based approaches. This approximating function is usually called *exponential penalty function* or *aggregation function*. Recently, Xu (2001) proposed such a sophisticated approach in combination with a quadratic programming gradient-based solver. Sequential quadratic programming (SQP) is a very popular approach in these cases.

In contrast to classical optimization methods, the unconventional form of the minimax problem does not raise any obstacle on the application of evolutionary algorithms, since objective function values can be directly obtained. PSO does not constitute an exception to this. Laskari *et al.* (2002) applied both the constriction coefficient and inertia weight PSO variants, denoted as $PSO^{[co]}$ and $PSO^{[in]}$, respectively, on a set of test problems and compared it with a classical optimization approach with SQP and Armijo line search applied on a smoothing function. The experimental setting of PSO is reported in Table 6.

Table 7 contains part of the results reported in Laskari *et al.* (2002) for the test problems TP_{MX-1}-TP_{MX-6}, defined in Appendix A of the book at hand. More specifically, the numbers of successful detections of the global minimizer with accuracy 10^{-4} in 30 independent experiments, as well as the mean required number of function evaluations, are reported for both PSO variants and SQP. As we observe, PSO was able to detect the optimum in almost all cases, with $PSO^{[co]}$ dominating $PSO^{[in]}$ in all problems. On the other hand, SQP was unpredictable, ranging from total failure to the most efficient performance. This is indicative of its dependence on the problem and initial conditions. However, we must note that the quality of solutions detected with PSO was in many cases better than that of SQP, since the latter employs penalty functions that often violate the constraints in implicitly defined minimax problems. The same conclusions hold for the whole set of test problems considered by Laskari *et al.* (2002). Counting furthermore the inherent complexity of the SQP approach, we conclude that PSO can serve as a very promising alternative for straightforwardly addressing minimax optimization problems.

The potential of PSO was also verified in a later work by Krohling *et al.* (2004). In this work, a co-evolutionary PSO variant that incorporates Gaussian distributions for solution fine-tuning was used on several benchmark problems with promising results. However, we must note that all test problems were given in explicit form and no comparisons with classical optimization methods were provided.

Table 7. Results reported in Laskari et al. (2002) for the minimax optimization problems

Problem	Algorithm	Success	Mean Func. Eval.
TP_{MX-1}	$PSO^{[in]}$	30/30	6012.0
	$PSO^{[co]}$	30/30	2348.0
	SQP	24/30	4044.5
TP_{MX-2}	$PSO^{[in]}$	30/30	5612.0
	$PSO^{[co]}$	30/30	1693.3
	SQP	18/30	8035.7
TP_{MX-3}	$PSO^{[in]}$	30/30	5124.0
	$PSO^{[co]}$	30/30	1142.6
	SQP	30/30	135.5
TP_{MX-4}	$PSO^{[in]}$	29/30	10526.6
	$PSO^{[co]}$	30/30	5150.0
	SQP	0/30	20000.0
TP_{MX-5}	$PSO^{[in]}$	30/30	5588.6
	$PSO^{[co]}$	30/30	1673.3
	SQP	30/30	140.6
TP_{MX-6}	$PSO^{[in]}$	30/30	15398.3
	$PSO^{[co]}$	30/30	10511.6
	SQP	30/30	611.6

CHAPTER SYNOPSIS

We presented PSO approaches for tackling three very interesting problem types. Due to the rich relative literature and the ongoing interest on new developments, we mainly considered multiobjective approaches. Different Pareto-based and non-Pareto algorithms were briefly described, providing the main concepts that constitute the latest developments today. Constrained optimization was also considered, especially under the use of penalty functions. Finally, minimax optimization and its tight ties with constrained problems were described, along with results and comparisons of PSO with classical optimization methods.

REFERENCES

Alvarez-Benitez, J. E., Everson, R. M., & Fieldsend, J. E. (2005). A MOPSO algorithm based exclusively on Pareto dominance concepts. [Berlin: Springer.]. *Lecture Notes in Computer Science, 3410,* 459–473.

Balling, R. (2003). The maximin fitness function: multiobjective city and regional planning. [Berlin: Springer.]. *Lecture Notes in Computer Science, 2632,* 1–15. doi:10.1007/3-540-36970-8_1

Bandler, J. W., & Charalambous, C. (1974). Nonlinear programming using minimax techniques. *Journal of Optimization Theory and Applications, 13*(6), 607–619. doi:10.1007/BF00933620

Bartz-Beielstein, T., Limbourg, P., Mehnen, J., Schmitt, K., Parsopoulos, K. E., & Vrahatis, M. N. (2003). Particle swarm optimizers for Pareto optimization with enhanced archiving techniques. In *Proceedings of the 2003 IEEE Congress on Evolutionary Computation (CEC'03), Canberra, Australia* (pp. 1780-1787).

Baumgartner, U., Magele, C., & Renhart, W. (2004). Pareto optimality and particle swarm optimization. *IEEE Transactions on Magnetics, 40*(2), 1172–1175. doi:10.1109/TMAG.2004.825430

Chow, C.-K., & Tsui, H.-T. (2004). Autonomous agent response learning by a multi-species particle swarm optimization. In *Proceedings of the 2004 IEEE Congress on Evolutionary Computation (CEC'04), Portland (OR), USA* (pp. 778-785).

Coello Coello, C. A. (1999). *A survey of constraint handling techniques used with evolutionary algorithms* (Tech. Rep. Lania-RI-99-04). Laboratorio Nacional de Informática Avanzada, Mexico.

Coello Coello, C. A., & Salazar Lechuga, M. (2002). MOPSO: A proposal for multiple objective particle swarm optimization. In *Proceedings of the 2002 IEEE Congress on Evolutionary Computation (CEC'02), Honolulu (HI), USA* (pp. 1051-1056).

Coello Coello, C. A., Toscano Pulido, G., & Salazar Lechuga, M. (2004). Handling multiple objectives with particle swarm optimization. *IEEE Transactions on Evolutionary Computation, 8*(3), 256–279. doi:10.1109/TEVC.2004.826067

Coello Coello, C. A., Van Veldhuizen, D. A., & Lamont, G. B. (2002). *Evolutionary algorithms for solving multi-objective problems*. New York: Kluwer Academic Publishing.

Deb, K. (1999). Multi-objective genetic algorithms: problem difficulties and construction of test problems. *Evolutionary Computation, 7*(3), 205–230. doi:10.1162/evco.1999.7.3.205

Deb, K., & Goldberg, D. E. (1989). An investigation of niche and species formation in genetic function optimization. In *Proceedings of the 3rd International Conference on Genetic Algorithms (ICGA'89), Fairfax (VA), USA* (pp. 42-50).

Deb, K., Pratap, A., Agarwal, S., & Meyarivan, T. (2002). A fast and elitist multiobjective genetic algorithm: NSGA-II. *IEEE Transactions on Evolutionary Computation, 6*(2), 182–197. doi:10.1109/4235.996017

Demyanov, V. F., & Molozemov, V. N. (1974). *Introduction to minimax*. New York: Wiley.

Du, D. Z., & Pardalos, P. M. (1995). *Minimax and applications*. Dordrecht, The Netherlands: Kluwer Academic Publishing.

Erickson, M., Mayer, A., & Horn, J. (2001). The niched Pareto genetic algorithm 2 applied to the design of groundwater remediation systems. [Berlin: Springer.]. *Lecture Notes in Computer Science, 1993*, 681–695.

Fieldsend, J. E., Everson, R. M., & Singh, S. (2003). Using unconstrained elite archives for multiobjective optimization. *IEEE Transactions on Evolutionary Computation, 7*(3), 305–323. doi:10.1109/TEVC.2003.810733

Fieldsend, J. E., & Singh, S. (2002). A multi-objective algorithm based upon particle swarm optimisation, an efficient data structure and turbulence. In *Proceedings of the 2002 UK Workshop on Computational Intelligence (UKCI-02), Birmingham, UK* (pp. 34-44).

Floudas, C. A., & Pardalos, P. M. (1990). *A collection of test problems for constrained global optimization algorithms*. Berlin: Springer.

Fonseca, C. M., & Fleming, P. J. (1993). Genetic algorithms for multiobjective optimization: formulation, discussion and generalization. In *Proceedings of the 5th International Conference on Genetic Algorithms (ICGA'93), Urbana-Champaign (IL), USA* (pp. 416-423).

He, Q., & Wang, L. (2007). An effective co-evolutionary particle swarm optimization for constrained engineering design problems. *Engineering Applications of Artificial Intelligence, 20*, 89–99. doi:10.1016/j.engappai.2006.03.003

Himmelblau, D. M. (1972). *Applied nonlinear programming*. Boston: McGraw-Hill.

Hock, W., & Schittkowski, K. (1981). *Test examples for nonlinear programming codes*. Berlin: Springer.

Homaifar, A., Lai, A. H.-Y., & Qi, X. (1994). Constrained optimization via genetic algorithms. *Simulation, 2*(4), 242–254. doi:10.1177/003754979406200405

Horn, J., Nafpliotis, N., & Goldberg, D. E. (1994). A niched Pareto genetic algorithm for multiobjective optimization. In *Proceedings of the 1st IEEE International Conference on Evolutionary Computation (ICEC'94), Orlando (FL), USA* (pp. 82-87).

Hu, X., & Eberhart, R. (2002). Multiobjective optimization using dynamic neighborhood particle swarm optimization. In *Proceedings of the 2002 IEEE Congress on Evolutionary Computation (CEC'02), Honolulu (HI), USA* (pp. 1677-1681).

Hu, X., Eberhart, R. C., & Shi, Y. (2003a). Particle swarm with extended memory for multiobjective optimization. In *Proceedings of the 2003 IEEE Swarm Intelligence Symposium (SIS'03), Indianapolis (IN), USA* (pp. 193-197).

Hu, X., Eberhart, R. C., & Shi, Y. (2003b). Engineering optimization with particle swarm. In *Proceedings of the 2003 IEEE Swarm Intelligence Symposium (SIS'03), Indianapolis (IN), USA* (pp. 53-57).

Jin, Y., Olhofer, M., & Sendhoff, B. (2001). Evolutionary dynamic weighted aggregation for multiobjective optimization: Why does it work and how? In *Proceedings of the 2001 Genetic and Evolutionary Computation Conference (GECCO'01), San Francisco (CA), USA* (pp. 1042-1049).

Joines, J. A., & Houck, C. R. (1994). On the use of non-stationary penalty functions to solve nonlinear constrained optimization problems with GA's. In *Proceedings of the 1st IEEE International Conference on Evolutionary Computation (ICEC'94), Orlando (FL), USA* (pp. 579-585).

Knowles, J. D., & Corne, D. W. (2000). Approximating the nondominated front using the Pareto archived evolution strategy. *Evolutionary Computation, 8*(2), 149–172. doi:10.1162/106365600568167

Krohling, R. A., Hoffmann, F., & Santos Coelho, L. (2004). Co-evolutionary particle swarm optimization for min-max problems using Gaussian distribution. In *Proceedings of the 2004 IEEE Conference on Evolutionary Computation (CEC'04), Portland (OR), USA* (pp. 959-964).

Krohling, R. A., & Santos Coelho, L. (2006). Coevolutionary particle swarm optimization using Gaussian distribution for solving constrained optimization problems. *IEEE Transactions on Systems, Man, and Cybernetics . Part B, 36*(6), 1407–1416.

Laskari, E. C., Parsopoulos, K. E., & Vrahatis, M. N. (2002). Particle swarm optimization for minimax problems. In *Proceedings of the 2002 IEEE Congress on Evolutionary Computation (CEC'02), Honolulu (HI), USA* (pp. 1576-1581).

Li, X. (2003). A non-dominated sorting particle swarm optimizer for multi-objective optimization. [Berlin: Springer.]. *Lecture Notes in Computer Science, 2723*, 37–48. doi:10.1007/3-540-45105-6_4

Li, X. (2004). Better spread and convergence: particle swarm multiobjective optimization using the maximin fitness function. [Berlin: Springer.]. *Lecture Notes in Computer Science, 3102*, 117–128.

Mahfouf, M., Chen, M.-Y., & Linkens, D. A. (2004). Adaptive weighted particle swarm optimisation for multi-objective optimal design of alloy steels. [Berlin: Springer.]. *Lecture Notes in Computer Science, 3242*, 762–771.

Mostaghim, S., & Teich, J. (2003a). Strategies for finding good local guides in multi-objective particle swarm optimization (MOPSO). In *Proceedings of the 2003 IEEE Swarm Intelligence Symposium (SIS'03), Indianapolis (IN), USA* (pp. 26-33).

Mostaghim, S., & Teich, J. (2003b). The role of ε-dominance in multi-objective particle swarm optimization methods. In *Proceedings of the 2003 IEEE Congress on Evolutionary Computation (CEC'03), Canberra, Australia* (pp. 1764-1771).

Mostaghim, S., & Teich, J. (2004). Covering Pareto-optimal fronts by subswarms in multi-objective particle swarm optimization. In *Proceedings of the IEEE 2004 Congress on Evolutionary Computation (CEC'04), Portland (OR), USA* (pp. 1404-1411). Washington, DC: IEEE Press.

Mostaghim, S., & Teich, J. (2006). About selecting the personal best in multi-objective particle swarm optimization. [Berlin: Springer.]. *Lecture Notes in Computer Science, 4193*, 523–532. doi:10.1007/11844297_53

Muñoz Zavala, A. E., Hernández Aguirre, A., & Villa Diharce, E. R. (2005). Constrained optimization via particle evolutionary swarm optimization. In *Proceedings of the 2005 Genetic and Evolutionary Computation Conference (GECCO'05), Washington (DC), USA* (pp. 209-216).

Parsopoulos, K. E., & Vrahatis, M. N. (2002a). Recent approaches to global optimization problems through particle swarm optimization. *Natural Computing, 1*(2-3), 235–306. doi:10.1023/A:1016568309421

Parsopoulos, K. E., & Vrahatis, M. N. (2002b). Particle swarm optimization method in multiobjective problems. In *Proceedins of the 2002 ACM Symposium on Applied Computing (SAC'02), Madrid, Spain* (pp. 603-607).

Parsopoulos, K. E., & Vrahatis, M. N. (2002c). Particle swarm optimization method for constrained optimization problems. In P. Sincak, J. Vascak, V. Kvasnicka, & J. Pospichal (Eds.), *Intelligent Technologies - Theory and Applications: New Trends in Intelligent Technologies (Frontiers in Artificial Intelligence and Applications series, Vol. 76)* (pp. 214-220). Amsterdam: IOS Press.

Parsopoulos, K. E., & Vrahatis, M. N. (2005). Unified particle swarm optimization for solving constrained engineering optimization problems. In [LNCS]. *Lecture Notes in Computer Science, 3612*, 582–591.

Parsopoulos, K. E., & Vrahatis, M. N. (2008). Multiobjective particle swarm optimization approaches. In L.T. Bui & S. Alam (Eds.), *Multi-Objective Optimization in Computational Intelligence: Theory and Practice, Chapter 2* (pp. 20-42). Hershey, PA: IGI Global.

Raquel, C. R., & Naval, P. C., Jr. (2005). An effecive use of crowding distance in multiobjective particle swarm optimization. In *Proceedings of the 2005 Genetic and Evolutionary Computation Conference (GECCO'05), Washington (DC), USA* (pp. 257-264).

Reyes-Sierra, M., & Coello Coello, C. A. (2005). Improving PSO-based multi-objective optimization using crowding, mutation and ε-dominance. [Berlin: Springer.]. *Lecture Notes in Computer Science, 3410*, 505–519.

Reyes-Sierra, M., & Coello Coello, C. A. (2006a). Multi-objective particle swarm optimizers: a survey of the state-of-the-art. *International Journal of Computational Intelligence Research, 2*(3), 287–308.

Reyes-Sierra, M., & Coello Coello, C. A. (2006b). On-line adaptation in multi-objective particle swarm optimization. In *Proceedings of the 2006 IEEE Swarm Intelligence Symposium (SIS'06), Indianapolis (IN), USA* (pp. 61-68).

Salazar Lechuga, M., & Rowe, J. E. (2005). Particle swarm optimization and fitness sharing to solve multi-objective optimization problems. In *Proceedings of the 2005 IEEE Congress on Evolutionary Computation (CEC'05), Edinburgh, UK* (pp. 1204-1211).

Schaffer, J. D. (1984). *Multiple Objective Optimization With Vector Evaluated Genetic Algorithms*. PhD thesis, Vanderbilt University, Nashville, TN, USA.

Srinivas, N., & Deb, K. (1994). Multiobjective optimization using nondominated sorting in genetic algorithms. *Evolutionary Computation, 2*(3), 221–248. doi:10.1162/evco.1994.2.3.221

Toscano Pulido, G., & Coello Coello, C. A. (2004). Using clustering tchniques to improve the performance of a particle swarm optimizer. [Berlin: Springer.]. *Lecture Notes in Computer Science, 3102*, 225–237.

Xu, S. (2001). Smoothing method for minimax problems. *Computational Optimization and Applications, 20*(3), 267–279. doi:10.1023/A:1011211101714

Yang, J.-M., Chen, Y.-P., Horng, J.-T., & Kao, C.-Y. (1997). Applying family competition to evolution strategies for constrained optimization. [Berlin: Springer.]. *Lecture Notes in Computer Science, 1213*, 201–211. doi:10.1007/BFb0014812

Yeniay, Ö. (2005). Penalty function methods for constrained optimization with genetic algorithms. *Mathematical and Computational Applications, 10*, 45–56.

Zitzler, E., & Thiele, L. (1999). Multiobjective evolutionary algorithms: a comparative case study and the strength Pareto approach. *IEEE Transactions on Evolutionary Computation, 3*(4), 257–271. doi:10.1109/4235.797969

Zuhe, S., Nuemaier, A., & Eiermann, M. C. (1990). Solving minimax problems by interval methods. *BIT, 30*, 742–751. doi:10.1007/BF01933221

Chapter 12
Afterword

In the previous chapters, we presented the fundamental concepts and variants of PSO, as along with a multitude of recent research results. The reported results suggest that PSO can be a very useful tool for solving optimization problems from different scientific and technological fields, especially in cases where classical optimization methods perform poorly or their application involves formidable technical difficulties due to the problem's special structure or nature. PSO was capable of addressing continuous and integer optimization problems, handling noisy and multiobjective cases, and producing efficient hybrid schemes in combination with specialized techniques or other algorithms in order to detect multiple (local or global) minimizers or control its own parameters.

All these properties have established PSO as one of the most popular intelligent optimization algorithms, adequately simple to be usable by non-expert researchers. Thus, one may arguably question how much room is left for further improvements and developments on PSO. The answer is: a lot. Although PSO has been shown to be a very successful algorithm, we are still far from the ultimate goal of introducing an intelligent optimization algorithm with the ability to self-adapt to the structure and nature of any given optimization problem. Indeed, as exposed in the largest part of the book, user intervention in selecting among alternative variants and/or parameter settings of the algorithm has a dominant position and impact on the successful culmination of the optimization procedure. Moreover, it is evident that, despite the large number of works on PSO applications, improvements targeted on specific problem types can further enhance its performance.

The aforementioned scope for improvement provides solid ground for further research on PSO. Actually, it opens a vast research horizon, where the ascertained plasticity of the algorithm can serve as the base material for the implementation of numerous ideas targeted at enhancing PSO's performance and promoting its more intelligent and autonomous operation. We anticipate the introduction of such new developments in

DOI: 10.4018/978-1-61520-666-7.ch012

Copyright © 2010, IGI Global. Copying or distributing in print or electronic forms without written permission of IGI Global is prohibited.

the form of novel research works and doctoral dissertations in the following years, maximally exploiting the available knowledge, of which, a large part has been presented in the previous chapters.

Of course, each research effort and achievement has a special (strong or weak) merit. Nevertheless, every effort must be recognized as a step towards the desirable direction. Unfortunately, due to the large number of researchers working on PSO worldwide, many of these efforts are inevitably uncoordinated, resulting in overlapping works and the weakening of some results that, if combined with other approaches, could be of great significance. In addition, some cases followed a wrong direction, resulting in fruitless efforts. One of the main reasons for this is the lack of an adequate number of sources where young researchers can find comprehensive presentations of the fundamental variants and the most important developments in PSO. We tried to address this problem, to the extent allowed by the limited space of a book, by writing the book at hand.

In mathematical research, we could say that, as compared to the past and present, the concept of *significance* is biased in favor of the future, with all the hopes and prospects it holds. Thus, we could not end this book without taking a glimpse at the research directions that are expected to significantly contribute in PSO in the following years, and distinguish the most promising ones:

a. Theoretical analysis.
b. Strategies and operators.
c. Self-adaptive models.
d. New variants suited to modern computation systems.
e. New and more fascinating applications.

In the following paragraphs, we briefly comment on what we believe are the most significant of these topics.

THEORETICAL ANALYSIS

As we already reported in Chapter Three, there have been some serious attempts to theoretically analyze PSO. However, the existing results do not fully describe the dynamics of the algorithm. This can be attributed to the specific approaches used, which impose the use of approximating models due to the inability of directly studying the original PSO model with the existing mathematical tools. Thus, a rigorous study of PSO's convergence properties on problems without nice mathematical properties is questionable. History reveals that although the existing theoretical studies have not completely succeeded in mathematically interpreting PSO, they have yielded significant contributions, providing a better understanding of the algorithm, as well as better control on its parameters and building blocks in general optimization problems. Our hope is that this knowledge will be enriched through similar theoretical studies in the future.

STRATEGIES AND OPERATORS

The handling of the swarms in PSO, as well as the operators applied on the particles, constitute significant choices that shall precede its application. The nature of the problem itself is the crucial factor that determines whether the existing velocity and particle update operators are adequate or modifications are needed. As we saw in the previous chapters, the existence of many local minimizers, discrete variables, noise or many objective functions, require the revision of existing PSO variants and their extension to more specialized schemes. Also, a question arising very often considers the effectiveness of using the same strategy for all particles (or swarms) against schemes that promote pluralism of behaviors and properties. Moreover, an ongoing discussion has been initiated regarding the combination of PSO with other (stochastic or deterministic) optimization algorithms, into hybrid methods, such as memetic algorithms.

Of course, the research on these topics is mostly problem-driven, since it is those inherent properties of the problem that dictate the use of the one approach or another. To date, the existing results are insufficient to establish sound methodologies. Our opinion is that research must be intensified towards this direction by encouraging young researchers to implement innovative ideas rather than use the same old recipes.

SELF-ADAPTIVE MODELS

As we already mentioned, the concept of an algorithm capable of self-adapting to any given problem is the Holy Grail of research in intelligent optimization. Nevertheless, there is an evident connection of this research field with others, such as artificial intelligence and artificial life, which deal with similar goals, probably from a different viewpoint. In order to achieve this goal, the development of proper PSO variants and operators is required, along with an efficient decision-making scheme that requires minor (or no) intervention by the user.

Under this prism, the two previously mentioned research issues, namely theoretical analysis and the study of strategies and operators, can offer an instrumental contribution. Today, we can clearly see that there is a long road ahead for the achievement of the most ambitious goals, although it seems that a solid background has been established for the further development and sophistication of the "intelligence" incorporated in the existing PSO approaches.

NEW VARIANTS SUITED TO MODERN COMPUTATION SYSTEMS

Modern technology has offered the potential to apply optimization algorithms in increasingly powerful computer systems. In many cases, the increase in computational capabilities is accompanied by new system architectures or new computation philosophies. Parallel processors with multiple cores and clusters of conventional machines constitute the most common available hardware. However, new fascinating ideas on computation arise, opening new roads in the future of computation. *Grid computing* and *quantum computers* can increase the available computational power enormously. Thus, the design of a new generation of algorithms that will be able to take full advantage of these developments is very interesting and important. Very significant contributions can be introduced by developing new PSO variants capable of benefiting from the new computation systems, exploiting the inherent capability of PSO for distributed and asynchronous operation. The amount of works towards this direction is still very limited.

NEW AND MORE FASCINATING APPLICATIONS

The high popularity of PSO is attributed to its ability to efficiently solve difficult optimization problems from different scientific and technological disciplines, requiring only minor implementation efforts. Thus, the large number of reported multidisciplinary applications of PSO was highly expected. Since, as previously mentioned, the optimization problems dictate the requirements and specifications in the development of new optimization algorithms, we anticipate new and more complex applications of PSO to drive the research to new and more fascinating developments. This can be a boost for all research directions described above.

Closing this short afterword, we would like to encourage new researchers to implement and assess new techniques or improve existing ones, even if the anticipated gain does not seem very important. It is very hard to accurately assess the influence and importance of a seemingly unimportant development, in future use.

Moreover, taking into consideration that intelligent optimization research is heavily based on experimental rather than theoretical results, we would like to encourage researchers to try to explicate and interpret their results, instead of simply reporting and presenting them in their works. It is far more important to understand what triggers a specific behavior of the algorithm, even based on a small amount of data, than to invest time and resources in conducting millions of computations without producing new knowledge or gaining further intuition on the algorithm. The proper statistical analysis of the results as well as the statistical design of the experiments, based on established methodologies, can produce high-quality research results. In this way, we will be able to design more efficient algorithms, increase our understanding on their workings, and retain the flow of new knowledge at high levels.

Appendix A

This appendix contains the benchmark problems used throughout this book. The problems are divided into categories based on the corresponding problem type. Information on each problem is reported when available, along with a relative reference.

UNCONSTRAINED OPTIMIZATION PROBLEMS

This section contains the unconstrained optimization problems used in the book at hand. Each problem is denoted with the general scheme, $TP_{UO\text{-}xx}$, where xx is its number. Unless otherwise stated, we assume that $x = (x_1, x_2, \ldots, x_n)^T$ is an n-dimensional vector.

$TP_{UO\text{-}1}$ (Sphere) (Storn & Price, 1997): This n-dimensional problem is defined as:

$$f(x) = x^T x = \sum_{i=1}^{n} x_i^2$$

and it has a global minimum $f^* = 0$ at $x^* = (0,0,\ldots,0)^T$.

$TP_{UO\text{-}2}$ (Rosenbrock) (Trelea, 2003): This n-dimensional problem is defined as,

$$f(x) = \sum_{i=1}^{n-1} [(1 - x_i)^2 + 100(x_{i+1} - x_i^2)^2]$$

and it has a global minimum $f^* = 0$ at $x^* = (1,1,\ldots,1)^T$. The 2-dimensional instance of the Rosenbrock function is also called *Banana Valley* function.

$TP_{UO\text{-}3}$ (Rastrigin) (Storn & Price, 1997): This n-dimensional problem is defined as:

Copyright © 2010, IGI Global, distributing in print or electronic forms without written permission of IGI Global is prohibited.

$$f(x) = \sum_{i=1}^{n} [x_i^2 - 10\cos(2\pi x_i) + 10]$$

and it has a global minimum $f^* = 0$ at $x^* = (0,0,\ldots,0)^{\mathrm{T}}$.

TP_{UO-4} (Griewank) (Storn & Price, 1997): This n-dimensional problem is defined as:

$$f(x) = \sum_{i=1}^{n} \frac{x_i^2}{4000} - \prod_{i=1}^{n} \cos\left(\frac{x_i}{\sqrt{i}}\right) + 1$$

and it has a global minimum $f^* = 0$ at $x^* = (0,0,\ldots,0)^{\mathrm{T}}$.

TP_{UO-5} (Schaffer's F6) (Trelea, 2003): This 2-dimensional problem is defined as:

$$f(x) = 0.5 + \frac{\left(\sin\left(\sqrt{x_1^2 + x_2^2}\right)\right)^2 - 0.5}{\left(1 + 0.001\left(x_1^2 + x_2^2\right)\right)^2}$$

and it has a global minimum $f^* = 0$ at $x^* = (0,0)^{\mathrm{T}}$.

TP_{UO-6} (Ackley) (Storn & Price, 1997): This n-dimensional problem is defined as:

$$f(x) = -20\exp\left(-0.02\sqrt{\frac{\sum_{i=1}^{n} x_i^2}{n}}\right) - \exp\left(\frac{\sum_{i=1}^{n} \cos(2\pi i)}{n}\right) + 20 + \exp(1),$$

and it has a global minimum $f^* = 0$ at $x^* = (0,0,\ldots,0)^{\mathrm{T}}$.

TP_{UO-7} (Corana) (Storn & Price, 1997): This 4-dimensional problem is defined as:

$$f(x) = \sum_{i=1}^{4} \begin{cases} 0.15(z_i - 0.05\mathrm{sign}(z_i))^2 d_i, & \text{if } |x_i - z_i| < 0.05, \\ d_i x_i^2, & \text{otherwise,} \end{cases}$$

where $x_i \in [-1000,1000]$, $d_1 = 1$, $d_2 = 1000$, $d_3 = 10$, $d_4 = 100$, and,

$$z_i = 0.2\left\lfloor \left|\frac{x_i}{0.2}\right| + 0.49999 \right\rfloor \mathrm{sign}(x_i).$$

All points with $|x_i^*| < 0.05$, $i = 1, 2, 3, 4$, are global minimizers with $f^* = 0$.

TP_{UO-8} (Lee & Yao, 2004): This n-dimensional problem is defined as:

$$f(x) = 0.1\left\{1 + \sin^2(3\pi x_1) + \sum_{i=1}^{n-1}(x_i-1)^2[1+\sin^2(3\pi x_{i+1})] + (x_n-1)^2[1+\sin^2(2\pi x_n)]\right\} + \sum_{i=1}^{n}u(x_i,5,100,4),$$

where,

$$u(z,a,k,m) = \begin{cases} k(z-a)^m, & z > a, \\ 0, & -a \le z \le a, \\ k(-z-a)^m, & z < -a, \end{cases}$$

and it has a global minimum $f^* = 0$ at $x^* = (1, 1,..., 1)^T$.

TP$_{UO-9}$ (Lee & Yao, 2004): This n-dimensional problem is defined as:

$$f(x) = \frac{\pi}{n}\left\{10 + \sin^2(\pi x_1) + \sum_{i=1}^{n-1}(x_i-1)^2[1+10\sin^2(\pi x_{i+1})] + (x_n-1)^2\right\} + \sum_{i=1}^{n}u(x_i,10,100,4),$$

where $u(x)$ is defined as in TP$_{UO-8}$. It has a global minimum $f^* = 0$ at $x^* = (1, 1,..., 1)^T$.

TP$_{UO-10}$ (Branin) (Michalewicz, 1999): This 2-dimensional problem is defined as:

$$f(x) = a_1(x_2 - a_2 x_1^2 + a_3 x_1 - a_4)^2 + a_5(1-a_6)\cos(x_1) + a_5,$$

where, $a_1 = 1$, $a_2 = 5.1/(4\pi^2)$, $a_3 = 5/\pi$, $a_4 = 6$, $a_5 = 10$, and $a_6 = 1/(8\pi)$. It has three global minimizers, $x_1^* = (-\pi, 12.275)^T$, $x_2^* = (\pi, 12.275)^T$, and $x_3^* = (9.42478, 2.475)^T$, with $f^* = 0.397887$.

TP$_{UO-11}$ (Six-hump camel) (Michalewicz, 1999): This 2-dimensional problem is defined as:

$$f(x) = (4 - 2.1x_1^2 + x_1^4/3)x_1^2 + x_1 x_2 + (-4 + 4x_2^2)x_2^2,$$

and it has two global minimizers, $x_1^* = (-0.0898, 0.7126)^T$, $x_2^* = (0.0898, -0.7126)^T$, with $f^* = -1.0316$.

TP$_{UO-12}$ (Freudenstein-Roth) (More *et al.*, 1981): This 2-dimensional problem is defined as:

$$f(x) = (-13 + x_1 + ((5-x_2)x_2 - 2)x_2)^2 + (-29 + x_1 + ((x_2+1)x_2 - 14)x_2)^2,$$

and it has a global minimum $f^* = 0$ at $x^* = (5, 4)^T$.

TP$_{UO-13}$ (Goldstein-Price) (Michalewicz, 1999): This 2-dimensional problem is defined as:

$$f(x) = [1 + (x_1+x_2+1)^2 (19-14x_1+3x_1^2-14x_2+6x_1x_2+3x_2^2)] \times [30 + (2x_1-3x_2)^2 (18-32x_1+12x_1^2+48x_2-36x_1x_2+27x_2^2)],$$

and it has a global minimum $f^* = 3$ at $x^* = (0,-1)^T$.

$TP_{UO\text{-}14}$ *(Levy no. 3)* (Levy *et al.*, 1981): This 2-dimensional problem is defined as:

$$f(x) = \sum_{i=1}^{5} i \cos((i-1)x_1 + i) \times \sum_{j=1}^{5} j \cos((j+1)x_2 + j),$$

and it has 18 global minimizers with $f^* = -176.542$.

$TP_{UO\text{-}15}$ *(Levy no. 5)* (Levy *et al.*, 1981): This 2-dimensional problem is defined as:

$$f(x) = \sum_{i=1}^{5} i \cos((i-1)x_1 + i) \times \sum_{j=1}^{5} j \cos((j+1)x_2 + j) + (x_1 + 1.42513)^2 + (x_2 + 0.80032)^2,$$

and it has a global minimum $f^* = -176.1375$ at $x^* = (-1.3068, -1.4248)^{\mathrm{T}}$.

$TP_{UO\text{-}16}$ *(Helical valley)* (More *et al.*, 1981): This 3-dimensional problem is defined as,

$$f(x) = \left(10(x_3 - 10\theta(x_1, x_2))\right)^2 + \left(10\left(\sqrt{x_1^2 + x_2^2} - 1\right)\right)^2 + x_3^2,$$

where,

$$\theta(x_1, x_2) = \begin{cases} \dfrac{1}{2\pi} \arctan(x_2 / x_1), & \text{if } x_1 > 0, \\[2ex] \dfrac{1}{2\pi} \arctan(x_2 / x_1) + 0.5, & \text{if } x_1 < 0, \end{cases}$$

and it has a global minimum $f^* = 0$ at $x^* = (1, 0, 0)^{\mathrm{T}}$.

$TP_{UO\text{-}17}$ *(Hyper-ellipsoid)* (Storn & Price, 1997): This *n*-dimensional problem is defined as,

$$f(x) = \sum_{i=1}^{n} i^2 x_i^2,$$

and it has a global minimum $f^* = 0$ at $x^* = (0, 0, \ldots, 0)^{\mathrm{T}}$.

$TP_{UO\text{-}18}$ *(Watson)* (More *et al.*, 1981): This *n*-dimensional problem is defined for $2 \leq n \leq 31$ as,

$$f(x) = \sum_{i=1}^{n} (f_i(x))^2,$$

where,

$$f_i(x) = \sum_{j=2}^{n} (j-1)x_j t_i^{j-2} - \left(\sum_{j=1}^{n} x_j t_i^{j-1}\right)^2 - 1,$$

with $t_i = i/29$ for $2 \leq i \leq 29$, while,

$f_{30}(x) = x_1$ and $f_{31}(x) = x_2 - x_1^2 - 1$.

For $n = 6$ it has a global minimum $f^* = 0.00228767$ at $x^* = (0, 0, \ldots, 0)^T$.

TP_{UO-19} *(Levy no. 8)* (Levy *et al.*, 1981): This n-dimensional problem is defined as:

$$f(x) = \sin^2(\pi y_1) + \sum_{i=1}^{n-1}(y_i - 1)^2[1 + 10\sin^2(\pi y_{i+1})] + (y_n - 1)^2,$$

where $y_i = 1 + (x_i - 1)/4$. For $n = 3$, it has a global minimum $f^* = 0$ at $x^* = (1, 1, 1)^T$.

TP_{UO-20} *(Quadric)* (Van den Bergh & Engelbrecht, 2002): This n-dimensional problem is defined as:

$$f(x) = \sum_{i=1}^{n}\left(\sum_{j=1}^{i}x_j\right)^2,$$

and it has a global minimum $f^* = 0$ at $x^* = (0, 0, \ldots, 0)^T$.

TP_{UO-21} *(Egg holder)* (Parsopoulos & Vrahatis, 2004): This 2-dimensional problem is defined as:

$f(x) = \cos^2(x_1) + \sin^2(x_2),$

and it has only global minimizers at the points $(k_1\pi/2, k_2\pi)^T$, with $k_1 = \pm1, \pm2,\ldots,$ and $k_2 = 0, \pm1, \pm2,\ldots$. In the range $[-5, 5]^2$ it has 12 global minimizers.

TP_{UO-22} *(Beale)* (More *et al.*, 1981): This 2-dimensional problem is defined as:

$f(x) = [y_1 - x_1(1-x_2)]^2 + [y_2 - x_1(1-x_2^2)]^2 + [y_3 - x_1(1-x_2^3)]^2,$

with $y_1 = 1.5$, $y_2 = 2.25$, and $y_3 = 2.625$. It has a global minimum $f^* = 0$ at $x^* = (3, 0.5)^T$.

NONLINEAR MAPPINGS

This section contains the nonlinear mappings used for assessing the performance of the presented algorithms on problems of detecting periodic orbits. Each mapping is denoted with the general scheme, TP_{NM-xx}, where xx is its number. Unless otherwise stated, we assume that $x = (x_1, x_2, \ldots, x_n)^T$ is an n-dimensional vector.

TP_{NM-1} *(Hénon 2-dimensional)* (Hénon, 1969): This is a 2-dimensional mapping defined as:

$$\Phi(x)=\begin{pmatrix}\cos a & -\sin a\\ \sin a & \cos a\end{pmatrix}\begin{pmatrix}x_1\\ x_2-x_1^2\end{pmatrix}\Leftrightarrow\begin{cases}\Phi_1(x)=x_1\cos a-(x_2-x_1^2)\sin a,\\ \Phi_2(x)=x_1\sin a-(x_2-x_1^2)\cos a,\end{cases}$$

where $a\in[0,\pi]$ is the rotation angle. The cases of $\cos a=0.8$ and $\cos a=0.24$ are very common.

TP$_{NM\text{-}2}$ (Standard) (Rasband, 1990): This is a discontinuous 2-dimensional mapping defined as:

$$\begin{cases}\Phi_1(x)=\left(x_1+x_2-\dfrac{k}{2\pi}\sin(2\pi x_1)\right)\bmod\dfrac{1}{2},\\ \Phi_2(x)=\left(x_2-\dfrac{k}{2\pi}\sin(2\pi x_1)\right)\bmod\dfrac{1}{2},\end{cases}$$

where $k=0.9$. The modulo function is defined as:

$$y\bmod\frac{1}{2}=\begin{cases}(y\bmod 1)-1, & \text{if }(y\bmod 1)>0.5,\\ (y\bmod 1)+1, & \text{if }(y\bmod 1)<-0.5,\\ y\bmod 1, & \text{otherwise.}\end{cases}$$

TP$_{NM\text{-}3}$ (Gingerbreadman) (Devaney, 1984): This is a nondifferentiable 2-dimensional mapping defined as:

$$\begin{cases}\Phi_1(x)=1-x_2+|x_1|,\\ \Phi_2(x)=x_1,\end{cases}$$

and it has a unique periodic orbit of period $p=1$.

TP$_{NM\text{-}4}$ (Predator-prey) (Maynard Smith, 1968): This is a 2-dimensional mapping defined as:

$$\begin{cases}\Phi_1(x)=ax_1(1-x_1)-x_1x_2,\\ \Phi_2(x)=bx_1x_2,\end{cases}$$

where $a=3.6545$ and $b=3.226$ (Henry *et al.*, 2000).

TP$_{NM\text{-}5}$ (Lorenz) (Lorenz, 1963): This is a 3-dimensional mapping defined as:

$$\begin{cases}\Phi_1(x)=\sigma(x_2-x_1),\\ \Phi_2(x)=rx_1-x_2-x_1x_3,\\ \Phi_3(x)=x_1x_2-bx_3,\end{cases}$$

where σ, r and b are the parameters of the system. Lorenz used the values $\sigma=10$ and $b=8/3$. Regarding r, the value $r=28$ is used (Henry *et al.*, 2000).

TP_{NM-6} (Rössler) (Rössler, 1976): This is a 3-dimensional mapping defined as:

$$\begin{cases} \Phi_1(x) = -(x_1 + x_2), \\ \Phi_2(x) = x_1 + ax_2, \\ \Phi_3(x) = b + x_3(x_1 - c), \end{cases}$$

where a, b, and c are parameters of the system. The values $a = b = 0.2$ and $c = 5.7$ are used (Henry *et al.*, 2000).

TP_{NM-7} (Hénon 4-dimensional) (Vrahatis, 1995): This 4-dimensional mapping is an extension of the 2-dimensional Hénon mapping defined previously as TP_{NM-1} and it is defined as:

$$\begin{pmatrix} \Phi_1(x) \\ \Phi_2(x) \\ \Phi_3(x) \\ \Phi_4(x) \end{pmatrix} = \begin{pmatrix} R(a) & O \\ O & R(a) \end{pmatrix} \begin{pmatrix} x_1 \\ x_2 - x_1^2 + x_3^2 \\ x_3 \\ x_4 - 2x_1x_3 \end{pmatrix},$$

where a is the rotation angle and $R(a)$, O, are matrices defined as:

$$R(a) = \begin{pmatrix} \cos a & -\sin a \\ \sin a & \cos a \end{pmatrix}, \quad O = \begin{pmatrix} 0 & 0 \\ 0 & 0 \end{pmatrix}.$$

The value $a = \cos^{-1}(0.24)$ is a common choice. The mapping can be also generalized to a symplectic map with two frequencies, a_1 and a_2:

$$\begin{pmatrix} \Phi_1(x) \\ \Phi_2(x) \\ \Phi_3(x) \\ \Phi_4(x) \end{pmatrix} = \begin{pmatrix} R(a_1) & O \\ O & R(a_2) \end{pmatrix} \begin{pmatrix} x_1 \\ x_2 + x_1^2 - x_3^2 \\ x_3 \\ x_4 - 2x_1x_3 \end{pmatrix}.$$

TP_{NM-8} (Kantz & Grassberger, 1988): This 6-dimensional mapping is the $n = 3$ case of the standard maps studied by Kantz and Grassberger (1988) and it is defined as:

$$\begin{cases} x_1' = x_1 + x_2' \\ x_2' = x_2 + \dfrac{k}{2\pi}\sin(2\pi x_1) - \dfrac{b}{2\pi}\left[\sin(2\pi(x_5 - x_1)) + \sin(2\pi(x_3 - x_1))\right] \\ x_3' = x_3 + x_4' \\ x_4' = x_4 + \dfrac{k}{2\pi}\sin(2\pi x_3) - \dfrac{b}{2\pi}\left[\sin(2\pi(x_1 - x_3)) + \sin(2\pi(x_5 - x_3))\right] \\ x_5' = x_5 + x_6' \\ x_6' = x_6 + \dfrac{k}{2\pi}\sin(2\pi x_5) - \dfrac{b}{2\pi}\left[\sin(2\pi(x_3 - x_5)) + \sin(2\pi(x_1 - x_5))\right] \end{cases} \quad (\text{mod } 1).$$

All variables of the mapping are taken (mod 1), hence $x_i \in [0,1)$, $i = 1, 2,\ldots, 6$. For $b = 0$, the mapping give three uncoupled standard maps, while for $b \neq 0$ the maps are coupled, influencing each other. In the present book, the values that were used in the experiments are $b = k = 1$.

INVENTORY OPTIMIZATION PROBLEMS

The continuous review inventory optimization problems used in the present book are based on the model of Chern *et al.* (2008):

$$TC(n,s_i,t_i) = \sum_{i=1}^{n} c_0 \exp(-r_1 t_i)$$

$$+ \sum_{i=1}^{n} c_p \exp(-r_2 t_i) \left(\int_{s_{i-1}}^{t_i} \beta(t_i - t) f(t) dt + \int_{t_i}^{s_i} \exp(\delta(t) - \delta(t_i)) f(t) dt \right)$$

$$+ \sum_{i=1}^{n} \sum_{j=1}^{2} c_{h_j} \int_{t_i}^{s_i} \exp(-r_j t) \int_{t}^{s_i} \exp(\delta(u) - \delta(t)) f(u) du dt$$

$$+ \sum_{i=1}^{n} \sum_{j=1}^{2} \frac{c_{b_j}}{r_j} \int_{s_{i-1}}^{t_i} \left(\exp(-r_j t) - \exp(-r_j t_i) \right) \beta(t_i - t) f(t) dt$$

$$+ \sum_{i=1}^{n} \sum_{j=1}^{2} c_{l_j} \int_{s_{i-1}}^{t_i} \exp(-r_j t) \left(1 - \beta(t_i - t) \right) f(t) dt,$$

with,

$$s_0 = 0, \quad s_{i-1} < t_i \leq s_i, \quad s_n = H.$$

The model can also admit constraints on the inventory size, resulting in the following global minimization problem:

$$\min_{n, t_i, s_i} TC(n, t_i, s_i) \quad such \ that \quad \int_{t_i}^{s_i} \exp(\delta(u) - \delta(t_i)) f(u) du \leq W,$$

$$s_0 = 0, \quad s_n = H, \quad s_{i-1} < t_i \leq s_i, \quad i = 1, 2, \ldots, n.$$

All test problems are defined as the minimization problem above (either constrained or unconstrained) but for different parameter settings.

TP_{INV-1} (Skouri & Papachristos, 2002): This problem is based on a simplified version of the model, with parameters:

$f(t) = 20 + 2t,$
$\beta(x) = \exp(-ax),$

$r_1 = r_2 = 0,$
$c_0 = 100, c_p = 0.2, c_{b_2} = 1.5, c_{l_2} = 0.5, c_{h_2} = 55, c_{h_1} = c_{b_1} = c_{l_1} = 0,$
$\theta(t) = 0.01.$

The problem was considered for three values of the parameter a, namely $a = 0.08, 0.05,$ and 0.02. In its constrained version, the maximum inventory size is set to the value, $W = 90$.

TP_{INV-2} (Chern *et al.*, 2008): In this problem, shortages are completely backlogged. Its parameters are defined as follows:

$f(t) = 200 + 50t,$
$\beta(x) = 1,$
$H = 10,$
$c_0 = 80, c_p = 9, c_{h_1} = 0.2, c_{h_2} = 0.4, c_{b_1} = 0.5, c_{b_2} = 0.4,$
$r = 0.2,$
$i_1 = 0.08, i_2 = 0.09,$
$\theta(t) = 0.01.$

In its constrained version, the maximum inventory size is set to the value, $W = 300$.
TP_{INV-3} (Chern *et al.*, 2008): In this problem, shortages are completely backlogged. Its parameters are defined as follows:

$f(t) = 200 + 50t - 3t^2,$
$\beta(x) = 1,$
$H = 10,$
$c_0 = 80, c_p = 15, c_{h_1} = 0.2, c_{h_2} = 0.4, c_{b_1} = 0.8, c_{b_2} = 0.6,$
$r = 0.2,$
$i_1 = 0.08, i_2 = 0.1,$
$\theta(t) = 0.01.$

In its constrained version, the maximum inventory size is set to the value, $W = 300$.

GAME THEORY PROBLEMS

The benchmark problems used for the detection of Nash equilibria were obtained through the state-of-the-art GAMBIT software suit (version 0.97.0.5) (Pavlidis *et al.*, 2005). GAMBIT is available through the web address: http://gambit.sourceforge.net/. All problems have more than one Nash equilibria and their complete list can be obtained through the GAMBIT routine "PolEnumSolve". For each test problem, its characteristics along with the corresponding defining GAMBIT file are reported.

TP_{NE-1}: This normal-form game has 4 players, with 2 pure strategies for each one. It has 3 Nash equilibria and corresponds to the GAMBIT file "2x2x2x2.nfg".

***TP**$_{NE-2}$*: This normal-form game has 4 players, with 2 pure strategies for each one. It has 5 mixed equilibria and corresponds to the GAMBIT file "g3.nfg".

***TP**$_{NE-3}$*: This normal-form game has 5 players, with 2 pure strategies for each one. It has 5 Nash equilibria and corresponds to the GAMBIT file "2x2x2x2x2.nfg".

***TP**$_{NE-4}$*: This normal-form game has 3 players, with 2 pure strategies for each one (McKelvey, 1991). The payoffs of the game are as follows:

s_{31}	s_{21}	s_{22}
s_{11}	9, 8, 12	0, 0, 0
s_{12}	0, 0, 0	9, 8, 2

s_{32}	s_{21}	s_{22}
s_{11}	0, 0, 0	3, 4, 6
s_{12}	3, 4, 4	0, 0, 0

It has 9 Nash equilibria (4 pure strategy, 3 mixed strategy, and 2 full support strategy equilibria). It corresponds to the GAMBIT file "2x2x2.nfg".

***TP**$_{NE-5}$*: This coordination game has 3 players, with 3 strategies for each one. It has 13 equilibria and it corresponds to the GAMBIT file "coord333.nfg".

***TP**$_{NE-6}$*: This game has 2 players, with 4 strategies for each one, and it has 15 Nash equilibria. Its payoff matrix is given as follows:

	s_{21}	s_{22}	s_{23}	s_{24}
s_{11}	3, 2	0, 0	0, 0	0, 0
s_{12}	0, 0	2, 2	0, 0	0, 0
s_{13}	0, 0	0, 0	1, 4	0, 0
s_{14}	0, 0	0, 0	0, 0	4, 7

The corresponding GAMBIT file is "coord4.nfg".

DATA SETS FROM BIOINFORMATICS

The following data sets were considered for performance assessment of SA-PNNs by Georgiou *et al.* (2006):

E.coli data set (Blake & Merz, 1998): The goal is to predict the cellular localization sites of *E.coli* proteins. There are 8 cellular sites:

1. Cytoplasm (cp).
2. Inner membrane without signal sequence (im).
3. Periplasm (pp).

4. Inner membrane with uncleavable signal sequence (imU).
5. Outer membrane (om).
6. Outer membrane lipoprotein (omL).
7. Inner membrane lipoprotein (imL).
8. Inner membrane with cleavable signal sequence (imS).

The attributes under consideration are the following:

1. McGeoch method for signal sequence recognition (mcg).
2. Von Heijne method for signal sequence recognition (gvh).
3. Presence of charge on N-terminus of predicted lipoproteins (chg).
4. Score of discriminant analysis of the amino acid content of outer membrane and periplasmic proteins (aac).
5. Score of the ALOM membrane spanning region prediction program (alm1).
6. Score of ALOM program after excluding putative cleavable signal regions from the sequence (alm2).

The size of the data set is 336 without any missing values and all its variables are continuous.

Yeast data set (Blake & Merz, 1998): The goal is to predict the cellular localization of yeast proteins. There are 10 sites:

1. CYT (cytosolic or cytoskeletal).
2. NUC (nuclear).
3. MIT (mitochondrial).
4. ME3 (membrane protein, no N-terminal signal).
5. ME2 (membrane protein, uncleaved signal).
6. ME1 (membrane protein, cleaved signal).
7. EXC (extracellular).
8. VAC (vacuolar).
9. POX (peroxisomal).
10. ERL (endoplasmic reticulum lumen).

The same attributes with the *E.coli* data set are considered, additionally including nuclear localization information. Thus, there are 8 continuous inputs of 1484 instances without missing values.

Breast Cancer data set (Prechelt, 1994): This data set was provided by the University of Wisconsin hospitals in 1992 and contains breast tumor records that can be categorized in two classes: benign and malignant. The input features are:

1. Uniformity of cell size and shape.
2. Bland chromatin.
3. Single epithelial cell size.
3. Mitoses.

There are 9 continuous inputs and 699 instances, without missing values.

Pima Indians Diabetes data set (Smith *et al.*, 1988): This data set was provided by the John Hopkins University in 1992 and concerns Pima Indians' diabetes. The goal is to determine whether someone suffers from diabetes or not; hence, there are two classification classes. It has the following input features:

1. Diastolic blood pressure.
2. Triceps skin fold thickness.
3. Plasma glucose concentration in a glucose tolerance test.
4. Diabetes pedigree function.

Classification is performed on the patient's exhibition of diabetes signs, based on criteria established by the World Health Organization. There are 8 inputs, all continuous without missing values, and 768 instances.

MULTIOBJECTIVE OPTIMIZATION PROBLEMS

The following widely used problems were used in experiments with multiobjective PSO variants:

TP_{MO-1} (Knowles & Corne, 2000; Zitzler *et al.*, 2000): This problem has two objective functions, and it is defined as follows:

$$f_1(x) = \frac{1}{n}\sum_{i=1}^{n} x_i^2, \qquad f_2(x) = \frac{1}{n}\sum_{i=1}^{n}(x_i - 2)^2.$$

It has a convex, uniform Pareto front.

TP_{MO-2} (Knowles & Corne, 2000; Zitzler *et al.*, 2000): This problem has two objective functions, and it is defined as follows:

$$f_1(x) = x_1, \qquad g(x) = 1 + \frac{9}{n-1}\sum_{i=2}^{n} x_i, \qquad f_2(x) = g(x)\left(1 - \sqrt{\frac{f_1(x)}{g(x)}}\right).$$

It has a convex, non-uniform Pareto front.

TP_{MO-3} (Knowles & Corne, 2000; Zitzler *et al.*, 2000): This problem has two objective functions, and it is defined as follows:

$$f_1(x) = x_1, \qquad g(x) = 1 + \frac{9}{n-1}\sum_{i=2}^{n} x_i, \qquad f_2(x) = g(x)\left(1 - \left(\frac{f_1(x)}{g(x)}\right)^2\right).$$

It has a concave Pareto front.

***TP*_{MO-4}** (Knowles & Corne, 2000; Zitzler *et al.*, 2000): This problem has two objective functions, and it is defined as follows:

$$f_1(x) = x_1, \qquad g(x) = 1 + \frac{9}{n-1}\sum_{i=2}^{n} x_i, \qquad f_2(x) = g(x)\left(1 - \sqrt[4]{\frac{f_1(x)}{g(x)}} - \left(\frac{f_1(x)}{g(x)}\right)^4\right).$$

It has neither a purely convex nor purely concave Pareto front.

***TP*_{MO-5}** (Knowles & Corne, 2000; Zitzler *et al.*, 2000): This problem has two objective functions, and it is defined as follows:

$$f_1(x) = x_1, \qquad g(x) = 1 + \frac{9}{n-1}\sum_{i=2}^{n} x_i, \qquad f_2(x) = g(x)\left(1 - \sqrt{\frac{f_1(x)}{g(x)}} - \frac{f_1(x)}{g(x)}\sin(10\pi f_1(x))\right).$$

Its Pareto front consists of separated convex parts.

CONSTRAINED OPTIMIZATION PROBLEMS

The following constrained optimization problems were used in experiments with constrained PSO approaches. The problems are divided in constrained benchmark and constrained engineering design problems.

Constrained Benchmark Problems

***TP*_{CO-1}** (Himmelblau, 1972): This 2-dimensional problem is defined as follows:

$$f(x) = (x_1\text{-}2)^2 + (x_2\text{-}1)^2,$$

subject to the constraints:

$$C_1(x): \qquad x_1 = 2x_2 - 1,$$

$$C_2(x): \qquad \frac{x_1^2}{4} + x_2^2 - 1 \le 0.$$

It has a solution with $f^* = 1.3934651$.

TP_{CO-2} (Floudas & Pardalos, 1990): This 2-dimensional problem is defined as follows:

$f(x) = (x_1-10)^3 + (x_2-20)^3,$

subject to the constraints:

$C_1(x)$: $100 - (x_1 - 5)^2 - (x_2 - 5)^2 \leq 0,$

$C_2(x)$: $(x_1 - 6)^2 + (x_2 - 5)^2 - 82.81 \leq 0,$

with $13 \leq x_1 \leq 100$ and $0 \leq x_2 \leq 100$. It has a solution with $f^* = -6961.81381$.

TP_{CO-3} (Hock & Schittkowski, 1981): This 7-dimensional problem is defined as follows:

$f(x) = (x_1-10)^2 + 5(x_2-12)^2 + x_3^4 + 3(x_4-11)^2 + 10x_5^6 + 7x_6^2 + x_7^4 - 4x_6x_7 - 10x_6 - 8x_7,$

subject to the constraints:

$C_1(x)$: $-127 + 2x_1^2 + 3x_2^4 + x_3 + 4x_4^2 + 5x_5 \leq 0,$

$C_2(x)$: $-282 + 7x_1 + 3x_2 + 10x_3^2 + x_4 - x_5 \leq 0,$

$C_3(x)$: $-196 + 23x_1 + x_2^2 + 6x_6^2 - 8x_7 \leq 0,$

$C_4(x)$: $4x_1^2 + x_2^2 - 3x_1x_2 + 2x_3^2 + 5x_6 - 11x_7 \leq 0,$

with $-10 \leq x_i \leq 10$, $i = 1, 2,..., 7$. It has a solution with $f^* = 680.630057$.

TP_{CO-4} (Hock & Schittkowski, 1981): This 5-dimensional problem is defined as follows:

$f(x) = 5.3578547x_3^2 + 0.8356891x_1x_5 + 37.293239x_1 - 40792.141,$

subject to the constraints:

$C_1(x)$: $0 \leq 85.334407 + 0.0056858T_1 + T_2x_1x_4 - 0.0022053x_3x_5 \leq 92,$

$C_2(x)$: $90 \leq 80.51249 + 0.0071317x_2x_5 + 0.0029955x_1x_2 + 0.0021813x_3^2 \leq 110,$

$C_3(x)$: $20 \leq 9.300961 + 0.0047026x_3x_5 + 0.0012547x_1x_3 + 0.0019085x_3x_4 \leq 25,$

with $78 \leq x_1 \leq 102$, $33 \leq x_2 \leq 45$, and $27 \leq x_i \leq 45$, $i = 3, 4, 5$, where $T_1 = x_2x_5$ and $T_2 = 0.0006262$. It has a solution $f^* = -30665.538$.

***TP**$_{CO-5}$* (Hock & Schittkowski, 1981): This problem is defined exactly as TP$_{CO-4}$ above, but with $T_1 = x_2 x_3$ and $T_2 = 0.00026$.

***TP**$_{CO-6}$* (Michalewicz, 1999): This 5-dimensional problem is defined as follows:

$$f(x,y) = -10.5x_1 - 7.5x_2 - 3.5x_3 - 2.5x_4 - 1.5x_5 - 10y - 0.5\sum_{i=1}^{5} x_i^2,$$

subject to the constraints:

$$C_1(x): \quad 6x_1 + 3x_2 + 3x_3 + 2x_4 + x_5 - 6.5 \leq 0,$$

$$C_2(x): \quad 10x_1 + 10x_3 + y \leq 20,$$

with $0 \leq x_i \leq 1$, $i = 1, 2,\ldots, 5$, and $y \geq 0$. It has a solution $f^* = -213.0$.

Constrained Engineering Design Problems

***TP**$_{ED-1}$ (Tension/compression spring)* (Arora, 1989): This problem concerns the minimization of weight for the tension/compression spring illustrated in Fig. 1, subject to constraints on minimum deflection, shear stress, surge frequency, diameter, and design variables. The design variables are the wire parameter, d, the mean coil diameter, D, and the number of active coils, N.

If $x = (d, D, N)^T$, the problem is defined as follows:

$$f(x) = (N + 2) D d^2,$$

subject to the constraints:

$$C_1(x): \quad 1 - \frac{D^3 N}{71785d^4} \leq 0,$$

$$C_2(x): \quad \frac{4D^2 - dD}{12566(Dd^3 - d^4)} + \frac{1}{5108d^2} - 1 \leq 0.$$

$$C_3(x): \quad 1 - \frac{140.45d}{D^2 N} \leq 0,$$

$$C_4(x): \quad \frac{D + d}{1.5} - 1 \leq 0,$$

with $0.05 \leq d \leq 2.0$, $0.25 \leq D \leq 1.3$, and $2.0 \leq N \leq 15.0$.

Figure 1. The tension/compression spring problem

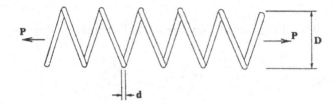

TP_{ED-2} (*Welded beam*) (Rao, 1996): This problem concerns the minimization of the cost of the welded beam illustrated in Fig. 2, subject to constraints on shear stress, τ; bending stress in the beam, σ; buckling load on the bar, P_c; end deflection of the bead, δ; and side constraints. There are 4 design variables, h, l, t, and b, henceforth denoted as x_1, x_2, x_3, and x_4, respectively.

If $x = (x_1, x_2, x_3, x_4)^T$, the problem is defined as follows:

$$f(x) = 1.10471x_1^2 x_2 + 0.04811 x_3 x_4 (14.0 + x_2),$$

subject to the constraints:

$C_1(x)$: $\tau(x) - \tau_{max} \leq 0,$

$C_2(x)$: $\sigma(x) - \sigma_{max} \leq 0,$

$C_3(x)$: $x_1 - x_4 \leq 0,$

$C_4(x)$: $0.10471x_1^2 + 0.04811 x_3 x_4 (14.0 + x_2) - 5.0 \leq 0,$

$C_5(x)$: $0.125 - x_1 \leq 0,$

$C_6(x)$: $\delta(x) - \delta_{max} \leq 0,$

$C_7(x)$: $P - P_c(x) \leq 0,$

where:

$$\tau(x) = \sqrt{\tau'^2 + 2\tau'\tau''\frac{x_2}{2R} + \tau''^2},$$

$$\tau' = \frac{P}{\sqrt{2}x_1 x_2}, \qquad \tau'' = \frac{MR}{J}, \qquad M = P\left(L + \frac{x_2}{2}\right),$$

$$R = \sqrt{\frac{x_2^2}{4} + \left(\frac{x_1 + x_3}{2}\right)^2}, \qquad J = 2\left[\sqrt{2}x_1 x_2 \left(\frac{x_2^2}{12} + \left(\frac{x_1 + x_3}{2}\right)^2\right)\right],$$

Figure 2. The welded beam problem

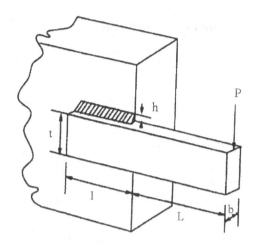

$$\sigma(x) = \frac{6PL}{x_4 x_3^2}, \qquad \delta(x) = \frac{4PL^3}{Ex_3^3 x_4}, \qquad P_c(x) = \frac{4.013E\sqrt{\dfrac{x_3^2 x_4^6}{36}}}{L^2}\left(1 - \frac{x_3}{2L}\sqrt{\frac{E}{4G}}\right),$$

with P = 6000 lb; L = 14 in; E = 30×10⁶ psi; G = 12×10⁶ psi; τ_{max} = 13600 psi; σ_{max} = 30000 psi; δ_{max} = 0.25 in; and $0.1 \le x_1, x_4 \le 2.0$, $0.1 \le x_2, x_3 \le 10.0$.

TP_{ED-3} *(Gear train)* (Sandgen, 1990): This problem concerns the minimization of the cost of the gear ratio of the gear train illustrated in Fig. 3. The gear ration, *gr*, is defined as follows:

$$gr = (n_B n_D) / (n_F n_A),$$

where n_j denotes the number of teeth of the *j*-th gearwheel, j = A, B, D, F. The design variables, n_A, n_B, n_D, and n_F, will be henceforth denoted as x_1, x_2, x_3, and x_4, respectively.

If $x = (x_1, x_2, x_3, x_4)^T$, the problem is defined as follows:

$$f(x) = \left(\frac{1}{6.931} - \frac{x_3 x_2}{x_1 x_4}\right)^2,$$

subject to the constraints: $12 \le x_i \le 60$, i = 1, 2, 3, 4.

TP_{ED-4} *(Pressure vessel)* (Sandgen, 1990): This problem concerns the minimization of the cost of the pressure vessel illustrated in Fig. 4. The design variables are shell thickness, T_s; thickness of head, T_h; inner radius, R; and the length, L, which will be henceforth denoted as x_1, x_2, x_3, and x_4, respectively. T_s and T_h are integer multiples of 0.0625, representing the available thicknesses of rolled steel plates.

If $x = (x_1, x_2, x_3, x_4)^{\mathrm{T}}$, the problem is defined as follows:

$$f(x) = 0.6224x_1x_3x_4 + 1.7781x_2x_3^2 + 3.1661x_1^2x_4 + 19.84x_1^2x_3,$$

subject to the constraints:

$C_1(x):\quad -x_1 + 0.0193x_3 \leq 0,$

$C_2(x):\quad -x_2 + 0.00954x_3 \leq 0,$

$C_3(x):\quad -\pi x_3^2 x_4 - (4/3)\pi x_3^3 + 1296000 \leq 0,$

$C_4(x):\quad x_4 - 240 \leq 0,$

with $1 \leq x_1, x_2 \leq 99,\ 10 \leq x_3,\ x_4 \leq 0.$

MINIMAX PROBLEMS

Minimax problems are intimately related to constrained optimization problems. As reported in Chapter Eleven of the book at hand, a minimax problem can be given in the explicit form:

$$\min_{x} \max_{1 \leq i \leq m} \{f_i(x)\},$$

where $i = 1, 2, \ldots, m$, denotes the number of functions. Also, a constrained problem of the form:

$$\min_{x} f(x) \text{ subject to the constraints } C_i(x) \geq 0, \quad i = 2, 3, \ldots, m,$$

can be given in an equivalent implicit minimax form:

Figure 3. The gear train problem

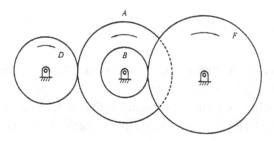

$$\min_{x} \max_{1 \le i \le m}\{f_i(x)\},$$

$$f_1(x) = f(x),$$
$$f_j(x) = f(x) - a_j C_j(x),$$
$$a_j > 0, \quad j = 2,3,...,m.$$

The following widely used minimax test problems were used for experimentation with PSO, given in either of the aforementioned forms:

TP_{MX-1} (Xu, 2001): This 2-dimensional problem is defined directly as a minimax problem, as follows:

$$\min_{x} \max_{1 \le i \le 3}\{f_i(x)\},$$

$$f_1(x) = x_1^2 + x_2^4,$$
$$f_2(x) = (2-x_1)^2 + (2-x_2)^2,$$
$$f_3(x) = 2\exp(-x_1+x_2).$$

TP_{MX-2} (Xu, 2001): This 2-dimensional problem is similar to the previous one, defined as follows:

$$\min_{x} \max_{1 \le i \le 3}\{f_i(x)\},$$

$$f_1(x) = x_1^4 + x_2^2,$$
$$f_2(x) = (2-x_1)^2 + (2-x_2)^2,$$
$$f_3(x) = 2\exp(-x_1+x_2).$$

TP_{MX-3} *(Rosen-Suzuki problem)* (Xu, 2001): This 4-dimensional problem is defined in implicit form, as follows:

Figure 4. The pressure vessel problem

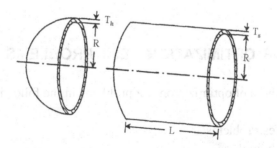

$$\min_{x} \max_{1 \le i \le 4} \{ f_i(x) \},$$

$f_1(x) = f(x) = x_1^2 + x_2^2 + 2x_3^2 + x_4^2 - 5x_1 - 5x_2 - 21x_3 + 7x_4,$
$f_i(x) = f(x) - a_i C_i(x), \quad i = 2, 3, 4,$

with,

$C_2(x) = -x_1^2 - x_2^2 - x_3^2 - x_4^2 - x_1 + x_2 - x_3 + x_4 + 8,$
$C_3(x) = -x_1^2 - 2x_2^2 - x_3^2 - 2x_4^2 + x_1 + x_4 + 10,$
$C_4(x) = -x_1^2 - x_2^2 - x_3^2 - 2x_1 + x_2 + x_4 + 5.$

***TP**$_{MX-4}$* (Xu, 2001): This 7-dimensional problem is defined in implicit form, as follows:

$$\min_{x} \max_{1 \le i \le 5} \{ f_i(x) \},$$

$f_1(x) = f(x) = (x_1\text{-}10)^2 + 5(x_2\text{-}12)^2 + 3(x_4\text{-}11)^2 + x_3^4 + 10x_5^6 + 7x_6^2 + x_7^4 - 4x_6 x_7 - 10x_6 - 8x_7,$
$f_i(x) = f(x) - a_i C_i(x), \quad i = 2, 3, 4, 5,$

with,

$C_2(x) = -2x_1^2 - 3x_3^4 - x_3 - 4x_4^2 - 5x_5 + 127,$
$C_3(x) = -7x_1 - 3x_2 - 10x_3^2 - x_4 + x_5 + 282,$
$C_4(x) = -23x_1 - x_2^2 - 6x_6^2 + 8x_7 + 196,$
$C_5(x) = -4x_1^2 - x_2^2 + 3x_1 x_2 - 2x_3^2 - 5x_6 + 11x_7.$

***TP**$_{MX-5}$* (Schwefel, 1995): This 2-dimensional problem is defined in explicit minimax form, as follows:

$$\min_{x} \max_{i=1,2} \{ f_i(x) \},$$

$f_1(x) = |x_1 + 2x_2 - 7|,$
$f_2(x) = |2x_1 + x_2 - 5|.$

***TP**$_{MX-6}$* (Schwefel, 1995): This 10-dimensional problem is defined in explicit minimax form, as follows:

$$\min_{x} \max_{1 \le i \le 10} \{ f_i(x) \},$$

$f_i(x) = |x_i|, \quad i = 1, 2, \ldots, 10.$

FURTHER LINKS FOR OPTIMIZATION TEST PROBLEMS

Researchers can find a plethora of optimization test problems in the following web sources:

http://titan.princeton.edu/TestProblems/
http://plato.la.asu.edu/bench.html

http://www-optima.amp.i.kyoto-u.ac.jp/member/student/hedar/Hedar_files/TestGO.htm
http://www.netlib.org/toms/667
http://www.netlib.org/uncon/data/
http://www.mat.univie.ac.at/~neum/glopt/test_results.html
http://www.mat.univie.ac.at/~neum/glopt/my_problems.html
http://infohost.nmt.edu/~borchers/optlibs.html
http://www.nada.kth.se/~viggo/wwwcompendium/wwwcompendium.html
http://www.ti3.tu-harburg.de/deutsch/optimierung.html
http://riot.ieor.berkeley.edu/riot/index.html
http://www.research.att.com/~mgcr/abstracts/guidelines.html

as well as in the *software for generation of classes of test functions* of Gaviano *et al.* (2003).

REFERENCES

Arora, J. S. (1989). *Introduction to optimum design*. New York: McGraw-Hill.

Blake, C. L., & Merz, C. J. (1998). *UCI repository of machine learning databases*. Retrieved from http://archive.ics.uci.edu/ml/

Chern, M. S., Yang, H. L., Teng, J. T., & Papachristos, S. (2008). Partial backlogging inventory lot-size models for deteriorating items with fluctuating demand under inflation. *European Journal of Operational Research, 191* (1), 127-141.

Devaney, R. L. (1984). A piecewise linear model for the zones of instability of an area preserving map. *Physica D: Nonlinear Phenomena, 10* (3), 387-393.

Floudas, C. A., & Pardalos, P. M. (1990). *A collection of test problems for constrained global optimization algorithms*. Berlin, Germany: Springer.

Gaviano, M., Kvasov, D. E., Lera, D., & Sergeyev, Y. D. (2003). Algorithm 829: Software for generation of classes of test functions with known local and global minima for global optimization. *ACM Transactions on Mathematical Software, 29* (4), 469-480.

Hénon, M. (1969). Numerical study of quadratic area-preserving mappings. *Quarterly of Applied Mathematics, 27*, 291–312.

Henry, B., Watt, S., & Wearne, S. (2000). A lattice refinement scheme for finding periodic orbits. *ANZIAM Journal, 42* (E), C735-C751.

Himmelblau, D. M. (1972). *Applied nonlinear programming*. New York: McGraw-Hill.

Hock, W., & Schittkowski, K. (1981). *Test examples for nonlinear programming codes*. Berlin, Germany: Springer.

Kantz, H., & Grassberger, P. (1988). Internal Arnold diffusion and chaos thresholds in coupled symplectic maps. *Journal of Physics A, 21* (3), L127-L133.

Knowles, J. D., & Corne, D. W. (2000). Approximating the nondominated front using the Pareto archived evolution strategies. *Evolutionary Computation, 8* (2), 149-172.

Lee, C.-Y., & Yao, X. (2004). Evolutionary programming using mutations based on the Lévy probability distribution. *IEEE Transactions on Evolutionary Computation, 8* (1), 1–13.

Levy, A. Montalvo, A., Gomez, S., & Galderon, A. (1981). *Topics in global optimization.* New York: Springer.

Lorenz, E. N. (1963). Deterministic nonperiodic flow. *Journal of the Atmospheric Sciences, 20* (2), 130-141.

Maynard Smith, J. (1968). *Mathematical ideas in biology.* London: Cambridge University Press.

McKelvey, R. D. (1991). *A Liapunov function for Nash equilibria* (Tech. Rep.). California, USA: California Institute of Technology.

Michalewicz, Z. (1999). *Genetic algorithms + data structures = evolution programs.* Berlin, Germany: Springer.

More, J. J., Garbow, B. S., & Hillstrom, K. E. (1981). Testing unconstrained optimization software. *ACM Transactions on Mathematical Software, 7* (1), 17-41.

Pavlidis, N. G., Parsopoulos, K. E., & Vrahatis, M. N. (2005). Computing Nash equilibria through computational intelligence methods. *Journal of Computational and Applied Mathematics, 175* (1), 113-136.

Prechelt, L. (1994). *Proben1: a set of neural network benchmark problems and benchmarking rules* (Tech. Rep. 21/94). Fakultät für Informatik, Universität Karlsruhe, Germany.

Rao, S. S. (1996). *Engineering optimization-Theory and practice.* New York: Wiley.

Rasband, S. (1990). *Chaotic dynamics of nonlinear systems.* New York: Wiley.

Rössler, O. (1976). An equation for continuous chaos. *Physics Letters A, 57* (5), 397-398.

Sandgen, E. (1990). Nonlinear integer and discrete programming in mechanical design optimization. *Journal of Mechanical Design (ASME), 112*, 223-229.

Schwefel, H.-P. (1995). *Evolution and optimum seeking.* New York: Wiley.

Skouri, K., & Papachristos, S. (2002). A continuous review inventory model, with deteriorating items, time-varying demand, linear replenishment cost, partially time-varying backlogging. *Applied Mathematical Modelling, 26* (5), 603-617.

Smith, J. W., Everhart, J. E., Dickson, W. C., Knowler, W. C., & Johannes, R. S. (1988). Using the ADAP learning algorithm to forecast the onset of diabetes mellitus. In *Proceedings of the 1998 Annual Symposium on Computer Application in Medical Care, Washington (DC), USA* (pp. 261-265).

Storn, R., & Price, K. (1997). Differential evolution - a simple and efficient heuristic for global optimization over continuous spaces. *Journal of Global Optimization, 11*, 341-359.

Trelea, I. C. (2003). The particle swarm optimization algorithm: convergence analysis and parameter selection. *Information Processing Letters, 85* (6), 317-325.

Van den Bergh, F., & Engelbrecht, A. P. (2002). A new locally convergent particle swarm optimizer. In *Proceedings of the 2002 IEEE International Conference on Systems, Man and Cybernetics (SMC'02), Hammamet, Tunisia* (Vol. 3, pp. 96-101).

Vrahatis, M. (1995). An efficient method for locating and computing periodic orbits of nonlinear mappings. *Journal of Computational Physics, 119*, 105-119.

Xu, S. (2001). Smoothing method for minimax problems. *Computational Optimization and Applications, 20* (3), 267-279.

Zitzler, E., Deb, K., & Thiele, L. (2000). Comparison of multiobjective evolution algorithms: empirical results. *Evolutionary Computation, 8* (2), 173-195.

Appendix B

This appendix contains a simple source code that implements the unified PSO (UPSO) approach, described in Chapter Four, in Matlab©. UPSO is implemented as a function with one input argument, namely the unification factor, u. We remind that UPSO for $u = 0$ corresponds to the standard local PSO variant, while for $u = 1$ we obtain the standard global PSO variant. Thus, the reported code implements both the standard PSO and UPSO variants. We analyze the source code separately in a following section, while comments are also provided in the program, in the lines starting with the symbol "%". The provided program uses test problem TP_{UO-1} to apply the algorithm for a number of experiments. For each experiment, the program provides the solution, its function value, as well as the required number of iterations. After the execution of all experiments, it also provides averaged statistics regarding the total number of successes, the mean number, and standard deviation of the required iterations and function evaluations.

A SIMPLE SOURCE CODE OF UNIFIED PSO

The following are the contents of a file named "UPSO.m". Line numbers are included to help analysis in the next section:

```
1     %%%%%%%%%%%%%%%%%%%%%%%%%%%%%%%%%%%%%%%
2     %%   A SIMPLE IMPLEMENTATION OF UNIFIED PSO (UPSO)   %%
3     %%%%%%%%%%%%%%%%%%%%%%%%%%%%%%%%%%%%%%%

4     function UPSO = UPSO(u)
5     clc;
6     rand('state',sum(100*clock));   % Different random generator seeds per execution

7     %===================
8     % PARAMETER SETTING
9     %===================
10    % PROBLEM-RELATED PARAMETERS
11    dim     = 2;          % Problem dimension
12    Xmax    = 5;          % Upper bound of particles per coordinate direction
13    Xmin    = -5;         % Lower bound of particles per coordinate direction
14    GM      = 0;          % Global minimum (used as stopping criterion)
```

Copyright © 2010, IGI Global, distributing in print or electronic forms without written permission of IGI Global is prohibited.

```
15   Acc       = 1e-3;        % Desired accuracy
16   nexp      = 100;         % Number of experiments
17   RunStats = [ ];          % Matrix of averaged statistics

18   % PSO-RELATED PARAMETERS
19   SS        = 10;                % Swarm size
20   MaxIt     = 1000;              % Maximum number of iterations
21   chi       = 0.729;             % Constriction coefficient
22   c1        = 2.05;              % Cognitive parameter
23   c2        = 2.05;              % Social parameter
24   NR        = 1;                 % Neighborhood radius (must be NR < SS)
25   vmax      = (Xmax-Xmin)/2;     % Maximum velocity

26   %====================================
27   % LOOP ON THE NUMBER OF EXPERIMENTS
28   %====================================
29   for T = 1:nexp,

30      % PARAMETER INITIALIZATION
31      success   = 0;        % Success flag
32      iter      = 1;        % Iteration counter
33      feval     = 0;        % Function evaluations counter
34      STOP      = 0;        % Experiment stopping flag

35      % SWARM, VELOCITIES & BEST POSITION INITIALIZATION
36      swarm     = rand(dim, SS)*(Xmax-Xmin) + Xmin;
37      vel       = rand(dim, SS)*2*vmax - vmax;
38      bestpos   = swarm;

39      % EVALUATE SWARM
40      for i=1:SS,
41        fswarm(i)   = f(swarm(:,i));
42        fbestpos(i) = fswarm(i);
43        feval       = feval+1;
44      end

45      % UPDATE OVERALL BEST PARTICLE INDICES
46      [fxopt, g_over] = min(fbestpos);
47      xopt = bestpos(:,g_over);

48      % UPDATE LOCAL BEST PARTICLE INDICES PER NEIGHBORHOOD
49      g_neig = 1:SS;
50      for i=1:SS,
51        for j=i-NR:i+NR,
52          if (j<=0)
53            jc = SS+j;
54          elseif (j>SS)
55            jc = j-SS;
56          else
57            jc = j;
58          end
59          if (fbestpos(jc)<fbestpos(g_neig(i)))
60            g_neig(i) = jc;
61          end
62        end
63      end

64      %=====================
65      % SWARM EVOLUTION LOOP
66      %=====================
67      while (STOP == 0)

68         % UPDATE ITERATION COUNTER
69         iter = iter+1;
```

```
70        % UPDATE VELOCITIES
71        for i=1:SS,

72            % Global PSO component
73            R1 = rand(dim,1);
74            R2 = rand(dim,1);
75            G  = chi .* (vel(:,i) + c1.*R1.*(bestpos(:,i)-swarm(:,i)) + c2.*R2.*(bestpos(:,g_over)-swarm(:,i)));

76            % Local PSO component
77            R1 = rand(dim,1);
78            R2 = rand(dim,1);
79            L  = chi .* (vel(:,i) + c1.*R1.*(bestpos(:,i)-swarm(:,i)) + c2.*R2.*(bestpos(:,g_neig(i))-swarm(:,i)));

80            % Standard UPSO velocity update
81            vel(:,i) = u.*G + (1-u).*L;
82        end

83        % CONSTRAIN VELOCITIES
84        vel(vel>vmax)  = vmax;
85        vel(vel<-vmax) = -vmax;

86        % UPDATE SWARM
87        swarm = swarm + vel;

88        % CONSTRAIN SWARM
89        swarm(swarm>Xmax) = Xmax;
90        swarm(swarm<Xmin) = Xmin;

91        % EVALUATE SWARM
92        for i=1:SS,
93            fswarm(i) = f(swarm(:,i));
94            feval        = feval+1;
95        end

96        % UPDATE BEST POSITIONS
97        bestpos(:,fswarm<fbestpos) = swarm(:,fswarm<fbestpos);
98        fbestpos(fswarm<fbestpos)  = fswarm(fswarm<fbestpos);

99        % UPDATE OVERALL BEST PARTICLE INDICES
100       [fxopt, g_over] = min(fbestpos);
101       xopt = bestpos(:,g_over);

102       % UPDATE LOCAL BEST PARTICLE INDICES PER NEIGHBORHOOD
103       g_neig = 1:SS;
104       for i=1:SS,
105         for j=i-NR:i+NR,
106           if (j<=0)
107              jc = SS+j;
108           elseif (j>SS)
109              jc = j-SS;
110           else
111              jc = j;
112           end
113           if (fbestpos(jc)<fbestpos(g_neig(i)))
114              g_neig(i) = jc;
115           end
116         end
117       end

118       % CHECK STOPPING CRITERION
119       if ((iter >= MaxIt)|(fxopt <= GM+Acc))
120          STOP = 1;
121          if (fxopt <= GM+Acc)
122             success = 1;
```

```
123        end
124      end

125    end % FINISH SWARM EVOLUTION

126    % UPDATE RUN STATISTICS VECTOR
127    RunStats(T,:) = [success iter feval];

128    % PRINT DETECTED SOLUTION ON SCREEN
129    fprintf('EXPERIMENT #%3d :: SUCCESS=%1d :: SOLUTION X*=(', T, success);
130    for i=1:dim
131       fprintf(' %15.5e', xopt(i));
132    end
133    fprintf(' ) :: F(X*)=%12.5e :: ITER=%5d\n', fxopt, iter);

134  end % FINISH LOOP ON EXPERIMENTS

135  %===========================
136  % PRINT AVERAGED STATISTICS
137  %===========================
138  fprintf('\nAVERAGED STATISTICS:\n');
139  SucRunStats = RunStats(RunStats(:,1)==1,:);
140  fprintf('SUCCESSES = %3d/%3d\n', sum(RunStats(:,1)),nexp);
141  fprintf('MEAN ITERATIONS = %8.2f\n', mean(SucRunStats(:,2)));
142  fprintf('StD  ITERATIONS = %8.2f\n', std(SucRunStats(:,2)));
143  fprintf('MEAN FUNC.EVAL. = %8.2f\n', mean(SucRunStats(:,3)));
144  fprintf('StD  FUNC.EVAL. = %8.2f\n', std(SucRunStats(:,3)));

145  %====================
146  % OBJECTIVE FUNCTION
147  %====================
148  function f = f(x)
149  f = x'*x;
```

ANALYSIS OF THE PROGRAM

The first part of the program consists of commands in lines 1-25. This part defines the basic problem- and algorithm-related parameters. The unification factor is an input argument provided by the user (line 4). We can distinguish two major blocks, namely those of lines 7-17 and 18-25. The first block defines problem-related parameters. Problem dimension is set to 2, while particle bounds are the same for all coordinate directions, restricting particles within $[-5,5]^2$. The global minimum is provided to serve as a stopping criterion along with the desired accuracy, which is set to 10^{-3}. The total number of experiments is set to 100, and a matrix is defined to store performance measures per run in line 17.

In the second block, the swarm size is set to 10 and the maximum number of iterations to 1000. The default constriction coefficient PSO parameter set, $\chi = 0.729$, $c_1 = c_2 = 2.05$, is defined in lines 21-23, along with the ring neighborhood radius for the local PSO components, which is set to 1 in line 24. Finally, the maximum velocity is set to half the dynamic range of particles in line 25.

The program continues with the commands in lines 26-63. At this stage, the execution of the specified number of experiments starts at line 29. First, a set of run parameters are initialized in lines 30-34. These include a success flag, which takes the value 1 if a run is successful, the initial iteration and function evaluation counter values, as well as a stopping flag. Then, swarm and velocities are randomly initialized and best positions are set to coincide with the initial swarm in lines 35-38. After that, particle evaluation takes place in lines 39-44, followed by the update of the best indices in lines 45-63.

There are two indices to be updated. The first one, denoted as "g_over", is the index of the overall best position and is updated in lines 45-47 by simply detecting the overall best position. The second index is actually an index vector denoted as "g_neig". Its *i*-th component corresponds to the index of the best position in the neighborhood of the *i*-th particle under the ring topology with the radius defined in line 24. Special care is taken in cases were indices recycle over the ring, with the commands in lines 52-58.

The initial iteration of the algorithm ends at line 63, and the main iterative procedure of UPSO takes place in lines 64-125. Initially, the iteration counter is updated in line 69, and velocities of all particles are updated according to the standard UPSO scheme, described in equations (1)-(4) of Chapter Four, in lines 71-82. Notice that random vectors are generated in lines 73-74 and 77-78 to avoid the common implementation mistake described in equations (3)-(4) and illustrated in Fig. 2 of Chapter Two. The global UPSO search direction of equation (1) in Chapter Four is computed in line 75 and the local one of equation (2) in line 79. Then, equation (3) of Chapter Four is implemented in line 81. Notice that if u = 1 was provided by the user, then line 81 computes the standard global PSO velocity update, while for $u = 0$ it computes the standard local PSO velocity update.

The execution continues with a check on maximum velocity in lines 83-85. Then, particles are updated and constrained within the search space in lines 86-90. Finally, the swarm is evaluated and best positions are updated, along with indices in lines 91-117. The iteration is completed by checking the stopping criterion and updating the corresponding flags in lines 118-124.

The commands in lines 126-133 are used to handle statistics of the last conducted experiment. The success flag, required number of iterations, and function evaluations, are stored in a matrix in line 127 and all information of the experiment is written on-screen with lines 128-133. After the execution of all experiments, averaged statistics are reported with lines 135-144 by analyzing the aforementioned matrix. The last lines 145-149 define the objective function under consideration. In our implementation, test problem TP_{UO-1} of Appendix A is used as an illustrative example.

The user can easily parameterize the provided source code and alter it suitably to render it in a desirable form for different use. The program can be easily extended to include the mutated UPSO variant by modifying line 81. It was not our intention to provide an optimal code but rather a comprehensive one. Thus, several parts can be re-written in matrix form to reduce the required execution time of the program.

The screen output of a simple execution of the provided source code for $u = 0.5$, using the Matlab© command:

```
» UPSO(0.5)
```

is provided below:

```
EXPERIMENT #  1 :: SUCCESS=1 :: SOLUTION X*=(  8.20493e-003   2.58746e-002 ) :: F(X*)=7.36814e-004 :: ITER=  25
EXPERIMENT #  2 :: SUCCESS=1 :: SOLUTION X*=(  3.83295e-003  -2.70715e-002 ) :: F(X*)=7.47557e-004 :: ITER=  19
EXPERIMENT #  3 :: SUCCESS=1 :: SOLUTION X*=(  1.24547e-002  -2.09935e-003 ) :: F(X*)=1.59527e-004 :: ITER=  25
EXPERIMENT #  4 :: SUCCESS=1 :: SOLUTION X*=(  3.61461e-003  -9.50201e-004 ) :: F(X*)=1.39683e-005 :: ITER=  19
EXPERIMENT #  5 :: SUCCESS=1 :: SOLUTION X*=( -2.86471e-003   4.42894e-003 ) :: F(X*)=2.78221e-005 :: ITER=  21
EXPERIMENT #  6 :: SUCCESS=1 :: SOLUTION X*=( -3.18816e-003  -7.84297e-004 ) :: F(X*)=1.07795e-005 :: ITER=  17
EXPERIMENT #  7 :: SUCCESS=1 :: SOLUTION X*=( -2.78762e-002  -4.91552e-004 ) :: F(X*)=7.77322e-004 :: ITER=  32
EXPERIMENT #  8 :: SUCCESS=1 :: SOLUTION X*=( -2.65138e-003   3.90574e-003 ) :: F(X*)=2.22846e-005 :: ITER=  21
EXPERIMENT #  9 :: SUCCESS=1 :: SOLUTION X*=(  3.12590e-002  -4.13055e-004 ) :: F(X*)=9.77293e-004 :: ITER=  13
EXPERIMENT # 10 :: SUCCESS=1 :: SOLUTION X*=( -9.93293e-003  -2.45458e-002 ) :: F(X*)=7.01157e-004 :: ITER=  28
EXPERIMENT # 11 :: SUCCESS=1 :: SOLUTION X*=(  1.27998e-002   1.62942e-002 ) :: F(X*)=4.29338e-004 :: ITER=  19
EXPERIMENT # 12 :: SUCCESS=1 :: SOLUTION X*=( -2.51056e-002  -1.18328e-003 ) :: F(X*)=6.31690e-004 :: ITER=  26
```

```
EXPERIMENT # 13 :: SUCCESS=1 :: SOLUTION X*=(  -1.51391e-002    9.35760e-003 ) :: F(X*)=3.16757e-004 :: ITER=  28
EXPERIMENT # 14 :: SUCCESS=1 :: SOLUTION X*=(  -2.20107e-002    1.88080e-002 ) :: F(X*)=8.38213e-004 :: ITER=   3
EXPERIMENT # 15 :: SUCCESS=1 :: SOLUTION X*=(  -2.35410e-002    1.00231e-002 ) :: F(X*)=6.54643e-004 :: ITER=  22
EXPERIMENT # 16 :: SUCCESS=1 :: SOLUTION X*=(  -1.84971e-002    1.13460e-002 ) :: F(X*)=4.70874e-004 :: ITER=  19
EXPERIMENT # 17 :: SUCCESS=1 :: SOLUTION X*=(  -2.02657e-002   -1.60819e-002 ) :: F(X*)=6.69326e-004 :: ITER=  29
EXPERIMENT # 18 :: SUCCESS=1 :: SOLUTION X*=(   2.01813e-003   -1.62399e-002 ) :: F(X*)=2.67808e-004 :: ITER=  17
EXPERIMENT # 19 :: SUCCESS=1 :: SOLUTION X*=(  -2.41084e-003   -1.97390e-002 ) :: F(X*)=3.95441e-004 :: ITER=  17
EXPERIMENT # 20 :: SUCCESS=1 :: SOLUTION X*=(   2.66296e-003   -1.15125e-002 ) :: F(X*)=1.39628e-004 :: ITER=  23
EXPERIMENT # 21 :: SUCCESS=1 :: SOLUTION X*=(  -3.59976e-003    7.04091e-003 ) :: F(X*)=6.25328e-005 :: ITER=  23
EXPERIMENT # 22 :: SUCCESS=1 :: SOLUTION X*=(   1.41340e-002   -1.64633e-002 ) :: F(X*)=4.70810e-004 :: ITER=  17
EXPERIMENT # 23 :: SUCCESS=1 :: SOLUTION X*=(   2.20253e-002   -1.39740e-002 ) :: F(X*)=6.80387e-004 :: ITER=  17
EXPERIMENT # 24 :: SUCCESS=1 :: SOLUTION X*=(  -5.95643e-003   -2.50471e-002 ) :: F(X*)=6.62837e-004 :: ITER=  26
EXPERIMENT # 25 :: SUCCESS=1 :: SOLUTION X*=(   7.12361e-004   -2.39330e-002 ) :: F(X*)=5.73296e-004 :: ITER=  23
EXPERIMENT # 26 :: SUCCESS=1 :: SOLUTION X*=(  -7.40934e-003   -2.39281e-002 ) :: F(X*)=6.27453e-004 :: ITER=  15
EXPERIMENT # 27 :: SUCCESS=1 :: SOLUTION X*=(   2.01486e-002    6.32761e-003 ) :: F(X*)=4.46004e-004 :: ITER=  23
EXPERIMENT # 28 :: SUCCESS=1 :: SOLUTION X*=(   8.49605e-004    2.45291e-003 ) :: F(X*)=6.73861e-006 :: ITER=  25
EXPERIMENT # 29 :: SUCCESS=1 :: SOLUTION X*=(  -1.49360e-003    2.40983e-002 ) :: F(X*)=8.03812e-004 :: ITER=  28
EXPERIMENT # 30 :: SUCCESS=1 :: SOLUTION X*=(  -6.59177e-003   -1.94126e-002 ) :: F(X*)=4.20302e-004 :: ITER=  21
EXPERIMENT # 31 :: SUCCESS=1 :: SOLUTION X*=(   8.89583e-003    3.28827e-003 ) :: F(X*)=8.99485e-005 :: ITER=  15
EXPERIMENT # 32 :: SUCCESS=1 :: SOLUTION X*=(   2.68264e-002    2.88856e-003 ) :: F(X*)=7.28002e-004 :: ITER=  14
EXPERIMENT # 33 :: SUCCESS=1 :: SOLUTION X*=(   1.64331e-002   -9.86652e-003 ) :: F(X*)=3.67395e-004 :: ITER=  16
EXPERIMENT # 34 :: SUCCESS=1 :: SOLUTION X*=(  -1.38726e-002   -1.44362e-002 ) :: F(X*)=4.00852e-004 :: ITER=  23
EXPERIMENT # 35 :: SUCCESS=1 :: SOLUTION X*=(  -2.72911e-002   -1.01103e-002 ) :: F(X*)=8.47022e-004 :: ITER=  15
EXPERIMENT # 36 :: SUCCESS=1 :: SOLUTION X*=(  -7.82163e-003    1.09136e-002 ) :: F(X*)=1.80285e-004 :: ITER=  26
EXPERIMENT # 37 :: SUCCESS=1 :: SOLUTION X*=(  -1.89804e-002    1.25057e-002 ) :: F(X*)=5.16650e-004 :: ITER=  16
EXPERIMENT # 38 :: SUCCESS=1 :: SOLUTION X*=(  -8.38985e-003    2.13726e-003 ) :: F(X*)=5.27179e-004 :: ITER=  20
EXPERIMENT # 39 :: SUCCESS=1 :: SOLUTION X*=(  -4.93147e-003   -1.58747e-003 ) :: F(X*)=2.68395e-005 :: ITER=  21
EXPERIMENT # 40 :: SUCCESS=1 :: SOLUTION X*=(   1.28486e-002    8.43406e-003 ) :: F(X*)=2.36220e-004 :: ITER=  24
EXPERIMENT # 41 :: SUCCESS=1 :: SOLUTION X*=(  -1.31404e-003   -1.69140e-002 ) :: F(X*)=2.87808e-004 :: ITER=  18
EXPERIMENT # 42 :: SUCCESS=1 :: SOLUTION X*=(   6.61844e-003   -2.57374e-002 ) :: F(X*)=7.06215e-004 :: ITER=  21
EXPERIMENT # 43 :: SUCCESS=1 :: SOLUTION X*=(  -3.41540e-003    2.54876e-002 ) :: F(X*)=6.61284e-004 :: ITER=  17
EXPERIMENT # 44 :: SUCCESS=1 :: SOLUTION X*=(  -2.54805e-003    1.23551e-002 ) :: F(X*)=1.59142e-004 :: ITER=  16
EXPERIMENT # 45 :: SUCCESS=1 :: SOLUTION X*=(   2.54059e-002    6.96628e-003 ) :: F(X*)=6.93987e-004 :: ITER=  16
EXPERIMENT # 46 :: SUCCESS=1 :: SOLUTION X*=(  -2.48786e-002   -1.51542e-002 ) :: F(X*)=8.45596e-004 :: ITER=  16
EXPERIMENT # 47 :: SUCCESS=1 :: SOLUTION X*=(  -1.05948e-002    5.72036e-003 ) :: F(X*)=1.44973e-004 :: ITER=  25
EXPERIMENT # 48 :: SUCCESS=1 :: SOLUTION X*=(  -6.27297e-003   -1.02131e-002 ) :: F(X*)=1.43657e-004 :: ITER=  18
EXPERIMENT # 49 :: SUCCESS=1 :: SOLUTION X*=(  -6.77770e-003   -3.05436e-002 ) :: F(X*)=9.78848e-004 :: ITER=  26
EXPERIMENT # 50 :: SUCCESS=1 :: SOLUTION X*=(  -1.79872e-002   -9.45686e-003 ) :: F(X*)=4.12971e-004 :: ITER=  14
EXPERIMENT # 51 :: SUCCESS=1 :: SOLUTION X*=(  -1.84248e-002   -1.07893e-002 ) :: F(X*)=4.55883e-004 :: ITER=  13
EXPERIMENT # 52 :: SUCCESS=1 :: SOLUTION X*=(  -1.08369e-002   -2.60657e-002 ) :: F(X*)=7.96860e-004 :: ITER=  18
EXPERIMENT # 53 :: SUCCESS=1 :: SOLUTION X*=(  -9.66945e-003   -1.49774e-003 ) :: F(X*)=9.57415e-005 :: ITER=  20
EXPERIMENT # 54 :: SUCCESS=1 :: SOLUTION X*=(  -1.18518e-002    2.58036e-002 ) :: F(X*)=8.06291e-004 :: ITER=  15
EXPERIMENT # 55 :: SUCCESS=1 :: SOLUTION X*=(  -1.47890e-003   -1.05713e-002 ) :: F(X*)=1.13940e-004 :: ITER=  23
EXPERIMENT # 56 :: SUCCESS=1 :: SOLUTION X*=(  -1.48266e-002    4.57483e-003 ) :: F(X*)=2.40758e-004 :: ITER=  17
EXPERIMENT # 57 :: SUCCESS=1 :: SOLUTION X*=(   7.42223e-003    1.24499e-003 ) :: F(X*)=5.66395e-005 :: ITER=  12
EXPERIMENT # 58 :: SUCCESS=1 :: SOLUTION X*=(   9.04777e-003   -2.89715e-003 ) :: F(X*)=9.02556e-005 :: ITER=  27
EXPERIMENT # 59 :: SUCCESS=1 :: SOLUTION X*=(  -6.11140e-003   -1.49777e-002 ) :: F(X*)=2.61681e-004 :: ITER=  22
EXPERIMENT # 60 :: SUCCESS=1 :: SOLUTION X*=(   2.22336e-002    1.89757e-002 ) :: F(X*)=8.54409e-004 :: ITER=  13
EXPERIMENT # 61 :: SUCCESS=1 :: SOLUTION X*=(   3.38245e-003    1.05181e-002 ) :: F(X*)=1.22072e-004 :: ITER=  23
EXPERIMENT # 62 :: SUCCESS=1 :: SOLUTION X*=(  -2.08610e-002    1.31809e-002 ) :: F(X*)=4.36918e-004 :: ITER=  18
EXPERIMENT # 63 :: SUCCESS=1 :: SOLUTION X*=(   7.03932e-003   -1.17298e-002 ) :: F(X*)=1.87140e-004 :: ITER=  19
EXPERIMENT # 64 :: SUCCESS=1 :: SOLUTION X*=(  -1.12538e-002   -1.08888e-002 ) :: F(X*)=2.45213e-004 :: ITER=  24
EXPERIMENT # 65 :: SUCCESS=1 :: SOLUTION X*=(  -2.74965e-002    1.08333e-002 ) :: F(X*)=8.73417e-004 :: ITER=  18
EXPERIMENT # 66 :: SUCCESS=1 :: SOLUTION X*=(  -3.88534e-004    1.37473e-002 ) :: F(X*)=1.89138e-004 :: ITER=  27
EXPERIMENT # 67 :: SUCCESS=1 :: SOLUTION X*=(  -2.49835e-002   -3.25187e-003 ) :: F(X*)=6.34751e-004 :: ITER=  21
EXPERIMENT # 68 :: SUCCESS=1 :: SOLUTION X*=(  -1.54809e-002    7.98801e-003 ) :: F(X*)=3.03467e-004 :: ITER=  21
EXPERIMENT # 69 :: SUCCESS=1 :: SOLUTION X*=(  -2.13055e-003   -2.65438e-002 ) :: F(X*)=7.09114e-004 :: ITER=  20
EXPERIMENT # 70 :: SUCCESS=1 :: SOLUTION X*=(   1.43669e-002    5.87321e-003 ) :: F(X*)=2.40902e-004 :: ITER=  16
EXPERIMENT # 71 :: SUCCESS=1 :: SOLUTION X*=(   1.35924e-002   -1.19171e-002 ) :: F(X*)=3.26771e-004 :: ITER=  18
EXPERIMENT # 72 :: SUCCESS=1 :: SOLUTION X*=(  -2.60697e-002    1.43166e-002 ) :: F(X*)=8.84594e-004 :: ITER=  13
EXPERIMENT # 73 :: SUCCESS=1 :: SOLUTION X*=(   1.29666e-002    1.31233e-002 ) :: F(X*)=3.40355e-004 :: ITER=  13
EXPERIMENT # 74 :: SUCCESS=1 :: SOLUTION X*=(  -1.48739e-002    2.36897e-002 ) :: F(X*)=7.82436e-004 :: ITER=  20
EXPERIMENT # 75 :: SUCCESS=1 :: SOLUTION X*=(  -1.30002e-002    7.04916e-003 ) :: F(X*)=2.18696e-004 :: ITER=  20
EXPERIMENT # 76 :: SUCCESS=1 :: SOLUTION X*=(   1.66187e-002    8.32411e-003 ) :: F(X*)=3.45473e-004 :: ITER=  24
EXPERIMENT # 77 :: SUCCESS=1 :: SOLUTION X*=(  -5.98262e-003    4.53823e-003 ) :: F(X*)=5.63873e-005 :: ITER=  27
EXPERIMENT # 78 :: SUCCESS=1 :: SOLUTION X*=(   1.31795e-002   -1.37656e-002 ) :: F(X*)=3.63191e-004 :: ITER=  22
EXPERIMENT # 79 :: SUCCESS=1 :: SOLUTION X*=(  -2.69209e-002   -1.36614e-002 ) :: F(X*)=9.11365e-004 :: ITER=  23
EXPERIMENT # 80 :: SUCCESS=1 :: SOLUTION X*=(   2.91334e-002    1.19672e-002 ) :: F(X*)=9.91965e-004 :: ITER=  16
EXPERIMENT # 81 :: SUCCESS=1 :: SOLUTION X*=(  -1.34526e-002   -1.78143e-002 ) :: F(X*)=4.98323e-004 :: ITER=  23
EXPERIMENT # 82 :: SUCCESS=1 :: SOLUTION X*=(   7.85566e-003    6.84412e-003 ) :: F(X*)=1.08553e-004 :: ITER=  20
EXPERIMENT # 83 :: SUCCESS=1 :: SOLUTION X*=(   2.01624e-003    9.88094e-003 ) :: F(X*)=1.01698e-004 :: ITER=  20
EXPERIMENT # 84 :: SUCCESS=1 :: SOLUTION X*=(  -2.49957e-002   -1.76589e-002 ) :: F(X*)=9.36623e-004 :: ITER=  17
EXPERIMENT # 85 :: SUCCESS=1 :: SOLUTION X*=(  -1.72408e-002   -2.29326e-002 ) :: F(X*)=8.23148e-004 :: ITER=  13
```

```
EXPERIMENT # 86 :: SUCCESS=1 :: SOLUTION X*=(  -1.79197e-002   -2.02930e-002 ) :: F(X*)=7.32920e-004 :: ITER=  15
EXPERIMENT # 87 :: SUCCESS=1 :: SOLUTION X*=(  -3.06098e-002    5.32501e-003 ) :: F(X*)=9.65313e-004 :: ITER=  19
EXPERIMENT # 88 :: SUCCESS=1 :: SOLUTION X*=(   1.13479e-002    1.65932e-002 ) :: F(X*)=4.04108e-004 :: ITER=  25
EXPERIMENT # 89 :: SUCCESS=1 :: SOLUTION X*=(   2.63679e-002   -5.34927e-003 ) :: F(X*)=7.23878e-004 :: ITER=  25
EXPERIMENT # 90 :: SUCCESS=1 :: SOLUTION X*=(   1.04269e-002   -6.64877e-003 ) :: F(X*)=1.52927e-004 :: ITER=  19
EXPERIMENT # 91 :: SUCCESS=1 :: SOLUTION X*=(  -9.36494e-003    1.73946e-002 ) :: F(X*)=3.90276e-004 :: ITER=  22
EXPERIMENT # 92 :: SUCCESS=1 :: SOLUTION X*=(  -2.43239e-002    1.08748e-002 ) :: F(X*)=7.09915e-004 :: ITER=  22
EXPERIMENT # 93 :: SUCCESS=1 :: SOLUTION X*=(   1.13855e-002   -9.39941e-003 ) :: F(X*)=2.17978e-004 :: ITER=  12
EXPERIMENT # 94 :: SUCCESS=1 :: SOLUTION X*=(  -2.04344e-002   -2.40546e-002 ) :: F(X*)=9.96190e-004 :: ITER=  17
EXPERIMENT # 95 :: SUCCESS=1 :: SOLUTION X*=(  -2.04062e-003   -5.49804e-003 ) :: F(X*)=3.43925e-005 :: ITER=  25
EXPERIMENT # 96 :: SUCCESS=1 :: SOLUTION X*=(   1.64662e-002    1.25430e-002 ) :: F(X*)=4.28463e-004 :: ITER=  14
EXPERIMENT # 97 :: SUCCESS=1 :: SOLUTION X*=(   1.16360e-002   -9.67004e-003 ) :: F(X*)=2.28906e-004 :: ITER=   7
EXPERIMENT # 98 :: SUCCESS=1 :: SOLUTION X*=(  -6.94740e-003   -1.18602e-002 ) :: F(X*)=1.88930e-004 :: ITER=  23
EXPERIMENT # 99 :: SUCCESS=1 :: SOLUTION X*=(  -1.49063e-002   -1.34528e-003 ) :: F(X*)=2.24008e-004 :: ITER=  23
EXPERIMENT #100 :: SUCCESS=1 :: SOLUTION X*=(  -1.00167e-002   -2.71101e-002 ) :: F(X*)=8.35291e-004 :: ITER=  20

AVERAGED STATISTICS:
SUCCESSES = 100/100
MEAN ITERATIONS =  19.82
StD  ITERATIONS =   4.91
MEAN FUNC.EVAL. =  198.20
StD  FUNC.EVAL. =   49.14
```

FURTHER SOURCES

The following sources can serve as good starting points for further information on PSO theory and implementations:

1. *Particle Swarm Central* (http://www.particleswarm.info/): This website is the best reference point for PSO. It contains extensive information on people, places, papers, programs, and problems related to PSO, as well as RSS feeds.

2. *PSO Resources* (http://clerc.maurice.free.fr/pso/): This webpage is maintained by Maurice Clerc and contains several problems, implementations, and preliminary theoretical results of Clerc.

3. *PSO Matlab Toolbox* (http://www.mathworks.com/matlabcentral/fileexchange/7506): This is a PSO toolbox for Matlab© developed by Brian Birge. It contains several objective functions and provides several interesting features, such as a neural network trainer, plot plug-ins, etc.

4. *Open Source PSO Toolbox* (http://psotoolbox.sourceforge.net/): This is an open source PSO toolbox in Matlab©, provided through Sourceforge.net under the GPL license. It is accompanied by a GUI that allows easy implementation of PSO solutions to different problems.

5. *JSwarm-PSO* (http://jswarm-pso.sourceforge.net/): This is an open source PSO implementation in Java, also provided through Sourceforge.net under the GPL license.

6. *CIlib* (Computational Intelligence Library) (http://cilib.sourceforge.net): This is an open source library developed by the research team of Professor A.P. Engelbrecht. It contains a huge collection of implemented and well tested computational intelligence algorithms, including PSO.

7. *Berkeley Lab* (http://gundog.lbl.gov/GO/jdoc/genopt/algorithm/PSOCC.html): This is a set of Java classes that implement the constriction coefficient PSO variant. Also, a PSO approach is incorporated in the GenOpt solver provided by the same laboratory (http://gundog.lbl.gov/GO/overview.html).

Numerous other contributions can be found by a simple search in any search engine.

About the Authors

Konstantinos E. Parsopoulos received his Diploma in Mathematics, with specialty in Computational Mathematics and Informatics, in 1998 from the Department of Mathematics, University of Patras, Greece, and his PhD degree in Intelligent Computational Optimization in 2005 from the Departments of Mathematics and Computer Engineering & Informatics, University of Patras, Greece. He served as a Lecturer with the Department of Mathematics, University of Patras, Greece, from May 2008 to September 2009. He is currently an Assistant Professor with the Department of Computer Science, University of Ioannina, Greece. His research is focused on Intelligent Computational Optimization algorithms with an emphasis on Swarm Intelligence and Evolutionary Computation approaches. His work consists of 16 papers in refereed international scientific journals, and 47 papers in books, edited volumes and conference proceedings, while it has received more than 990 citations from other researchers. He has been a visiting researcher at the Department of Computer Science, University of Dortmund, Germany (2001), as well as at the Institut de Recherche en Informatique et en Automatique (INRIA), Sophia-Antipolis, France (2003 and 2006). He has served as a member of the editorial board of 2 international scientific journals, as well as of the technical and program committee in 13 international conferences. Also, he has served as reviewer for 24 scientific journals, as well as for the Portuguese Foundation for Science and Technology (FCT). He has participated in national and international research projects, and he is a member of ACM and IEEE since 2005 and 1999, respectively.

Michael N. Vrahatis received his PhD in Computational Mathematics (1982) from the University of Patras, Greece. His work includes topological degree theory, systems of nonlinear equations, numerical and intelligent optimisation, data mining and unsupervised clustering, cryptography and cryptanalysis, bioinformatics and medical informatics, as well as computational and swarm intelligence. He is a Professor of Computational Mathematics in the University of Patras, since 2000 and serves as Director of the Computational Intelligence Laboratory of the same Department (since 2004). Several of his students serve as Lecturers, Assistant Professors, Associate Professors or Full Professors in Greece, England and Japan. He was a visiting research fellow at the Department of Mathematics, Cornell University, Ithaca, NY, USA (1987–1988) and a visiting professor to the INFN (Istituto Nazionale di Fisica Nucleare), Bologna, Italy (1992, 1994, and 1998); the Department of Computer Science, Katholieke Universiteit Leuven, Belgium (1999); the Department of Ocean Engineering, Design Laboratory, MIT, Cambridge, MA, USA (2000); and the Collaborative Research Center "Computational Intelligence" (SFB 531) at the Department of Computer Science, University of Dortmund, Germany (2001). He was a visiting researcher at CERN (European Organization of Nuclear Research), Geneva, Switzerland (1992) and at INRIA (Institut National de Recherche en Informatique et en Automatique), Sophia-Antipolis, France

Copyright © 2010, IGI Global, distributing in print or electronic forms without written permission of IGI Global is prohibited.

(1998, 2003, 2004 and 2006). He has participated in the organisation of over 80 conferences serving at several positions, and participated in over 200 conferences, congresses and advanced schools as a speaker (participant as well as invited) or observer. He is/was a member of the editorial board of eight international journals. His work consists of over 330 publications (125 papers in refereed international scientific journals, and 211 papers in books, edited volumes and conference proceedings) that has been cited by researchers over 4000 times (co-authors citations excluded).

Index

Symbols

ε-dominance 255
λ-fold cross-validation (λ-CV) 209

A

acceleration constant 28
additive noise 224
aggregation function 262
algorithm design 64, 65, 70, 72, 73, 80
analysis of variance 257
another MOPSO (AMOPSO) 256
ANOVA 257
ant colony optimization 17, 18
apparent urgency 186
artificial neural networks 150, 204
average relative percent deviation 188

B

balanced UPSO 98
bang-bang weighted aggregation (BWA) 249
batch training 151
binary representations 16
bioinformatics 204
black-box optimization 6
black-box PSO variant 121
Brownian motion 25
bunching effect 79

C

Camberra 152
center of area 159
characteristic bisection methods 169
characteristic diagram 180
Chebychev 152

chi-square 152
classification and regression trees (CART) 64
cognitive only model 120
collective intelligence 17, 24, 41
colored noise 225
comma-strategy 11
composite particle swarm optimization 88, 110
computational statistics 42, 63, 64, 85, 86
concave 247
constrained optimization (CO) 3, 43, 105, 142, 258
constraint satisfaction problem 187
constriction coefficient 38, 48, 52, 60, 61, 62, 63, 69, 70, 79, 109
context vector 117
continuous-based PSO 144
continuous inventory optimization problem 189
continuous optimization 6, 10
continuous review inventory 190
continuous-time 58
control strategy 76
conventional weighted aggregation (CWA) 233, 249
convex 247
convexity 92
convex optimization 4
cooperative co-evolutionary GA (CCGA) 119
cooperative particle swarm optimization 88, 116, 117
correlation 152
corresponding objective 176
corresponding optimization 227
corresponding parameters 139
corresponding potential 177

Copyright © 2010, IGI Global, distributing in print or electronic forms without written permission of IGI Global is prohibited.